To

Louise and Conor Taaffe

We are all in their debt

Contents

Preface

It is appropriate, on the republication of the first two books about soft systems methodology (SSM), to add new material written from the perspective of 1999. It would not be appropriate to change the text of the books themselves; they stand, telling the first part of the story. It seems appropriate to bring the story up to date by reflecting upon the 30 years of research on, and use of, SSM, and to offer a summing up of what is now a mature and much-tested process of inquiry into problem situations in human affairs. That is what is done in this chapter, which extends the earlier books.

The chapter examines: the emergence of soft systems thinking as systems engineering failed in problems which were not technically defined; the evolving expression of the methodology as a whole from 1972 to the 1990s; and then its parts, covering all the significant developments in the intellectual 'apparatus' of SSM as these have been honed in action. Next, in the light of these developments the chapter returns to look again at the methodology as a whole, first examining the implications of 'methodology' not being the same as 'method' (something almost completely ignored in the secondary literature), then re-examining the question of what must be done if a claim to be 'using SSM' is to be sustainable. Finally, what happens as the methodology is internalized and use moves from naïvety to sophistication is discussed. The last section then relates SSM to its larger context of social inquiry and research.

This structure is itself a small reflection of the cyclic learning which has always been involved in the development of SSM, and will no doubt continue.

Very many people have been involved in the development of SSM, and the acknowledgements made in *Systems Thinking, Systems Practice* (1981), *Soft Systems Methodology in Action* (1990) and *Information, Systems and Information Systems* (1998) are reiterated, with renewed thanks. For discussions relevant to the new material in this chapter I am greatly indebted to a particular group of people. *Primus inter pares* is Sue Holwell, research collaborator *par excellence*, with whom I have had a rich ongoing conversation over 10 years. We quickly learned always to have a notebook handy, and I have a collection of notes on our deliberations with headings like: 'M6, Feb 92' or 'Bullet train to Osaka, July 94'. Others in this special group include academic colleagues Nimal Jayaratna and Mark Winter, together with an exceptional group of reflective practitioners: Steve Clarke, Mike Haynes, Kees van der Heiden, Luc Hobeke, Jaap Leemhuis, John Poulter, Peter Wood.

Finally, for their ever-willing professional help in the production of this chapter, I am very grateful to Jenny Seddon and Martin Lister.

Peter Checkland
Lancaster, April 1999

INTRODUCTION

Although the history of thought reveals a number of holistic thinkers—Aristotle, Marx, Husserl among them—it was only in the 1950s that any version of holistic thinking became institutionalized. The kind of holistic thinking which then came to the fore, and was the concern of a newly created organization, was that which makes explicit use of the concept of 'system', and today it is 'systems thinking' in its various forms which would be taken to be the very paradigm of thinking holistically. In 1954, as recounted in Chapter 3 of *Systems Thinking, Systems Practice*, only one kind of systems thinking was on the table: the development of a mathematically expressed general theory of systems. It was supposed that this would provide a meta-level language and theory in which the problems of many different disciplines could be expressed and solved; and it was hoped that doing this would help to promote the unity of science.

These were the aspirations of the pioneers, but looking back from 1999 we can see that the project has not succeeded. The literature contains very little of the kind of outcomes anticipated by the founders of the Society for General Systems Research; and scholars in the many subject areas to which a holistic approach is relevant have been understandably reluctant to see their pet subject as simply one more example of some broader 'general system'!

But the fact that general systems theory (GST) has failed in its application does not mean that systems thinking itself has failed. It has in fact flourished in several different ways which were not anticipated in 1954. There has been development of systems ideas as such, development of the use of systems ideas in particular subject areas, and combinations of the two. The development in the 1970s by Maturana and Varela (1980) of the concept of a system whose elements generate the system itself provided a way of capturing the essence of an autonomous living system without resorting to use of an observer's notions of 'purpose', 'goal', 'information processing' or 'function'. (This contrasts with the theory in Miller's *Living Systems* (1978), which provides a general model of a living entity expressed in the language of an observer, so that what makes the entity autonomous is not central to the theory.) This provides a good example of the further development of systems ideas as such. The rethinking, by Chorley and Kennedy (1971), of physical geography as the study of the dynamics of systems of four kinds, is an example of the use of systems thinking to illuminate a particular subject area.

The two books to which this chapter is an adjunct provide an example of the third kind of development: a combination of the two illustrated above. We set out to see if systems ideas could help us to tackle the messy problems of 'management', broadly defined.

In trying to do this we found ourselves having to develop some new systems concepts as a response to the complexity of the everyday problem situations we encountered, the kind of situations which we all have to deal with in both our professional and our private lives. The aim in the research process we adopted was to make neither the ideas nor the practical experience dominant. Rather the intention was to allow the tentative ideas to inform the practice which then became the source of enriched ideas—and so on, round a learning cycle. This is the action research cycle whose emergence is described in *Systems Thinking, Systems Practice* and whose use and further development is the subject of *SSM in Action*.

The action research programme at Lancaster University was initiated by the late Gwilym Jenkins, first Professor of Systems at a British university, and Philip Youle, the perspicacious manager in ICI who saw the need for the kind of collaboration between universities and outside organizations which the action research programme required. Thirty years later that programme still continues, and with the same aim: to find ways of understanding and coping with the perplexing difficulties of taking action, both individually and in groups, to 'improve' the situations which day-to-day life continuously creates and continually changes. Specifically, the programme explores the value of the powerful bundle of ideas captured in the notion 'system', and they have not been found wanting, though both the ideas themselves and the ways of using them have been extended as a result of the practical experiences.

The progress of the 30 years of research has been chronicled and reflected upon since 1972 in about 100 papers and four books—which will be referred to in the remainder of this chapter by the initials of their titles. The nature of the books is summarized briefly below.

Systems Thinking, Systems Practice (*STSP*) (Checkland 1981) makes sense of systems thinking by seeing it as an attempt to avoid the reductionism of natural science, highly successful though that is when investigating natural phenomena; it describes early experiences of trying to apply 'systems engineering' outside the technical area for which it was developed, the rethinking of 'systems thinking' which early experience made necessary, and sets out the first developed form of SSM as a seven-stage process of inquiry.

Systems: Concepts, Methodologies and Applications (*SCMA*) (Wilson 1984, 2nd Edn 1990) describes the response of a professional control engineer to experiences in the Lancaster programme of action research; less concerned with the human and social aspects of problem situations, it cleaves to the functional logic of engineering and presents an approach which Holwell (1997) argues is best viewed as classic systems engineering with the transforming addition of human activity system modelling.

Soft Systems Methodology in Action (*SSMA*) (Checkland and Scholes 1990) describes the use of a mature SSM in both limited and wide-ranging situations in both public and private sectors; it moves beyond the 'seven-stage' model of the methodology (still useful for teaching purposes and—occasionally—in some real situations) to see it as a sense-making approach, which, once internalized, allows exploration of how people in a specific situation create for themselves the meaning of their world and so act intentionally; the book also initiates a wider discussion of the concept of 'methodology', a discussion which will be extended below.

Information, Systems and Information Systems (*ISIS*) (Checkland and Holwell 1998) stems from the fact that in very many of the Lancaster action research projects the creation of 'information systems' was usually a relevant, and often a core, concern; it attempts some conceptual cleansing of the confused field of IS and IT, treating IS as being centrally concerned with the human act of creating meaning, and relates experiences based on a mature use of SSM to a fundamental conceptualization of the field of IS/IT; it carries forward the discussion of SSM as methodology but less explicitly than will be attempted here.

It is important to understand the nature of these books if the aim of this chapter is itself to be properly understood. The less than impressive but nevertheless sprawling literature of 'management' caters in different ways for several different audiences. There is an apparent insatiable appetite for glib journalistic productions, offering claimed insights for little or no reader effort—*Distribution Management in an Afternoon*: that kind of thing. Such books are more often purchased than actually read. There is also a need for textbooks which systematically display the conventional wisdom of a subject for aspiring students. These need to be updated periodically in new editions. And also, more austerely, there are books which carry the discussion which is the real essence of any developing subject, and try to extend the boundaries of our knowledge. The books described above are of this kind. It is not usually appropriate—as it is with textbooks—to update them in new editions. They are 'of their time'. But it is useful on republication, as here, to offer reflections on the further development of the ideas as new experiences have accumulated since the books were written. That is what is done here for *STSP* and *SSMA*.

A particular structure is adopted. First, the emergence of soft systems thinking is briefly revisited. Then the methodology as a whole is considered, since the way in which it is thought about now is very different from the view of it in the 1970s, when it was a redefined version of systems engineering. This consideration of the methodology as a whole frames reflection on the separate parts which make up the whole (Analyses One, Two, Three; CATWOE; rich pictures; the three Es, etc.). This in turn yields a richer understanding of both the whole and its context. Such a structure, in which an initial consideration of the whole leads to an understanding of the parts, which in turn enables a richer understanding of the whole to be gained, is itself an example of Dilthey's 'hermeneutic circle' (Mueller-Vollmer 1986; Morse 1994, Chs 7 and 8). Here, it is a modest reflection of the same process through which SSM was itself developed, a process which tried to ensure that both whole and parts were continually honed and refined in cycles of action.

THE EMERGENCE OF SOFT SYSTEMS THINKING

The Starting Position

In the culture of the UK the word 'academic' is more often than not used in a pejorative sense. To describe something as 'academic' is usually to condemn it as unrelated to the rough and tumble of practical affairs. This was certainly the outlook of Gwilym Jenkins when he moved to Lancaster University in the mid-1960s to found the first systems department in a UK university. He did not want a department which could be dismissed as 'academic'. He rejected the idea that the name of the department should be Systems Analysis, in favour of a Department of Systems Engineering. 'Analysis is not enough', he used to say heretically. 'Beyond analysis it is important to put something together, to create, to "engineer" something.' Given this attitude it was not surprising that he initiated the programme of action research in real-world organizations outside the university. The intellectual starting point was Optner's concept (1965) that an organization could be taken to be a system with functional sub-systems—concerned with production, marketing, finance, human resources, etc. Jenkins' idea was that the real-world experiences would enable us gradually to build up knowledge of systems of various kinds: production systems, distribution systems, purchasing systems, etc. and that this knowledge would support the better design and operation of such systems in real situations. History did not, however, unfold in this way. Instead, the practical experiences led us to reject the taken-as-given assumption underlying the initial expectation, so taking the thinking in a very different direction. In doing this we had to distinguish between two fundamentally different stances within systems thinking: the two outlooks now known as 'hard' and 'soft' systems thinking.

At the outset, by formulating a research aim to uncover the fundamental characteristics of systems of various kinds, we were making the unquestioned assumption that the world contained such systems. Along with this went a second assumption that such systems could be characterized by naming their objectives. It seems obvious, for example, that 'a production system' will have objectives which can be expressed as: to make product X with a certain quality, at a certain rate, with a certain use of resources, under various constraints (budgetary, legal, environmental, etc.). Given such an explicit definition of an objective, then a system can in principle be 'engineered' to achieve that end. This is the stance of classic systems engineering (as described in Chapter 5 of *STSP*). This was what constituted 'systems thinking' at the time our research started, and its origins, as far as application to organizations goes, lie in the great contribution to management science made by Herbert Simon in the 1950s and 1960s (Simon 1960, 1977), which propounded the clarifying (but ultimately limited) concept that managing is to be thought of as decision-taking in pursuit of goals or objectives.

The Learning Experience

We found that although we were armed with the methodology of systems engineering and were eager to use its techniques to help engineer real-world systems to achieve their objectives, the management situations we worked in were always too complex for straightforward application of the systems engineering approach. The difficulty of answering such apparently simple questions as: What is the system we are concerned with? and What are its objectives? was usually a reason why the situation in question had come to be regarded as problematical. We had to accept that in the complexity of human affairs the unequivocal pursuit of objectives which can be taken as given is very much the occasional special case; it is certainly not the norm. A current long-running example of the surprising difficulty in using the language of 'objectives' in human affairs is provided by the arguments which wax and wane over the Common Agricultural Policy (CAP) of the EEC. The Treaty of Rome boldly declares that the CAP has three equally important objectives: to increase productivity in the agricultural industry; to safeguard jobs in the industry; and to provide the best possible service to the consumer. No wonder the CAP is a constant source of never resolved issues: progress towards any one of its (equally important) objectives will be at the expense of the other two! This is typical of the complexity we meet in human affairs as soon as we move out of the more straight-forward area in which problems can be technically defined: e.g. 'increase as much as possible the productivity of this phthalic anhydride plant', or 'make a device to produce radio waves with a 10 cm wavelength'. (If you insisted on using the language of 'objectives', you would have to conclude that the objective of the CAP is constantly to maintain and adjust a balance between the three incompatible objectives which is politically acceptable—which is not a very useful definition for 'engineering' purposes.)

It was having to abandon the classic systems engineering methodology which caused us to undertake the fundamental thinking described in Chapters 2–4 of *STSP*. And it was this rethink which led ultimately to the distinction between 'hard' and 'soft' systems thinking.

Four Key Thoughts

The process of learning by relating experience to ideas is always both rich and confusing. But as long as the interaction between the rhetoric and the experienced 'reality' is the subject of conscious and continual reflection, there is a good chance of recognizing and pinning down the learning which has occurred. Looking back at the development of SSM with this kind of reflective hindsight, it is possible to find four key thoughts which dictated the overall shape of the development of SSM and the direction it took (Checkland 1995).

Firstly, in getting away from thinking in terms of some real-world systems in need of repair or improvement, we began to focus on the fact that, at a higher level, every situation in which we undertook action research was a human situation in which people were attempting to take purposeful action which was meaningful for them. Occasionally, that purposeful action might be the pursuit of a well-defined objective, so that this broader concept *included* goal seeking but was not restricted to it. This led to the idea of modelling purposeful 'human activity systems' as sets of linked activities which together could exhibit the emergent property of purposefulness. Ways of building such models were developed.

Secondly, as you begin to work with the idea of modelling purposeful activity—in order to explore real-world action—it quickly becomes obvious that many interpretations of any declared 'purpose' are possible. Before modelling can begin choices have to be made and declared. Thus, given the complexity of any situation in human affairs, there will be a huge number of human activity system models which *could* be built; so the first choice to be made is of which ones are likely to be most relevant (or insightful) in exploring the situation. That choice made, it is then necessary to decide for each selected purposeful activity the perspective or viewpoint from which the model will be built, the *Weltanschauung* upon which it is based. Thus when David Farrah, a director of the then British Aircraft Corporation asked us to use our systems engineering approach to see how the Concorde project might be improved, possible relevant systems might have included 'a system to manage relations with the British Government' (since they were funding it) or 'a system to sustain a European precision engineering industry' (since Concorde would help to stimulate such activity). Thinking like systems engineers at the time (What is the system? What are its objectives?) Dave Thomas and I in fact proceeded only with the most basic and obvious of possible choices: 'a system to carry out the project'. Neither did the second choice give us pause: how would we conceptualize that project? Again, with our systems engineering blinkers firmly in place, it did not occur to us to think of it as anything other than an engineering project. But given its origins, at a time when President de Gaulle of France was vetoing British entry into the European Common Market, a defensible alternative world-view would be to treat it as a *political* project. On the day the Concorde project agreement was signed the British Government let it be known that it expected Britain to join the European Community within a year, while de Gaulle a few weeks later told a press conference that it was probable that negotiations for British entry might not succeed; in fact he made the supersonic aircraft project a touchstone of Britain's sincerity in applying for membership (Wilson 1973, pp. 31–32). So a model of the project based on a political world-view might be as useful as—or perhaps more useful than—the more obvious one based on a technical world-view.

The learning here was that in making the idea of modelling purposeful activity a usable concept, we had to accept that it was necessary to declare both a world-view which made a chosen model relevant, and a world-view which would then determine the model content. Equally, because interpretations of purpose will always be many and various, there would always be a number of models in play, never simply one model purporting to describe 'what is the case'.

This moved us a good way away from classic systems engineering, and the next key thought in understanding our experience recognized this. It was the thought which can now be seen to have established the shape of SSM as an inquiring process. And that in turn established the 'hard/soft' distinction in systems thinking, though that too was not immediately recognized at the time.

We had moved away from working with the idea of an 'obvious' problem which required solution, to that of working with the idea of a *situation* which some people, for various reasons, may regard as problematical. We had developed the idea of building models of concepts of purposeful activity which seemed relevant to making progress in tackling the problem situation. Next, since there would always be many possible models it seemed obvious that the best way to proceed would be to make an initial handful of models and—conscious of them as embodying only pure ideas of purposeful activity rather than being descriptions of parts of the real world—to use them as a source of questions to ask of the real situation. SSM was thus inevitably emerging as an organized *learning system*. And since the initial choice of the first handful of models, when used to question the real situation, led to new knowledge and insights concerning the problem situation, this leading to further ideas for relevant models, it was clear that the learning process was in principle ongoing. What would bring it to an end, and lead to action being taken, was the development of an accommodation among people in the situation that a certain course of action was both desirable in terms of this analysis and feasible for these people with their particular history, relationships, culture and aspirations.

SSM thus gradually took the form shown in Figure 1.3 of *SSMA* (p. 7), repeated with some embellishment here as Figure A1. This was the form of representation of SSM which eventually took hold, and is the one now normally used. The initial version of it was the 'seven-stage model' which is shown in Figure 6 in *STSP*, p. 163 and Figure 2.5 in *SSMA*, p. 27. This version, though still often used for initial

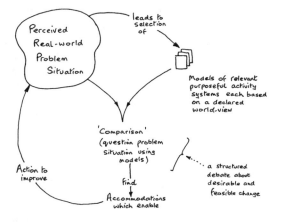

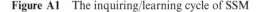

Principles

o real world : a complexity of relationships

o relationships explored via models of purposeful activity based on explicit world-views

o inquiry structured by questioning perceived situation using the models as a source of questions

o 'action to improve' based on finding accommodations (versions of the situation which conflicting interests can live with)

o inquiry in principle never-ending; best conducted with wide range of interested parties; give the process away to people in the situation

Figure A1 The inquiring/learning cycle of SSM

teaching purposes, has a rather mechanistic flavour and can give the false impression that SSM is a prescriptive process which has to be followed systematically, hence its fall from favour.

These three key thoughts capture succinctly the learning which accumulated with experience of using SSM, and they make sense of its development. The fourth such thought, that models of purposeful activity can provide an entry to work on information systems (which are less than ideal in virtually every real-world situation) is not our concern here, this aspect of SSM's use being the detailed subject of *ISIS*.

Hard and Soft Systems Thinking

Our final concern in this section is the major thought which came from these parti-cular experiences of relating systems thinking to systems practice: the 'hard'–'soft' distinction. This was first sharply expressed in a paper written two years after the publication of *STSP* in 1981 (Checkland 1983). It took some time for this idea to sink in!

In systems engineering (and also similar approaches based on the same fundamental ideas, such as RAND Corporation systems analysis and classic OR) the word 'system' is used simply as a label for something taken to exist in the world outside ourselves. The taken-as-given assumption is that the world can be taken to be a set of interacting systems, some of which do not work very well and can be engineered to work better. In the thinking embodied in SSM the taken-as-given assumptions are quite different. The world is taken to be very complex, problematical, mysterious. However, our coping with it, the process of inquiry into it, it is assumed, can itself be organized as a learning *system*. Thus the use of the word 'system' is no longer applied to the world, it is instead applied to the process of our dealing with the world. It is this shift of systemicity (or 'system-ness') from the world to the process of inquiry into the world which is the crucial intellectual distinction between the two fundamental forms of systems thinking, 'hard' and 'soft'.

In the literature it is often stated that 'hard' systems thinking is appropriate in well-defined technical problems and that 'soft' systems thinking is more appropriate in fuzzy ill-defined situations involving human beings and cultural considerations. This is not untrue, but it does not *define* the difference between 'hard' and 'soft' thinking. The definition stems from how the word 'system' is used, that is from the attribution of systemicity.

Experience shows that this distinction is a slippery concept which many people find it very hard to grasp; or, grasped one week it is gone the next. Probably this is because very deeply embedded within our habits is the way we use the word 'system' in everyday language. In everyday talk we constantly use it as if it were simply a label-word for a part of the world, as when we talk about the legal system, health care systems, the education system, the transport system, etc. even though many of these things named as systems do not in fact exhibit the characteristics associated with the word 'system' when it is used properly. This day-by-day use unconsciously but steadily reinforces the assumptions of the 'hard' systems paradigm; and the speaking habits of a lifetime are hard to break!

As the thinking about SSM gradually evolved, the formulation of this precise definition of 'hard' and 'soft' systems thinking did not arrive in the dramatic way events unfold in adventure stories for children ('With one bound, Jack was free!'). Rather the ultimate definition is the result of our feeling our way to the difference between 'hard' and 'soft', as experience accumulated, via a number of different formulations. These have been spotted and extracted by Holwell (1997, Table 4.2, p. 126) who collects eight different ways of discussing the hard/soft distinction between 1971 and 1990. These begin unpromisingly—judged by today's criteria—by assuming that 'hard' and 'soft' systems (roughly, determinate and indeterminate respectively) exist in the world. The shift in thinking comes between the publication of *STSP* and *SSMA*, its very first explicit appearance being in Checkland (1983), a paper which can now also be seen as part of the developments which have made the phrase 'soft OR' meaningful.

The eventual definition of the hard–soft distinction is succinctly expressed in Figure 2.3 of *SSMA* (p. 23), but this diagram is over-rich for many, and so here it is supplemented by Figure A2, a further attempt to make clear the difference between hard systems thinking and soft systems thinking. Understanding this idea is the crucial step in understanding SSM.

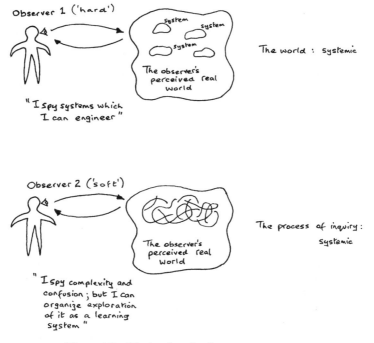

Figure A2   The hard and soft systems stances

SOFT SYSTEMS METHODOLOGY—THE WHOLE

Learning from books or lectures is relatively easy, at least for those with an academic bent, but learning from experience is difficult for everyone. Everyday life develops in all of us trusted intellectual structures which to us seem good enough to make sense of our experiences, and in general we are reluctant to abandon or modify them even when new experience implies that they are shaky. Even professional researchers, who ought to be ready to welcome change in taken-as-given structures of thinking, show the same tendency to distort perceptions of the world rather than change the mental structures we use to give us our bearings. So we were lucky in our research programme that the failure of classic systems engineering in rich 'management' problem situations, broadly defined, was dramatic enough to send us scurrying to examine the adequacy of the systems thinking upon which systems engineering was based. (The early experiences are described in *STSP*, Chapter 7.) But in spite of this it is still the case that the story of our learning is also the story of our gradually managing to shed the blinkered thinking which we started out with as a result of taking classic systems engineering as given.

Holwell (1997) has an appendix to her thesis which collects four different representations of SSM between 1972 and 1990 and correctly suggests that these 'show how the methodology has become less structured and broader as it has developed' (p. 450). It is useful briefly to review this changing perception of the methodology as a whole before moving on to a consideration of its parts.

1972—Blocks and Arrows

The first studies in the research programme were carried out in 1969, and the first account of what became SSM (though that phrase was not used at the time) was published three years later in a paper: 'Towards a systems-based methodology for real-world problem solving' (Checkland 1972). The paper argues the need for methodology 'of practical use in real-world problems' [*sic*](p. 88), reviews the context provided by the systems movement, introduces the case for action research as the research method, describes three projects in detail, refers to six others, and describes the emerging methodology. It finishes with the very important argument that any methodology which will be used by human beings cannot, as methodology, be *proved* to be useful:

> Thus, if a reader tells the author 'I have used your methodology and it works', the author will have to reply 'How do you know that better results might not have been obtained by an *ad hoc* approach?' If the assertion is: 'The methodology does not work' the author may reply, ungraciously but with logic, 'How do you know the poor results were not due simply to your incompetence in using the methodology?' (p. 114)

With reference to human situations, neither of these questions can be answered. Methodology, as such, remains undecidable.

Nearly 30 years later the paper has a somewhat quaint air, though not embarrassingly so. Apart from the reference noted above to 'real-world problems', rather than problem *situations*, the main inadequacy now is in the legacy of hard systems thinking which leads to reference being made to both 'hard systems' and 'soft systems' as existing in the real world; thus we find a few remarks of the kind: 'In soft systems like those of the three studies under discussion . . .' (p. 96). Such statements would not have been made a few years later. Also the methodology is presented as a sequence of stages with iteration back to previous stages, the sequence being: analysis; root definition of relevant systems; conceptualization; comparison and definition of changes; selection of change to implement; design of change and implementation; appraisal.

The focus on implementing *change* rather than introducing or improving *a system* is a signal that the thinking was on the move as a result of these early experiences, even if the straight arrows in the diagrams and the rectangular blocks in some of the models do now cause a little pain!

1981—Seven Stages

By the time the first book about SSM was written (*STSP*, 1981) the engineering-like sequence of the 1972 paper was being presented as a cluster of seven activities in a circular learning process: the 'seven-stage model', versions of which are Figure 6 in *STSP* (p. 163) and Figure 2.5 in *SSMA* (p. 7). In this model the first two stages entail entering the problem situation, finding out about it and expressing its nature. Enough of this has to be done to enable some first choices to be made of relevant activity systems. These are expressed as root definitions in stage three and modelled in stage four. The next stages use the models to structure the further questioning of the situation (the stage five 'comparison') and to seek to define the changes which could improve the situation, the changes meeting the two criteria of 'desirable in principle' and 'feasible to implement' (stage six). Stage seven then takes the action to improve the problem situation, so changing it and enabling the cycle to begin again.

The arrows which link the seven stages simply show the logical structure of the mosaic of actions which make up the overall process; it has always been emphasized that the work done in a real study will not slavishly follow the sequence from stage one to stage seven in a flat-footed or dogged way. Thus, to give one example, the stage five 'comparison' cannot but enhance the finding out about the situation, leading to new ideas for 'more relevant' systems to model. Similarly, the process can take a real-world change being implemented to be an example of stage seven; you can then work backwards to construct the notional 'comparison' which would lead to *this* change being selected, thus teasing out what world-views are being taken as given by people in the situation.

The seven-stage model of SSM has proved resilient, not least because it is easy to understand as a sequence which unfolds logically. This makes it easy to teach, and that too helps explain its resilience. Certainly it has three virtues worth noting before we begin to undermine it in what follows.

Firstly—an intangible, aesthetic point, but an important one—its fried-egg shapes and curved arrows begin to undermine the apparent *certainty* conveyed by straight arrows and rectangular boxes. These are typical of work in science and engineering, and the style conveys the implication: 'this is the case'. The more organic style of the seven-stage model (and of the rich pictures and hand-drawn models in *SSMA*) is meant to indicate that the status of all these artefacts is that they are working models, currently relevant *now* in *this* study, not claiming permanent ontological status. They are also meant to look more human, more natural than the ruled lines and right angles of science and engineering.

Secondly, it is a happy chance that the learning cycle of this model of the process has seven stages. Miller's well-known account of laboratory experiments on perception (1956) suggests that the channel capacity of our brains is such that we can cope with about seven items or concepts at once, hence the title of his famous paper: 'The magical number seven, plus or minus two: some limits on our capacity for processing information'. (He reminds us that there are seven days of the week, seven wonders of the world, seven ages of man, seven levels of hell, seven notes on the musical scale, seven primary colours ...) Irrespective of whether or not seven is truly a crucial number in human culture, the comfortable size of the model of the SSM process does mean that you can easily retain it in your mind. You do not have to look it up in a book, and this is very useful when using it flexibly in practice.

Another feature of the seven-stage model worthy of note is that the stages of forming most definitions and building models from them (stages three and four) were separated from the other stages by a line which separates the 'systems thinking world' below the line from the everyday world of the problem situation above the line. This distinguishing between the everyday world and the systems thinking *about* it was intended to draw attention to the conscious use of systems language in developing the intellectual devices (the activity models) which are consciously used to structure debate. The purpose of the line was essentially heuristic, and its elimination from the 1990 model of SSM will be discussed later in this chapter.

Finally, as far as the 1981 model is concerned, it was important at that stage of development to think about what it was you had to do in a systems study if you wished to claim to be using SSM. This problem was first addressed by Naughton (1977). He was tackling the problem of teaching SSM to Open University students, and for the sake of clarity in teaching, distinguished between 'Constitutive Rules' which had to be obeyed if the SSM claim was to be made, and 'Strategic Rules' which allowed a number of options among which the user could choose. Versions of these rules endorsed in *STSP* are given in Table 6 (p. 253). This was a very useful development in its time, though this is another area which will be further discussed in the light of current thinking.

In summary, formulation of SSM in the 1981 book was at least rich enough to enable it to be taught and used; accounts began to appear of uses of SSM by people other than its early developers. See, for example, Watson and Smith (1988) for an account of 18 studies carried out in Australia between 1977 and 1987.

1988—Two Streams

All of the action research which developed and used SSM was carried out in the spirit of Gwilym Jenkins' remark quoted earlier, that 'Analysis is not enough'. The overall aim in all the projects undertaken was to facilitate action, and it was always apparent that making things happen in real situations is a complex and subtle process, something which will not happen simply because some good ideas have been generated or a sophisticated analysis developed. Ideas are not usually enough to trigger action and that is why industrial companies value highly their 'shakers and movers': they are a much rarer breed than intelligent analysts. So, although a debate structured by questioning perceptions of the real situation by means of purposeful activity models was always insightful, moving on to action entailed broader considerations.

In the very first research in the programme, for example, in the failing textile company described in Checkland and Griffin (1970) and in *STSP* (p. 156), we were brought into the situation by a recently appointed marketing director. He had been brought into the company because the crisis due to falling revenues and disappearing profitability had at last been recognized by a relatively unsophisticated and rather inbred group of managers. This was the first instance in that company's history of appointing a senior manager from outside. The newcomer was thus not part of what had become a closed tribe, and though his previous experience gave him many ideas relevant to improving company performance, his effectiveness was profoundly affected by suspicion of the 'off-comer'. Understanding that, and taking it into account in influencing thinking in the company was crucial to initiating action.

It was thus important always to gain an understanding of the culture of the situations in which our work was done. For some years this was done informally, but—we hoped—with insights from experience, since all the original action researchers developing SSM were ex-managers rather than career academics— who are often naïve about life in unsubsidized organizations.

During those years much reflection went on concerning how we went about 'reading' situations culturally and politically, and it was a significant step forward when SSM was presented as an approach embodying not only a logic-based stream of analysis (via activity models) but also a cultural and political stream which enabled judgements to be made about the accommodations between conflicting interests which might be reachable by the people concerned and which would enable action to be taken. This two-stream model of SSM (*SSMA*, Figure 2.6, p. 29) was first expounded at a plenary session of the Annual Meeting of the International Society for General Systems Research in 1987, and was published the following year (Checkland 1988).

This version of SSM as a whole recognizes the crucially important role of history in human affairs. It is their history which determines, for a given group of people, both what will be noticed as significant and how what is noticed will be judged. It reminds us that in working in real situations we are dealing with something which is both perceived differently by different people and is continually changing.

Also, it is worth noting that this particular expression of SSM as a whole omits the dividing line between the world of the problem situation and the systems thinking world. It had served its heuristic purpose.

1990—Four Main Activities

Published in 1981, *STSP* covered broadly the first decade of development of SSM. The seven-stage model gave a version of the approach which was by then sufficiently well founded to be applied in new real-world situations, large and small, in both the public and the private sector. That was what happened during the second decade of development, some of those experiences being described in *SSMA*. They cover action research in different organizational settings (industry, the Civil Service, the NHS) and include involvements which took from a few hours (ICL, Chapter 6, pp. 164–171) to more than a year (Shell, Chapter 9).

When it came to expressing the shape of the methodology in the 1990 book, the seven-stage model was no longer felt able to capture the now more flexible use of SSM; and even the two-streams model was felt to carry a more formal air than mature practice was now suggesting characterized SSM use, at least by those who had internalized it. The version presented was the four-activities model (*SSMA*, Figure 1.3, p. 7) of which Figure A1 in this chapter is a contemporary form. This is iconic rather than descriptive, and subsumes the cultural stream of analysis in the four activities, which it implies rather than declares.

The four activities are, however, capable of sharp definition:

1. Finding out about a problem situation, including culturally/politically;
2. Formulating some relevant purposeful activity models;
3. Debating the situation, using the models, seeking from that debate both
   (a) changes which would improve the situation and are regarded as both desirable and (culturally) feasible, *and*
   (b) the accommodations between conflicting interests which will enable action-to-improve to be taken;
4. Taking action in the situation to bring about improvement.

((a) and (b) of course are intimately connected and will gradually create each other.)

A decade after *SSMA* was published this iconic model of SSM is still relevant. Why that is so will be discussed when we return to discussing the methodology as a whole. But first it is useful to review the evolving thinking about the parts which make up the whole.

SOFT SYSTEMS METHODOLOGY—THE PARTS

The gradual change in the way SSM as a whole has been thought about, described above, has been paralleled by more substantive changes to some of the separate parts which make up the whole. Many of these represent conscious attempts to improve and enrich such things as model building, or the uses to which rich pictures are put; some have entailed dropping earlier ways of doing things, for example the shift away from using 'structure/process/climate' as a framework for initial finding out about a situation (*STSP*, pp. 163, 164, 166), or the deliberate dropping of the 'formal system model' (*STSP*, Figure 9; *SSMA* pp. 41, 42). But whether the changes to the parts were additions or deletions, they were never made by sitting at desks being 'academic'. They have always been made as a result of experiences in using the approach in a complex world, and they have played their part in changing perceptions of SSM as a whole. This section will review the changes to the parts of SSM, the review being structured by the four activities which underpin the mature icon for SSM which is Figure A1 here.

Finding Out about a Problem Situation

Rich Picture Building

Making drawings to indicate the many elements in any human situation is something which has characterized SSM from the start. Its rationale lies in the fact that the complexity of human affairs is always a complexity of multiple interacting relationships; and pictures are a better medium than linear prose for expressing relationships. Pictures can be taken in as a whole and help to encourage holistic rather than reductionist thinking about a situation.

Producing such graphics is very natural for some people, very difficult for others. If it does not come naturally to you, it is a skill worth cultivating, but experience suggests that its *formalization* via use of ready-made fragments, such as is advocated by Waring (1989) is not usually a good idea, except perhaps as a way of making a start. Users need to develop skill in making 'rich pictures' in ways they are comfortable with, ways which are as natural as possible for them as individuals.

As far as use of such pictures is concerned, we have found them invaluable as an item which can be tabled as the starting point of exploratory discussion with people in a problem situation. In doing so we are saying, in effect 'This is how we see this situation at present, its main stakeholders and issues. Have we got it right from your perspective?' For example, when researching the subtle relationship between a health authority and one of its acute hospitals a few years ago (during the short-lived experiment with 'contracting' in the NHS) we assembled from a great many semi-structured interviews a somewhat large and complicated picture—though even very elaborate pictures are of course selections. (Bryant (1989) is correct to emphasize that 'Selection of the key features of a situation is a crucial skill in developing a picture' (p. 260).) The picture in question became known as 'the briar patch', since that was the impression it gave at first glance! Nevertheless it was found extremely useful, in a second round of interviews, to talk people through it and ask them for both their comments about things we had got wrong, as they saw it, and for their views on what were the main issues concerning contracting (Duxbury 1994). Their responses not only improved the picture, and hence our holistic view of the situation, but also contributed to our understanding of the social and cultural features of the situation—the subject, in SSM of Analyses One, Two and Three (discussed below).

In recent work in the Health Service a new role for rich-picture-like illustrations has emerged. In December 1997 the Government White Paper *The New NHS* (HMSO 1997) described a new concept of the NHS, which was to exhibit such features as: led from the front line of health care ('primary care' by family doctors and other local services); founded on evidence-based medicine, with national standards and guidelines; and supported by modern information systems. Achieving this, according to the Minister of Health responsible for it, involved 'a demanding ten year programme' of development (p. 5). In 1998 the necessary information strategy to support this vision was published, the two documents being coherently linked (Burns 1998). Together, these two publications represent the best conceptual thinking about the NHS for 20 years, though realizing the vision will be an immense and difficult task for medics who are usually not very interested in thinking deeply about *managing* their work (as opposed to its professional execution) and for an organization in which sophisticated 'informatics' skills are scarce.

The White Paper and the information strategy are documents of 86 and 123 pages respectively; absorbing their message is not an easy task for people as busy as health care professionals and Health Service managers. We have found it exceptionally useful, in work commissioned by the centre of the NHS on the information system implications of the new concept for acute hospitals, to turn these excellent but overwhelming documents into picture form. (The documents themselves, being products of a Government service in which prose rules, contain only a handful of rather unadventurous diagrams.) For the White Paper, Figure A3 gives the basic shape of the concept, while Figure A4 adds much more detail to this simple picture. The information strategy, more complicated at a detailed level than the White Paper,

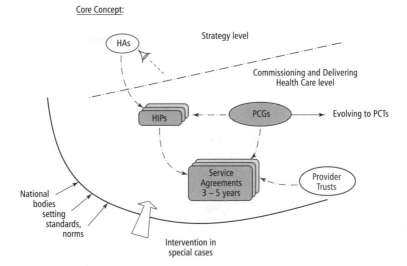

Model of the White Paper – The New NHS – Modern : Dependable

Core Concept:

HAs

Strategy level

Commissioning and Delivering Health Care level

HIPs ← PCGs → Evolving to PCTs

Service Agreements 3 – 5 years ← Provider Trusts

National bodies setting standards, norms

Intervention in special cases

Figure A3   The core concept of the NHS White Paper 1997 (HA = health authority; HIP = health improvement plan; PCG = primary care group; PCT = primary care trust)

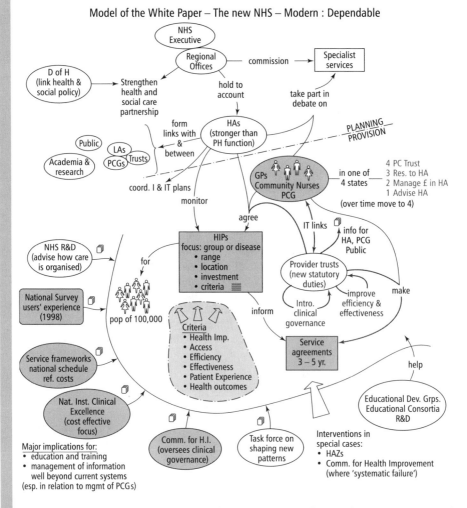

Model of the White Paper – The new NHS – Modern : Dependable

Figure A4   The White Paper concept of the New NHS 1997 (D of H = Department of Health; HAZ = health action zone; HI = health improvement; PH = public health)

was converted into a suite of eight pictures covering its core processes and structures, as well as the intended technical solution: electronic patient records which gradually evolve into each person's lifelong electronic health record. These picture versions of long documents have been very useful in conceptualizing our work, and no NHS audience sees them without asking for copies. This experience does suggest that there is a useful role for pictures of this kind wherever there is detailed written exposition of plans and strategies—at least until the happy times when such documents will themselves use seriously the medium of pictures as well as words.

Figures A3 and A4 can be seen as representations of combined structures and processes which enable the relation between the two elements to be debated. But the use of 'structure', 'process' and 'the relation between them' as a *formal* framework for 'finding out' in SSM, emphasized in the 1972 paper and in *STSP* (pp. 163, 164, 166), has not survived. I believe personally that I still use that framework *mentally*, without giving it much focused thought, but its more formal use, as described in 1972, has fallen into disuse. This seems to be because when you are faced with the energy and confusion which greet you whenever you enter any human problem situation, that particular framework seems highly abstract, a long way away from enabling you to grapple with pressing issues. However, as always with methodology, if it seems useful to you, then use it!

Analyses One, Two and Three

In addition to rich picture building, other frameworks which help to make the grasp of the problem situation as rich as possible are provided by Analyses One, Two and Three (*STSP*, pp. 194–198, 229–233; *SSMA*, pp. 45–53). Analysis One is an examination of the intervention itself, and its development was a direct result of our experience of research for the late Kenneth Wardell, a respected mining engineering consultant in that industry. (He is the 'Mr Cliff' of *STSP* (pp. 194–198).) This analysis is now a deeply embedded part of the thinking. The rich pictures will draw attention to the (usually) many people or groups who could be seen as stakeholders in any human situation, and Analysis One's list of possible, plausible 'problem owners', selected by the 'problem solver', is always a main source of ideas for 'relevant systems' which might usefully be modelled.

The freedom of the person or group intervening in a problem situation to answer the question: 'Who could I/we take the problem owner to be?' is important in achieving a grasp of the situation which is as holistic as possible. Thus in work which helped a community centre in Liverpool to rethink its role in a run-down part of that city, it was relevant to consider Liverpool Social Services Department as one among many possible problem owners, even though at the time the relationship between the centre and the department had not surfaced as an issue for anyone in the department. This kind of choice is what trying to be 'as holistic as possible' entails— even though *the* whole will always remain an unreachable grail. To adopt the counter-view suggested by Bryant (1989) that to be a problem owner you have to be *aware* of owning the problem, would put a completely unnecessary constraint on interventions founded on soft systems thinking.

Analyses Two and Three, comprising a framework for the social and political analyses, are also now thoroughly embedded in praxis. Some commentators have suggested that they are less highly developed than some of the other parts of SSM, such as model building, but that is to misunderstand them. The roles/norms/values framework and the ongoing analysis of 'commodities which embody power' are certainly simply expressed. That is the point of them. You can keep them in your head, and they can constantly guide all of the thinking which goes on throughout an intervention. But though they are simple in expression they reflect one of the main underlying conclusions from the whole 30 years of SSM development: that to make sense of it you have to adopt the view argued in Chapter 8 of *STSP* (pp. 264–285), namely that social reality is no reified entity 'out there', waiting to be investigated. Rather, it is to be seen as continuously socially constructed and reconstructed by individuals and groups (the latter never perfectly coherent). This represents an intellectual stance defined by such features as: deriving from the work of Max Weber; articulated, for example, in the sociology of Alfred Schutz; and underpinned by the philosophy of Edmund Husserl (Luckmann 1978, pp. 7–13). In practical terms, the usable framework which underpins Analyses Two and Three was found in the autopoietic model teased out of the work of Vickers on 'appreciative systems' (Checkland and Casar 1986). That will be discussed further towards the end of this chapter.

Analysis Three moves beyond the model of an appreciative system but is compatible with it. (The appreciative system model describes a social process; Analysis Three covers one of the main determinants of the *outcomes* of that process: the distribution of power in the social situation.) This analysis is avowedly practical, a highly significant contribution to the development of SSM from the action research carried out by Stowell in a light engineering company and in an educational publisher (Stowell 1989). He reviews the extensive social science literature on 'power', but his main aim is not to add to that literature—which is strong on words, less interested in action—but to find practical ways of enabling open discussions to take place on topics which are usually taboo, or emerge only obliquely in the local organizational jokes. These are discussions focused on power, its manifestations and the pattern of its distribution.

Analysis Three is not based on an answer to the question: What is power? It works with the fact that everyone who participates in the life of any social grouping quickly acquires a sense of what you have to do to influence people, to cause things to happen, to stop possible courses of action, to significantly affect the actions the group or members of it take. The metaphor of the 'commodities' which embody power is used to encourage discussion of these matters. Views can be elicited on what you have to possess to *be* powerful in this group or this organization. Is it knowledge, a particular role, skills, charisma, experience, clubbability, impudence, commitment, insouciance ... etc? Recent history of the organization or group can be questioned and/or illustrated in these terms, all with the aim of finding out as deeply as possible how this particular culture 'works', what change might be feasible and what difficulties would attend that change. Stowell (1989) describes the use of the metaphor 'commodity' thus:

'Commodity', then, is the proposed means of providing organisational members involved in change with a practical means of addressing power. Acknowledging, with Giddens, 'that speech and language provide us with useful clues as to how to conceptualise processes of social production and re-production', what has been suggested within this thesis is an idea by which the notion of power can be articulated in terms which are appropriate to a given organizational culture and which can be understood by those most affected (p. 246).

The aspiration of openness here is admirable, but do not be surprised if Analysis Three has to be carried out with great sensitivity and tact. In many human situations there is not the confidence necessary for open discussion of issues hinging on power.

Before moving on from the 'finding out' activity, it is worth reiterating that 'finding out' is never finished; it goes on *throughout* a study, and must never be thought of as a preliminary task which can be completed before modelling starts.

Building Purposeful Activity Models

The Role of Modelling in SSM

The purposeful activity models used in SSM are *devices*—intellectual devices—whose role is to help structure an exploration of the problem situation being addressed. This is not an easy thought to absorb for many people, since the normal connotation of the word 'model', in a culture drenched in scientific and technological thinking, is that it refers to some representation of some part of the world outside ourselves. This is the case, for example, for models as used within classic operational research. If an operational researcher builds a model of a production facility, then there is a need, before experimenting on the model to obtain results which can be used to improve the real-world performance, to first show that the model is a 'valid' representation. This might be done by showing that the model, fed with the last six months' input, can generate something which is close to the actual output produced over that period. But models in SSM are not at all like this. They do not purport to be representations of anything in the real situation. They are accounts of concepts of pure purposeful activity, based on declared world-views, which can be used to stimulate cogent questions in debate about the real situation and the desirable changes to it. They are thus not models *of* ... anything; they are models *relevant to debate* about the situation perceived as problematical. They are simply devices to stimulate, feed and structure that debate.

In the early stages of SSM's development, devices of only one type were built. Blinkered by our starting position in systems engineering, we tended only to make models whose (systems) boundaries corresponded to real-world organizational boundaries. This self-imposed limitation derived, we can now see, from the systems engineers' view that the world consists of interacting systems. Thus, working in, say, a manufacturing company with a conventional functional organization structure, we would make models of a production system, an R&D system, or a marketing system. These would map on to Production, R&D and Marketing departments. But organizations have to carry out, corporately, many more purposeful activities than the handful which can be institutionalized in an organization structure. For example, suppose the manufacturing company in question to be in the petrochemicals business. Such companies, in order to survive in a science-based international business full of very smart competitors, have to be technological innovators. In a systems study carried out in just such a company (the study being concerned with improving relations between R&D and other functions) it was found very useful to make a model based on the core idea of *innovating* in that industry. That model a ... system ... to innovate ... in the petrochemical industry ...) had a boundary which did not coincide with the organizational boundaries of the (functionally defined) existing departments. Not surprisingly, many of the activities in that model were actually taking place in the company: some in R&D, some in Production, some in Marketing. Also, some of the activities in the model were missing in the real situation. The great value of the model was that its boundary *cut across* the organizational boundaries of the actual departments. This was very helpful in stimulating discussion and debate within the company, when the model was used to question the existing situation.

Models which map existing organization structures (such as 'a system to carry out R&D' in this example) are thought of as 'primary task' models; models like that of the innovation system are 'issue-based'—the notional issue here being that some-how or other this particular company has to ensure that it has the ongoing capability to innovate. This primary task/issue-based distinction (*SSMA*, p. 31) has been found to be a source of confusion for many. This is probably because the distinction is not absolute. The petrochemical company, if its thinking had been a little differ-ent, *might* well have brought together people with the appropriate skills and exper-tise to staff an Innovation Department. Had they done so the issue-based model here would then have been a primary task model. Pragmatically, to make sure that the useful provocation provided by models whose boundaries cut across existing orga-nization boundaries is not neglected, the rule from experience is simple: make sure that you do not think only in terms of models which map existing structures. This will help ensure that the modelling fulfils its intended role in SSM: to lift the thinking in the situation out of its normal, unnoticed, comfortable grooves.

Root Definitions, CATWOE and Multi-level Thinking

To build a model of a concept of a complex purposeful activity for use in a study using SSM, you require a clear definition of the purposeful activity to be modelled. These definitional statements, SSM's 'root definitions', are constructed around an expression of a purposeful activity as a transformation process T. *Any* purposeful activity can be expressed in this form, in which an entity, the input to the transform-ing process, is changed into a different state or form, so becoming the output of the process. A bold sparse statement of T could stand as a root definition, for example 'a system to make electric toasters', but this would necessarily yield a very general model. Greater specificity leads to more useful models in most situations, so the T is elaborated by defining the other elements which make up the mnemonic CATWOE, as described in *STSP* (Chapters 6, 7, Appendix 1) and *SSMA* (Chapter 2; illustrated *passim*).

These are not abstruse ideas; the skills required for model building are not arcane: logical thought and an ability to see the wood *and* the trees; also, any model should be built in about 20 minutes. Nevertheless there are classic errors which recur time and time again. The most common error, often found in the literature, is to confuse the input which gets transformed into the output with the resources needed to carry out the transformation process. This conflates two different ideas: input and resources, which coherency requires be kept separate. Also, when people realize that there is a formula (an abstract one) which will always produce a formulation which is at least technically correct, namely: 'need for X' transformed into 'need for X met', they seize on this with glee. Unfortunately, they then often slip into writing down such transformations as 'need for food' transformed into 'food'. What a fortune you could make in the catering industry if you knew how to bring off that remarkable transformation! It is evidently not easy to remember that in a transfor-mation what comes out is the same as what went in, but in a changed (transformed) state.

In recent years experience has shown the value of not only including CATWOE elements in definitions but also casting root definitions in the form: do P by Q in order to contribute to achieving R, which answers the three questions: What to do (P), How to do it (Q) and Why do it (R)? [This formulation was, alas, initially given in terms of XYZ rather than PQR (*SSMA*, p. 36). Using P, Q and R avoids the chance that Y may be confused with why?] The simplest possible definition is of 'a system to do P'. 'Do P by Q' is richer, answering the question: how? And also forcing the model builder to be sure that there is a plausible theory as to why Q is an appropriate means of doing P. For example: 'communicate (P) by letter writing (Q)' is certainly plausible, but would provoke examination of the reasons for doing this communication (i.e. the R question) by this chosen means. In this particular case, the question of required timing would have to be thought about. This could lead to examining, for example, whether there was a case for replacing the cultural resonance which goes with writing a letter to someone by the more brutal but quicker e-mail.

The formal aim of this kind of thinking prior to building the model is to ensure that there is clarity of thought about the purposeful activity which is regarded as relevant to the particular problem situation addressed. The idea of levels, or layers (or 'hierarchy', though that word tends to carry connotations of authoritarianism which are not relevant here) is absolutely fundamental to systems thinking. Much human conversation is dogged by the confusion which follows from the common inability to organize thoughts and expression consciously in several layers. Thus, the Chair of the Tennis Club Social Committee opens a meeting by saying that the committee needs to decide whether or not to organize the club fête this year, given the wet day last July and the unfortunate arrival of a gang of unruly bikers! As you begin to think about this, sitting in committee, you are surprised (but should not be) that the first member to speak says: 'My sister and I will do the cake stall as usual'. Systems thinkers are adept at consciously separating 'whether' from 'what' and 'how'.

In selecting some hopefully relevant systems to model, there are in principle always a number of levels available, and it is necessary to decide for each root definition which level will be that of 'the system', the level at which will sit the T of CATWOE. This makes the next lower level the 'sub-system' level: that of the individual activities which, linked together, meet the requirements of the definition. The next higher level is then defined automatically as that of the 'wider system': the system of which the system defined by T is itself only a sub-system. In SSM this higher level is the level at which a decision to stop the system operating would be taken: it is the level of the system 'owner', i.e. the O of CATWOE. Thus, this intellectual apparatus of T, CATWOE, root definition and PQR, ensures that the thinking being done covers at least three levels, those of system, sub-systems and wider system. It prevents the thinking from being too narrow, and stimulates thoughts about whether or not to build other models. For example it might be decided also to model at the wider-system level, or to expand some of the individual activities in the initial model by making them sources of further root definitions. (Figure 8.14 in *SSMA*, p. 231, for example, shows a model in which activity 1 has been expanded into four more detailed activities. Similar structuring is shown in Figure 5.6, p. 136, and Figure 7.8 of *ISIS*, p. 209 shows a simple model in which most of the activities have been expanded in this way.) Figure A5 summarizes the importance of thinking consciously at several different levels, and also makes the point that different people might well make different judgements about which level to take as that of 'the system'. 'What' and 'how', 'system' and 'sub-system' are relative, not absolute concepts.

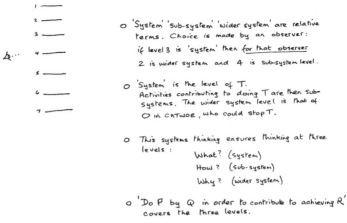

o 'System' 'sub-system' 'wider system' are relative
  terms. Choice is made by an observer:
  if level 3 is 'system' then for that observer
  2 is wider system and 4 is sub-system level.

o 'System' is the level of T.
  Activities contributing to doing T are then sub-
  systems. The wider system level is that of
  O in CATWOE, who could stop T.

o This systems thinking ensures thinking at three
  levels :
           What? (system)
           How ? (sub-system)
           Why ? (wider system)

o 'Do P by Q in order to contribute to achieving R'
  covers the three levels.

But the choice of level is always observer-dependent :

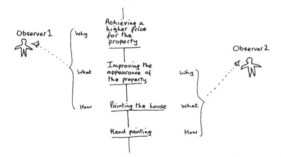

Figure A5   Systems thinking entails thinking in layers defined by an observer

Measures of Performance

It is obvious from the form of SSM (as in Figure A1) that it would be possible to use the approach without creating *systems* models as the devices used to shape the exploration of the situation addressed. It would be possible to use instead models based on theories of, say, aesthetics, psychology, religion, or even, if you were foolish enough to abandon rationality completely, astrology! We use systems models because our focus is on coping with the complexity in everyday life, and that complexity is always, at least in part, a complexity of interacting and overlapping relationships. Systems ideas are intrinsically concerned with relationships, and so systems models seem a sensible choice; and since they have been found, time after time, to lead to insights, they have not been abandoned.

Now, the core systems image is that of the whole entity which can adapt and survive in a changing environment. So our models, to use systems insights, need to be cast in a form which in principle allows the system to adapt in the light of changing circumstances. That is why models of purposeful activity are built as sets of linked activities (an operational system to carry out the T of CATWOE) together with another set of activities which monitor the operational system and take control action if necessary. Since there is no such thing as completely neutral monitoring, it is necessary to define the criteria by which the performance of the system as a whole will be judged. Hence the core structure of the monitoring and control sub-system is always the same: a 'monitor' activity contingent upon definition of the criteria by which system performance will be judged, and an activity rendered as 'take control action' which is contingent upon the monitoring. This can of course be augmented if justified in particular cases—as in the model in *STSP*, p. 291. The basic structure is seen in many of the models in *SSMA* and *ISIS*.

For many years the concept of 'measures of performance' was felt to be sufficient for use in models, but was then enriched by an analysis which flows from the consideration that SSM's models are simply logical machines for carrying out a purposeful transformation process expressed in a root definition. Measuring the performance of a logical machine can be expressed through an instrumental logic which focuses on three issues: checking that the output is produced; checking whether minimum resources are used to obtain it; and checking, at a higher level, that this transformation is worth doing because it makes a contribution to some higher level or longer-term aim. This gives definition of the '3Es' which will be relevant for every model: the criteria of efficacy (E_1), efficiency (E_2), and effectiveness (E_3), first developed in 1987 (Forbes and Checkland 1987; Checkland *et al* 1990; *SSMA*, pp. 38, 39). This core set of criteria can be extended in particular cases—for example by adding E_4 for ethicality (is this transformation morally correct?) and E_5 for elegance (is this an aesthetically pleasing transformation?). Since it will not be possible to name the criteria for effectiveness without thinking about the aspirations of the notional system owner (O in CATWOE), this analysis is another contribution which prevents the modeller's thinking being restricted only to one level, that of the system itself.

Model Building

Given the preliminary thinking expressed in root definition, CATWOE, the three Es and PQR, assembling an activity model ought not to be too difficult: simply a matter of assembling the activities required to obtain the input to T, transform it, and dispose of the output, ensuring that activities required by the other CATWOE elements are also covered; then link the activities according to whether or not they are dependent upon other activities. And the task ought not to be an elaborate one either, given the oft-proved value of the heuristic rule that the overall activity of the operational sub-system should be captured in Miller's (1968) 'magical number' 7 ± 2 individual activities (any of which can if necessary be made the source of a more detailed model). Nevertheless some people manage to make model building a task fraught with difficulty. This is probably because there are in fact subtle features of the process which are masked in the simple account just given.

These subtleties are illustrated by the fact that, for example, the distribution manager of a manufacturing company is probably not the person who will find it very easy to build a model from a root definition of a system to distribute manufactured products. The difficulty for such a person is to focus only on unpacking and displaying the *concept in the root definition*; the tendency will be to slip into describing the real-world arrangements for distributing products in his or her own company. Equally, inexperienced users, fresh-from-school undergraduates especially, find all such models difficult to build because they know so little about real-world arrangements. The fact is—and this is where the unobvious difficulties of modelling lie—it is not usually possible to construct a model *exclusively* on the basis of root definition, CATWOE, three Es and PQR; real-world knowledge *does* inform model building, but, crucially, must not dominate it. The craft skill is to build a model using a background of real-world knowledge without including features of typical practice which are not justified by the root definition, CATWOE, 3Es and PQR. As always with craft skills, practice, practice, practice is the watchword.

Because most practitioners initially 'feel their way' to a method of modelling comfortable for them, it may be helpful to provide some templates which derive purely from the logic of the process and which may provide help for those just starting to use the process of SSM. Two such templates are provided here; they are meant to be abandoned as experience grows.

Figure A6 sets out a logical procedure for modelling purposeful activity systems in a series of steps; Figure A7 expresses the process in Figure A6 as a partial activity model. These are self-explanatory, though it may be remarked with reference to Figure A6 that although the stages can be carried out on a computer screen, there is a good case, as long as you can manage it in good visual style, for producing the final model in hand-drawn form. The reason for this is psychological, and is the

Given : definition of T , $E_{1,2,3}$, CATWOE, Root Definition (PQR)

(1) Using verbs in the imperative ('obtain raw material X') write down activities necessary to carry out T (obtain I , transform it, dispose of Output). Aim for 7 ± 2 activities.

(2) Select activities which could be done at once (ie not dependent on others) :

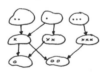

(3) Write these out on a line , then those dependent on these first activities on a line below ; continue in this fashion until all activities are accounted for.

Indicate the dependencies:

(4) Redraw to avoid overlapping arrows where possible and add monitoring and control

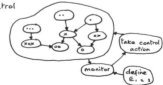

Figure A6   A logical procedure for building activity models

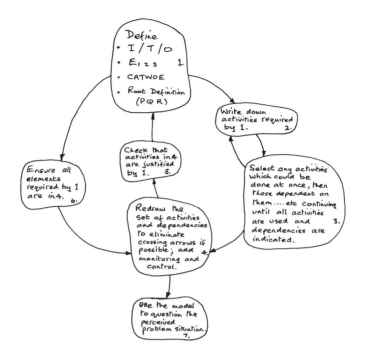

Figure A7 The process of modelling in SSM, embodying the logic of Figure A6

same as that for drawing egg or cloud shapes rather than rectangular boxes: it acknowledges the models' role as pragmatic devices, not definitive once-and-for-all statements. In Figure A7 the process form emphasizes the exercise of judgement during modelling. Iteration around activities 2, 3, 4 continues until it is felt that the minimum but necessary cluster of activities has been assembled; the wider iterations around activities 1 to 6, and around 1–6–4–5 represent the checks that the model is *defensible* in relation to the concept being expressed.

Once a model is constructed by such a process, the golden rule for 'reading' a model—something which the many people unconsciously straitjacketed in linear thinking find difficult—is always to start from the activities which are not dependent upon other activities but have others dependent upon them, i.e. those which have arrows from them but none to them.

Finally, on modelling, a few remarks about the formal system model are in order (*STSP*, pp. 173–177; *SSMA*, pp. 41–42). As formulated in Figure 9 of *STSP* (p. 175) this was useful when we were acquiring a sense of what is meant to treat purposeful activity seriously as a systems concept. In *SSMA* it is said that it can now be 'cheerfully dropped' (p. 92). Its language was the problem. Since it was built using concepts such as boundary, sub-systems, decision-takers, resources, etc. the unfortunate effect of its use was to reinforce the wrong impression that the devices called 'human activity systems' are in some way to be thought of as would-be descriptions of real-world purposeful action. Since that is a main source of misunderstanding about SSM, and since what it offers conceptually is captured in CATWOE, the 3Es and PQR, it can indeed be 'cheerfully dropped'. The same argument speaks against the phrase 'human activity system', but that is probably too deeply embedded to be prised out of SSM and ditched. The best antidote to these dangerous phrases is undoubtedly to encourage the use of Arthur Koestler's neologism for the abstract concept of a whole, namely 'holon' (Checkland 1988). That is what models in SSM are: holons for use in structuring debate.

Exploring the Situation and Taking Action

As human beings experience the unrolling flux of happenings and thoughts which make up day-to-day life, both professional and private, they are all the time likely to see parts of that flux as 'situations', and certain features of it as 'problems', or 'issues'. These concepts and this kind of language—of 'situations', 'issues', 'problems'—are very commonly used in everyday talk, but they are subtle concepts, and we need to beware of giving them a status they do not deserve. We must not reify them; they do not exist 'out there', beyond ourselves, as we can assume 'that beech tree' and 'that dog scratching itself' do. 'This situation', and 'this problem' indicate dispositions to think about (parts of) the flux in particular ways, and they are themselves *generated by human beings*; also, no two people will see them in *exactly* the same way. If, for example, the senior managers of a company all agree in discussion that they have a problem due to the failure of a new product to build up sales following its launch, no two of them will have precisely the same perception of this situation and/or this problem. What is more, some among those who 'agree' about the situation/problem may privately be seeking to ensure the failure of the new product in order that more resources can then come their way! (Remember, we can never know for sure what is going on inside the head of another person; and we cannot assume that their words necessarily reveal it.)

These are bleak thoughts, but necessary ones if applied social science is to be pursued with adequate intellectual rigour. They mean that neither problem situations nor problem types can be classified and made the basis of pigeon holes into which particular examples may be slotted, for one person's 'major issue' or 'serious problem' may well be another's unruffled normality. Both the existence of a problem situation and its interpretation are human judgements, and human beings are not like-thinking automata.

A result of this is that the later stages of a study using SSM cannot be pinned down and as sharply defined as the early stages, in which a situation can be tentatively defined and explored, plausible 'problem owners' named, 'relevant' systems selected and models built. The many uses of SSM described in *STSP, SSMA, SCMA* and *ISIS*, as well as the many accounts in the literature from people outside the Lancaster group, reveal the variety which is possible. This ranges from quick, short, tactical studies to much longer ones oriented to strategy. Because of this, comments on the later parts of SSM are bound to be generalizations from experiences which are very diverse, those generalizations being themselves subject to change as the flux of experience rolls on. Nevertheless a few very broad generalizations from a rich mass of experience can be entertained.

The initial ways of using models, described in *STSP* (pp. 177–180) and *SSMA* (pp. 42–44 and *passim* in the cases described), are still the most common way of initiating the 'comparison' stage of SSM, in which well-structured debate about possible change is sought. Most common, at least as an *initiator* of debate, is the completion of charts in which questions derived from the models brought to the debate are answered from perceptions of current reality on the part of people in the situation addressed. But do not expect the debate to be tidy or predictable; be deft, light on your feet, ready to follow where the debate leads, unready to follow any dogmatic line.

Looking back over experiences in the last decade, an emerging pattern can be discerned in which there are two common foci of the later stages of SSM, during which the driving principle is to bring the study to some sort of conclusion. The first of these is the original one: SSM as an action-oriented approach, seeking the accommodations which enable 'action to improve' to be taken. This is exemplified in the work in 'Index Publishing and Printing Company', described in *STSP*, pp. 183–189. Here action was taken to improve the working relationship between publishers and printers, who represent two very different cultures. A new process to deal with issues surrounding the decision 'where to print' was established, and a new unit to carry out the work was set up. The second focus, very prevalent in the great complexity which characterizes the public sector, is on SSM as a sense-making approach. This is exemplified in recent work in the NHS, and is discussed below.

In the first (action-oriented) case the change sought can usefully be thought about in terms of structural change, process change and changes of outlook or attitude. Normally in human affairs any explicitly organized change will entail all three, and the relationship and interactions between the three need careful thought. Of course the easy option to take—in the public sector for Government or, in other organizations, for senior managers—is to impose *structural* change; and that is often done without serious attention to the other two dimensions: process and attitude. The long series of changes imposed by the UK Government on the Health Service, for example, give a good illustration of imposed *structural* change with relatively little attention to the process and attitudinal change also required (Ham 1992; Rivett 1998; Webster 1998). [It has been significant, recently, that an experienced commentator on the Health Service, Chris Ham, has detected that 'the obsession with structural change that has dominated health policy in recent years has given way to a focus on how staff and services can be developed...' (Ham, 1996). That is a much-needed change.]

In general, thinking about desirable and feasible change can initially be structured in the way shown in Figure A8. A most important feature of this is the need in human affairs to think not only about the substance of the intended change itself but also about the additional things you normally have to do in human situations to *enable* change to occur. (In introducing a clinical information system in a big acute hospital, for example, a project described in *ISIS* (pp. 192–198), its instigator Peter Wood, Chairman of the District Health Authority had already spent several years preparing the ground with hospital consultants inevitably suspicious that such systems could lead to greater control of their clinical activity by hospital managers.)

The second broad category of use to which SSM-style activity models can be put is to use them to make sense of complex situations (though that sense making may of course also lead on to action being taken). It is significant that this category of use has grown markedly in the last decade of SSM development, as concepts such as 'organization', 'function', 'profession' and 'career' have all become more fluid.

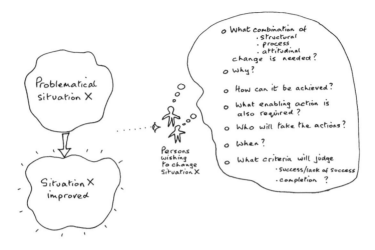

Figure A8   Thinking about desirable and feasible change

Sense-making use of models is well illustrated in recent research in the NHS. The work has been described in some detail in Checkland (1997) and in *ISIS*, pp. 165–172 and will only be sketched in here. Setting out to research the new 'contract'-based relationship between purchasers of health care for a given population and providers of that care (such as acute hospitals, for example) a research team from Lancaster University Management School, using SSM, first built an activity model of the contracting process (Figure 6.2 in *ISIS*). The concept expressed in this model did not rely at all upon observation of the NHS. It reflected simply the interests of the multi-disciplinary research team: information support, organization change, etc. This model was used as a source of structure for open-ended interviews with more than 60 NHS professionals. This produced a daunting mass of interview material. This was analysed by extracting from it the nouns and verbs used by NHS professionals in describing the contracting process and their expectations of it during its first year. These nouns and verbs were fashioned into the elements of an activity model, and these elements were combined to make an activity model relevant to the contracting process as it was initially being interpreted by both purchasers and providers of health care. (This 'backwards' modelling—not based on a root definition but teased out of the interview material—represented an innovation within SSM. See *ISIS*, pp. 165–172.) The difference between the first model (based on the researchers' world-views) and the second one (based on the world-views of NHS professionals) defined the learning achieved in this first phase of the research. This led to 10 pieces of action research in the NHS, and eventually to another sense-making model which helped to unpack and illuminate the purchaser–provider relationship.

This second sense-making model sought to flesh out coherently the complex interactions between a particular purchaser (a health authority) and a particular provider (an acute hospital), interactions to which we had had access over a two-year period. In order to find our way to a model which would richly express all we had observed, 47 previous models relevant to NHS purchasing/providing were first examined (Duxbury 1994). (These came from earlier SSM-based work in the NHS.) This established what language had been found relevant to describing purchasing or providing. This language, together with the recorded observations of what had happened in the present experience between the collaborating hospital and their local health authority, yielded an activity model which makes sense of all that had been observed. The derivation of the model [Figure 8.4 in the book about research in the NHS edited by Flynn and Williams (1997)] was a subtle process. The guide to that process was the question: What activity model could generate all the happenings observed over the two-year period? Its role was to provide a coherent frame for the 10 further pieces of action research at NHS sites.

This completes the necessarily tentative discussion of the variegated later stages of SSM-based studies or projects. Enough has been said to illustrate that the just described sense-making use of activity models calls for rather more than a slavish adherence to the apparently prescriptive seven-stage model of SSM! It also illustrates the fact that the role of methodology, properly interpreted as a set of guiding principles, is not to produce 'answers': that it can never do on its own; it is to enable you, the user, to produce better outcomes than you could without it.

This examination of the parts of SSM is now complete, and we can return to a re-examination of the methodology as a whole, a re-examination which we may hope is made richer by this examination of the parts.

SOFT SYSTEMS METHODOLOGY—THE WHOLE REVISITED

In the earlier section which examined SSM as a whole the focus was on the way in which its representation changed as experience of use accumulated and the different parts of it gradually became more sophisticated. This indicated a shift from the rather biff-bang 'engineering' atmosphere of the 1982 paper to the 'four-activities' model of Figure A1, with its deliberate reticence about the 'hows' and its avoidance of any implication of a prescription to be followed. Having now examined the parts of SSM in their developed form, a re-examination of the whole can try to address the question of what it is which characterizes the approach, making it more than the sum of its parts. This requires an examination of three things: the fundamental notion of *methodology*, as opposed to *method*; the question of what *constitutes* SSM (what you must do if you wish to claim to be guided by it in a particular study); and what happens to SSM when it is internalized in the practice of experienced users—at which point it is apparently a world away from the original formulation in the 1972 paper.

Methodology and Method: the LUMAS Model

The word 'methodology' was originally used to mean 'the science of method', which technically makes the concept of '*a* methodology' meaningless. I remember clearly the day in the early 1970s when my colleague, the late Ron Anderton, said to me: 'You're misusing the word "methodology"; you can't have *a* methodology, the word refers to the whole body of knowledge about method', to which I replied: 'We'll have to change the way the word is used, then.' The deplorable arrogance of that reply stemmed from the fact that I was at that time just becoming aware that, outside the study of social facts, as Durkheim (1895) advocated, the normal scientific method is inadequate as a way of inquiring into human situations; and I was starting to see systems thinking as a holistic reaction against the reductionism of natural science. This meant that the principles of scientific investigation, as used to underpin investigation of natural phenomena, would not adequately support our work. We needed a different methodology, that is a different set of principles. Happily for me, the way that the word 'methodology' is now used has indeed changed, and in the late 1990s Oxford dictionaries of current English now define it not only as 'the science of method' but also as 'a body of methods used in a particular activity' (*Concise Oxford Dictionary of Current English*, 1996). This latter definition makes the crucial distinction between 'methodology' and 'method', and it is the failure to understand this which characterizes much of the secondary literature on SSM.

As the structure of the word indicates, methodology, properly considered, is 'the logos of method', the *principles of method*. When those principles are used to underlie, justify and inform the things which are actually done in response to a particular human problem situation, those actions are at a different level from the overarching principles. Methodology in that situation leads to 'method', in the form of the specific approach adopted, the specific things the methodology user chooses to do in that particular situation. If the user is competent then it will be possible to relate the approach adopted, the specific 'method', to the general framework which is the methodology. And if the methodological principles are well thought out and clearly expressed, then a repertoire of regularly used methods which are found to work will emerge over time as experience is gained. (And of course some methods, over time, in some fields of study, acquire the status that they can—if skilfully employed— guarantee a particular result; they become *techniques*. Examples are the simple algebraic technique which enables you to solve any pair of simultaneous equations, or the physical technique which will cause a cricket ball to 'swing' (move sideways) in mid-air as it is bowled, this latter being a rather more difficult technique to master! Given the multiple perceptions which define and characterize human situations, it is extremely unlikely that any of the methods used within a methodology like SSM could become techniques in the sense used here.)

Since methodology is at a meta level with respect to method (i.e. *about* method) this argument means that no generalizations about methodology-in-use can ever be taken seriously. Thus to read commentators who declare that SSM is 'managerialist', or 'radical', or 'conservative', or 'emancipatory', or 'authoritarian' tells you something about the writer—that they have confused methodology with method—but it tells you nothing about SSM. SSM may exhibit any of these characteristics, as *method*, when it is used by particular users in particular situations. In fact whenever a user knowledgeable about a methodology perceives a problem situation, and uses the methodology to try to improve it, the three elements in Figure A9 are intimately linked: user; methodology as words on paper, and situation as perceived by the user. Any analysis of what happens, carried out by an outsider, would have to embrace all three elements and the interactions between them. This would include converting the methodology (as a set of principles) into a specific approach or 'method' which the user felt was appropriate for *this* particular situation at a particular moment in its history. What happens whenever a methodology is

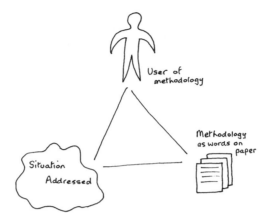

Figure A9 Three interacting elements always present in methodology use

used is shown in the LUMAS model which is Figure A10. Here a user, U, appreciating a methodology M as a coherent set of principles, and perceiving a problem situation S, asks himself (or herself): *What can I do?* He or she then tailors from M a specific approach, A, regarded as appropriate for S, and uses it to improve the situation. This generates learning L, which may both change U and his or her appreciations of the methodology: future versions of all the elements LUMAS may be different as a result of each enactment of the process shown. All the systems studies described in *STSP*, *SSMA* and *ISIS* can be seen as enactments of this process, which accepts that what the user can do depends upon the nexus consisting of U, U's perceptions of M, and U's perceptions of S (Tsouvalis 1995). Never imagine that any methodology can itself lead to 'improvement'. It may, though,

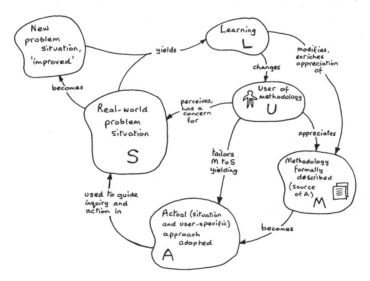

Figure A10 The LUMAS model: Learning for a User by a Methodology-informed Approach to a problem Situation

help *you* to achieve better 'improvement' than you would without its guidelines. But different users tackling the same situation would achieve different outcomes, and an outside observer can form sensible judgements not about M, as if it could be isolated and judged on its own, but about LUMAS as a whole. The model in fact pictures the process by which SSM was developed.

SSM's Constitutive Rules

In the early 1980s Atkinson researched SSM in use. His work included a very detailed examination of three completed systems studies in which different people had made use of SSM as their guiding methodology. He found their uses to show interesting differences. His shorthand summary for the three modes of use he observed were: 'liberal' (eclectic, problem-oriented); 'professional' (SSM as a management consultant's expertise, not necessarily shared with clients); and 'ideological' (the work dominated by an ideological commitment to help cooperatives become more effective). This kind of observation supports the argument developed in the previous section, that methodology use will always be user-dependent. But at the same time that he is noting these differences Atkinson (1984) also observes that the studies all show 'a family resemblance', which raises the questions: What then *is* SSM, the source of this resemblance? and What must a user do if he or she wishes to claim to be 'using SSM'? In *SSMA* the statement is made (p. 58) that

> ... *mouldability* by a *particular user* in a *particular situation* is the point of methodology...

which prompts us to ask what it is that gets shaped into the different forms which different users and different situations evoke.

This question had been addressed before Atkinson did his research, being raised initially by Naughton (1977) in the context of *teaching* SSM. In his commentary on SSM written for Open University students, Naughton argued that there were 'Constitutive Rules' which had to be followed if a claim to be using SSM was to be accepted as valid, and 'Strategic Rules' which 'help one to select among the basic moves'; for example the user might choose (or not) to use the structure/process/climate model in doing the initial exploration of the problem situation. These rules, deriving from the seven-stage model of SSM, were very helpful at the time, and were endorsed in *STSP* (pp. 252, 253). By the time that SSMA was written, however, the seven-stage model was no longer the preferred expression of SSM as a whole, and a new set of constitutive rules were proposed (*SSMA*, pp. 284–289). These defined five characteristics of uses of SSM and set out its epistemology (rich pictures, CATWOE, etc.). A use of SSM was one which could be described using these concepts and language.

In 1997, in the most cogent exegesis of SSM carried out so far, Holwell found these 1990 rules to be at the same time 'both too loose and not extensive enough' (p. 398). They are too loose because they allow people who have done no more than draw a rich picture to claim they are using SSM (the literature contains such examples!). And they are too restrictive, in the sense of being not extensive enough, because they are silent on some basic assumptions which SSM always takes as given. To correct this, Holwell (1997) argues that the answer to the question:

what is SSM? has to be made at three levels: the taken-as-given assumptions; the process of inquiry; and the elements used within that process. She writes:

> ... there are three necessary statements of principle or assumption:
>
> (1) you must accept and act according to the assumption that social reality is socially constructed, continuously;
> (2) you must use explicit intellectual devices consciously to explore, understand and act in the situation in question; and
> (3) you must include in the intellectual devices 'holons' in the form of systems models of purposeful activity built on the basis of declared worldviews.
>
> Then there are the necessary elements of process. The activity models ... are used in a process informed by an understanding of the history of the situation, the cultural, social and political dimensions of it ... (the process being) about learning a way, through discourse and debate, to accommodations in the light of which either 'action to improve' or 'sense making' is possible . Such a process is necessarily cyclical and iterative. Finally, while not limited to this pool ... a selection from Rich Picture, Root Definition, CATWOE ... etc may be used in the process.

These arguments are well made, and this work gives us a solid basis for definitive constitutive rules for SSM. We need rules which are oriented to practice rather than teaching, and which can encompass the wide range of sophistication brought to the use of SSM. At one end of the spectrum is a naïve following of the seven stages in sequence. This is not necessarily wrong, simply something users quickly grow out of as the ideas take root in their thinking. Once internalized, SSM's concepts lead to the deft, light-footed and flexible use which characterizes the other end of the spectrum of sophistication. The two 'ideal types' of SSM use which define the spectrum are termed Mode 1 and Mode 2 in *SSMA* (pp. 280–284). The difference between them is very relevant to the question of SSM as a whole, and is discussed in the next section.

Prescriptive and Internalized SSM: Mode 1 and Mode 2

SSM grew out of the failure of systems engineering—excellent in technically defined problem situations—to cope with the complexities of human affairs, including management situations. As systems engineering failed we were naturally interested in discovering what kind of approach *could* cope with problems of managing. So the research programme which yielded SSM was initially rather methodology-oriented. Then what happened was that as the shape of SSM emerged, as its assumptions became clearer, and its process and elements became firm, so the whole methodology became, for its pioneers, internalized. SSM became the way we thought about coping with complexity in real situations, and the research itself could become more problem-oriented. The process of internalization is a very real one for those for whom it is happening, but it is not an easy process to describe, certainly not as a series of steps recognized at the time they occur, for the steps are often not so recognized. The descriptions of the two ideal types of SSM use in *SSMA* enabled the 10 studies described in the book to be (subjectively) placed relative to each other on the spectrum between Mode 1 and Mode 2 (Figure 10.3 in *SSMA*; see also *ISIS*, pp. 163–172). This implicitly invited the reader to get a feel for what internalizing the methodology means, and to see whether he or she agrees with the placings.

Certain dimensions may be used to differentiate the two ideal types, recognizing that actual studies will never exactly match either of the two idealized concepts, but will reflect elements of both. Such dimensions are:

Mode 1		Mode 2
Methodology-driven	versus	situation-driven
Intervention	versus	interaction
Sometimes sequential	versus	always iterative
SSM an external recipe	versus	SSM an internalized model

and it follows from these that there will never be a generic version of what happens in 'near-Mode 2' studies precisely because they are situation-driven. Perhaps the best approach to understanding internalized SSM in action is through examples. One was given in the previous section (in which an activity model was teased out from the nouns and verbs used by Health Service professionals in talking about the then mandatory contracting process between purchasers and providers of health care). Another is now briefly described.

This example of near-Mode 2 use of SSM occurred at a one-day conference on 'Mergers in the NHS'. This was a topic of interest because the Health Service has seen many mergers in recent years—between district health authorities joining to form bigger purchasers of health care, between hospitals, and more recently between health authorities; and ministers have indicated that more such mergers will occur. In the morning the conference heard a number of talks from people who had been involved in mergers, in industry as well as the NHS, including in the case of the latter, examples from both a health authority and a hospital perspective. After lunch the participants split into small groups for discussion, this to be followed by a final plenary session to summarize the day. The organizers were anxious to avoid the usual problem in such circumstances: small-group discussions generate flip charts containing long unstructured lists of points made, usually covering several different (unstated) levels; and so everyone ends up unable to see any patterns which would help the audience to see and retain important lessons. To do better than this the people chairing the small groups were asked to structure the discussion by following an explicit agenda written out for them. Three of us spent the discussion period touring the various groups, trying to get a feel for the content and tone of the group discussions.

Alas for the well-laid plans, and in spite of the best efforts of those in the chair, what happened was what always happens when health professionals meet on occasions like this: uncontrollable discussion broke out and anecdotes were exchanged! The problem now to be solved during the afternoon tea-break was to prepare for the final plenary presentation and discussion in the absence of the hoped-for coherent responses from the groups. This is where SSM was helpful.

To provide a recognizable context for talk of mergers, a simple model relevant to the Health Service was jotted down, as shown in Figure A11. Here the public (who are occasionally patients of the NHS) both elect a Government and—in the UK— provide resources through direct taxation. Those resources are disbursed via NHS structures so that appropriate configurations of services can be made available to the

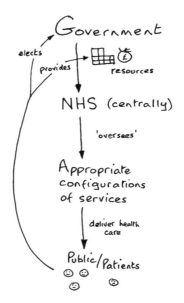

Figure A11   The simple model built to explore mergers in the NHS

public. Talk of 'mergers' can be thought about as talk about changes which will affect those configurations of services, changes which will involve any or all of: Health Authorities, hospitals, community service providers, family doctors and local authorities. The three of us who had spent the small-group discussion period touring the groups now annotated the model with our generic impressions of either the issues which were being discussed, or the issues which underlay the stories being told. These consolidated into five main points, and the final plenary session was opened by my displaying the model of Figure A11 and then adding the five main discussion points, as shown in Figure A12. This served to structure the final discussions. Feedback from delegates about the coherence of the day was good.

I can guarantee that this near-Mode 2 use of SSM was problem-oriented, not methodology-oriented. The fact that we had only the half-hour tea-break to prepare for the final discussion session concentrated the mind. Figures A11 and A12 represent the only explicit output from the work done in the tea-break, but I could retrospectively produce a conventional SSM-style model, together with root definition, CATWOE, E_1, E_2, E_3, etc., which would map Figure A11, as well as an issue-based root definition and model relevant to 'talk of mergers' (a system to decide the structural and service entailments of a configuration of health services considered desirable for population x in area y . . . etc.). None of that work was done at the time— or has been done since, for that matter; the internalization of SSM enabled the practical response to the 'tea-break problem' to be generated. Reconstructing this Mode 2-like use of SSM after the event, we can see that the small-group discussions three of us had dipped into were the source of a holistic impression of the work done in the small groups. We then made sense of that overall impression—for the purpose, on the day, of exposition—by means of the models in Figures A11 and A12.

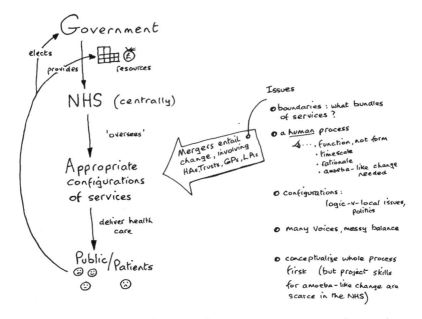

Figure A12   The NHS model annotated to structure presentation of merger issues

It is inevitable that users of SSM will internalize its guidelines and use them in an increasingly sophisticated way. This is akin to learning physical skills: beginners at rock climbing treat each hold as a new problem, appearing clumsy as they make their jerky progress up a rock face. Experienced climbers who have internalized their skill, at whatever level they have attained, put together sequences of moves and appear to 'flow' up a climb. They are likely to believe that you cannot be said to be *truly* rock climbing until this internalization has occurred, and so it is with use of SSM. The more subtle nature of human situations will be revealed to sophisticated users while the novice is struggling to remember what Analysis Two is, and what CATWOE and E_1, E_2, E_3 mean. So the disappearance of near-Mode 1 use is to be welcomed, apart, that is, from the fact that it has its virtues for initial teaching purposes. Just as novice climbers need to be taken up easy climbs, and to have the next hold, and how to use it, pointed out to them, so people coming to SSM for the first time *need* to treat it as a series of stages, each with a definite output, just as Naughton declared in the original constitutive rules.

Finally, though, we cannot advise inexperienced users simply to seek out straight-forward problem situations to tackle, since *all* human situations have their subtleties!

SOFT SYSTEMS METHODOLOGY—THE CONTEXT

Before concluding, two aspects of the context of SSM's development are worth attention, since they have emerged as virtually inseparable from SSM as a way of conducting inquiry in human affairs. The two are: the 'action research' mode; and the assumptions about the nature of the social process which underpin SSM as a whole.

Action Research

The fact that the research which produced SSM started out from a base in systems *engineering* indicates that it was part of the strand of research which concentrates on situations in which people are trying to take action. From the start the researchers tried not simply to observe the action as external watchers but to *take part* in the change process which the action entailed; this made change, and how to achieve it, the object upon which research attention fastened. This puts the research into the 'action research' tradition stemming from Kurt Lewin's views, developed in the 1940s, that real social events could not be studied in a laboratory. This mode of research is discussed in *STSP*, pp. 146–154 and illustrated throughout *SSMA*. Here I shall focus only briefly on what experience and reflection have shown to be an important requirement of this kind of research, a requirement which is, surprisingly, almost completely neglected in the literature of action research. (It is discussed in *ISIS*, pp. 18–28, and Checkland and Holwell 1998a).

The point is this. For findings to be accepted as part of the body of 'scientific knowledge' they have to be repeatable, time and again, by scientists other than those who first discovered them. If you announce that you have discovered the 'inverse square law of magnetism', working in Berlin, then that finding has to be repeatable in Brazil, Barnsley, Brisbane and Bournemouth if the happenings in your experiments are to be accepted as 'scientific knowledge'. Apparent findings in human situations, however, no one of which is ever either static or exactly the same as any other human situation, cannot match this strong criterion. It is the public testability which makes 'scientific knowledge' different from other kinds of knowledge; though do not expect unanimity on any *interpretation* of the findings, since the interpreting is a human act, and can in principle be as various as the people who make the interpretations.

In the human domain, in the province of 'social science', the findings are of a different nature, as are the criteria by which they can be judged. Emile Durkheim (1895), who made up the word 'sociology', suggested that the concern of this new 'science of society' should be 'social facts'. 'Treat social facts as things' is his best-known dictum. Social facts refer to aggregates, and are defined by an observer: for example, the fraction of marriages which end in divorce in a given society, or the rate at which people commit suicide—which was the subject of a famous study by Durkheim himself. But action research in local situations is concerned not with social facts but with study of the myths and meanings which individuals and groups attribute to their world and so make sense of it. This is part of that other great strand within sociology, the interpretive tradition stemming from Max Weber (1904). This is relevant to SSM since the meaning attribution by individuals and groups leads to their forming particular intentions and undertaking particular purposeful action.

The question of the criteria by which findings of this kind can be judged is obviously a tricky one. I have heard sociologists argue that the criterion by which their findings can be judged can be no more than mere plausibility: do these findings make a believable story? But if this weak criterion is accepted there seems to be virtually no difference between writing novels and doing social research. Surely we can do better than that?

In between the strong criterion of repeatability (of the happenings) and the weak criterion of plausibility, we argue (Checkland and Holwell 1998a) that action research should be conducted in such a way that the whole process is subsequently *recoverable* by anyone interested in critically scrutinizing the research. This means declaring explicitly, at the start of the research, the intellectual frameworks and the process of using them which will be used to define what counts as knowledge in this piece of research. By declaring the epistemology of their research process in this way, the researchers make it possible for outsiders to follow the research and see whether they agree or disagree with the findings. If they disagree, well-informed discussion and debate can follow. Also, the learning gained in a piece of organization-based action research may concern any or all of: the area focused on in the research; the methodology used; or the framework of ideas embodied in the methodology. SSM is itself the result of 30 years of this kind of learning in real-world situations.

The Social Process: Appreciative Systems

Once a systems thinker has taken on board the idea of conceptualizing the world and its structures in terms of a series of layers, with any layer being justified by definable emergent properties at that level (see *STSP*, Chapter 3), it is always appropriate to think at more than one level. As discussed earlier, the 'apparatus' of SSM ensures that whatever level is taken by an observer or researcher to be that of 'system', the level above ('wider system') and that below ('sub-system') will always be taken into account, as Figure A5 illustrates. But the systems thinker also accepts that an observer, investigator or researcher will not only select the level which is to be that of 'system' but will also interpret the nature of 'system' according to his or her own *Weltanschauung* or world-view (or, in SSM, deliberately select multiple world-views whose adoption might yield insights into the problem situation). These ideas of 'layers' and 'world-views' mean that developers of SSM could not avoid taking a position on both the nature of the methodology and the higher-level assumptions which it takes as given.

The methodology is taken to be a process of social inquiry which aims to bring about improvement in areas of concern by articulating a learning cycle (based on systems concepts) which can lead to action. This raises the question of what higher-level assumptions about the nature of social reality SSM implicitly takes as given: hence the discussion in Chapter 9 of *STSP*. The conclusion there is that in order to make sense of the research experiences it is necessary to take 'social reality' to be

> the ever-changing outcome of the social process in which human beings, the product of their genetic inheritance and previous experiences, continually negotiate and re-negotiate with others their perceptions and interpretations of the world outside themselves (pp. 283, 284).

This makes SSM in harmony with the sociology of Alfred Schutz and the philosophy of Edmund Husserl; but in practical terms it was Geoffrey Vickers' work on what he calls 'appreciative systems' which mapped most completely our experiences.

Vickers' theoretical work was done in his retirement after 40 years in what he always referred to as 'the world of affairs'. (He was City lawyer, a civil servant, a member of the National Coal Board responsible for management, training and personnel issues, and a member of many public bodies—as well as a young subaltern who won the Victoria Cross on the day after his twenty-first birthday at the Battle of Loos in 1915.) In his retirement he set himself the task of making sense of all his experience, and wrote a series of books in which he developed his account of 'the social process' in terms of his theory of 'appreciation'. SSM's debt to Vickers is recorded in *STSP* (Chapter 8) but more work has been done since then, and is here summarized in the Appendix.

In essence: Vickers discovered systems thinking in his retirement, found it very helpful for sense-making purposes, and was amazed that the greatest use of systems ideas seemed to be made in a technical context, whereas he saw them as richly relevant to 'human systems'. In a taped interview at the Open University in 1982 he said:

> While I was pursuing these thoughts, everyone else who was responding at all was busy with man-made systems for guided missiles and getting to the moon or forcing the most analogic mental activities into forms which would go on digital computers. 'Systems' had become embedded in faculties of technology and the very word had become dehumanized (quoted in Blunden 1984, p. 21).

In his thinking he rejected first the 'goal-seeking' model of human life (the core of Simon's great contribution to management science) and then the cybernetic model because in it the course to which the Steersman steers is a 'given' from outside the system whereas in human affairs the course being followed is continuously generated and regenerated from inside the system. This led him to his notion of 'appreciation' in which, both individually and in groups, we all do the following: selectively perceive our world; make judgements about it, judgements of both fact (what is the case?) and value (is this good or bad, acceptable or unacceptable?); envisage acceptable forms of the many relationships we have to maintain over time; and act to balance those relationships in line with our judgements. [The Appendix contains our model of what Vickers meant by 'an appreciative system' (Checkland and Casar 1986), and links his work to SSM.]

In summary, SSM can be seen as a systemic learning process which articulates the working of 'appreciative systems' in Vickers' sense.

CONCLUSION

The saxophonist John Coltrane was the greatest innovator in the jazz idiom since Charlie Parker reminted the coinage of jazz expression in the mid-1940s. Playing with the Miles Davis Quintet, Coltrane took to playing long long solos which might last for 20 minutes or more. On one occasion at the Apollo in Harlem, when he eventually finished a very lengthy solo he was asked why he had gone on so. He replied 'I couldn't find nothing good to stop on', whereupon Davis said, 'You only have to take the horn out of your mouth.' Authors too face the problem of finding 'something good to stop on', and obviously all they have to do is lift the pen from the page. But that would not satisfy a systems thinker, who would want to effect some kind of closure. Hence this conclusion, which adds some final comments on what has been an enthralling 30-year research experience for this writer.

SSM has been ill-served by its commentators, many of whom demonstrably write on the basis of only a cursory knowledge of the primary literature. However, both life and this chapter are too short to expend time and energy on correcting these nonsenses; but it is probably worth illustrating the size of the problem by recording the spectacular example which Holwell found during her masterly exegesis of the secondary literature (1997, p. 335). It is from a book on information systems published in 1995. The authors refer to *STSP* but—all too typically—do not mention *SSMA*, even though it had been published for five years when they wrote their book.

> This methodology stems from the work of Checkland (1981) who took a radically different approach to the *analysis and design of information human activity systems.* Starting from the premise that *organizations* (and therefore their *subsystem information systems*) are *open systems that interact with their environment,* he *includes the human activity subsystems as part of* his modelling process. The methodology starts by taking a particular view *of the system* and incorporating subjective and objective impressions into a 'rich picture' *of the system* that includes the people involved, the problem areas, sources of conflict and other 'soft' aspects *of* the overall system. A 'root definition' is then formed *about the system* which *proposes improvements* to *the system* to *tackle the problems identified in the rich picture.* Using *the* root definition, *various conceptual models of the new system* can be built, compared and evaluated against the *problems in the rich picture.* A *set of recommendations* is then suggested to deal with the specific changes that are necessary to *solve the problems.*

The italics here are used to highlight fundamental errors: nearly 20 in less than 200 words! Cheerful stoicism seems to be the necessary response to a lack of understanding as profound as this. Pity the poor students.

Although the secondary literature often creates a barrier, it is not the only reason that *teaching* SSM is not straightforward. In teaching such a methodology you are teaching not what to think, but a *way of thinking* which the user can consciously reflect upon. Many people coming to SSM for the first time in a classroom have never before consciously thought about their own thinking, and there is some rearrangement of mental furniture entailed in this which many find strange. Certainly the biggest difficulty in understanding SSM is to absorb its shift from assuming 'systems' exist in the world (as in everyday language) to assuming that the process of inquiry into the world can be a consciously organized learning system.

This is to say that *process thinking* is very unfamiliar for many people, and there is no doubt that teaching a way of thinking is harder than teaching substantive factual material—which is why many MBA courses, which ought to focus on teaching 'how to think in problem situations', instead opt for current factual material about marketing, finance, and other common organizational functions. How strange process thinking is for some people was illustrated recently by a journalist, Matthew Parris, who described in *Literary Review* (December 1998) how much he hated a training week in Brussels to which he was sent as a junior Foreign Office employee. He found a suffocating respect for questions of process combined with a carefree disregard for questions of substance. They kept telling me how a policy was steered into being. I kept wanting to know whether the policy was any good. They looked at me as though I was missing the point.

Of course, he *was* missing the point. A systems thinker would know that the process of policy creation and policy content are entirely complementary, the process itself conditioning what might emerge as content. Both need to be thought about together; but this is not yet a familiar concept.

The other difficulty faced by teachers of SSM is overcoming the shock some people feel when they discover the rigour involved in building purposeful models, thinking out their measures of performance, and so on. (Perhaps there is a tendency for newcomers to equate 'soft' with sloppy or casual: as if anything will do.) But the rigour helps clear the mind, as well as ensuring that the devices which will structure debate are themselves defensible.

In the first heady days of the Gorbachev reforms in the USSR the Institute for Systems Analysis in Moscow wanted to hear about SSM, since the Institute's researchers were intrigued by the idea of undertaking action research projects in Soviet industry. At the end of a week of lectures and seminars, the Director of the Institute, J. M. Gvishiani, said to me that he saw SSM as 'a rigorous approach to the subjective'. This struck me as a very insightful phrase. Both the primacy of the subjective in human affairs *and* the rigour in the thinking about process are important.

Oddly enough, the difficulties of teaching a systemic way of thinking in a classroom disappear when people learn it by using SSM in a real situation. But the situation has to be real for this to happen. There is a huge gap between real uses of SSM and 'pretend' uses on case studies in the classroom. Pretending to invest £10m., or deciding who to make redundant, in a case study, costs you nothing; doing it in real life is a world away from the pretence. But, when the use is real, our experience is that SSM is quickly grasped, and seems 'natural' to those using it. This adds weight to the argument in *SSMA* (p. 300) that the process of SSM reflects the everyday process we all engage in whenever we form sentences and entertain alternative predicates, comparing them with each other and with the perceived world in order to make judgements about action. SSM simply makes a special kind of predicate, in the form of models of purposeful activity, each of which expresses a pure world-view. It is a more organized, more holistic form of what we do when we engage in serious conversation.

But in observing that SSM, in use, seems natural, we need also to remind ourselves that its concern is with would-be purposeful action; and we should never forget how easy it is to overestimate the role of the purposeful in human life. Being able to act with intention, purposefully, is an important part of what makes us human. But it is only a part, and maybe not the most important part.

It has been argued above (and that argument is extended in the Appendix) that SSM can be seen as articulating 'the social process', in the form of what Vickers calls an 'appreciative system'. If, thinking systemically, we ask: what is the level above that of 'the social process'? then we are moving into very abstract realms indeed: in this case into the level at which the concern could be defined as 'being human'. This is two levels above that at which the concern is 'use of SSM', but it provides the ultimate context in which SSM is used.

This suggests several self-admonitory instructions for the user of SSM. We should be rigorous in thinking but circumspect in action. We should remember that many people painfully find their way unconsciously to world-views which enable them to be comfortable in their perceived world. Coming along with a process which challenges world-views and shifts previously taken-as-given assumptions, we should remember that this can hurt. So what right do we have to cause such pain? None at all unless we do it with respect and in the right spirit: no lofty hauteur. And we must remember, feet on ground, that all we can do with our 'natural' but intellectually sophisticated process of inquiry is learn our way to improved purposeful action, which is a ubiquitous part of human life but only a limited part of it, not the whole.

And so, to complete this chapter, let us remind ourselves—using a true story—of what it means to be *fully* human, and end with that image.

In 1993 in south London a black teenager, Stephen Lawrence, was fatally stabbed in a racist attack by a group of white youths. Six years later, with no one found guilty of the murder, Sir William Macpherson delivered to the Home Secretary his report on the incompetence of the criminal investigation, precipitating national soul searching and debate about institutionalized racism in British society. A writer, Richard Norton-Taylor, brilliantly crafted a play—*The Colour of Justice*—from the transcript of the Lawrence tribunal; this was shown on BBC television in February 1999. The production contained one of those moments, exceptionally rare on television, when the viewer is transfixed and transformed. A witness described how he and his wife, returning home from a church meeting, came upon Stephen as he lay bleeding on the pavement. The wife cradled Stephen, as the young man's life ebbed away. Knowing that hearing is the last sense to go, she whispered in his ear 'You are loved'. When he got home, the man washed his bloodied hands into a container and poured the water on to the roots of a favourite rosebush. He said that he supposed that in some way Stephen lived on.

We should never entertain the idea, even for a moment, that a mere 'systems approach', or any 'systems methodology' could ensure that we behave as Louise and Conor Taaffe did on that April night in Eltham in 1993.

APPENDIX: SYSTEMS THEORY AND MANAGEMENT THINKING

Two inquiring systems developed since the 1960s—Vickers's concept of the appreciative system and the soft systems methodology, are highly relevant to the problems of the 21st century. Both assume that organizations ane more than rational goal-seeking machines and address the relationship-maintaining and Gemeinschaft aspects of organizations, characteristically obscured by functionalist and goal-seeking models of organization and management. Appreciative systems theory and soft systems methodology enrich rather than replace these approaches.

Two rich metaphors provide a useful frame within which any consideration of the problems facing us in the late 20th century can, with advantage, be placed. As a result of the first industrial revolution, based on energy, and the current second one, based on information, the world is increasingly Marshall McLuhan's 'global village'. More and more problems need to be examined in a global rather than a local context and, as we do so, we need to remember that we are all of us, in, Buckminster Fuller's great phrase, 'the crew of Spaceship Earth'.

Thanks to the material successes of the two industrial revolutions we are a crew with rising expectations of high living standards. But we are increasingly aware that the wealth-generating machine may not be able to meet those expectations without doing unacceptable damage to Spaceship Earth, which, together with the free supply of energy from our sun, is the only given resource we have.

This triangle—of expectations, wealth generation, and protection of the planet—will have to be managed with great care at many different levels as we enter the 21st century if major disasters are to be avoided. Unfortunately, our current ideas on *management* are rather primitive and are probably not up to the task. They stem from the technologically oriented thinking of the 1960s, and they now need to be enlarged and enriched. This may well be possible from the systems thinking of the 1970s and 1980s, which has placed that body of thought more firmly within the arena of human affairs.

This article will examine the legacy of thinking about management and organizations that we get from the 1960s and develop a richer view that stems from more recent systems thinking, especially Vickers's work on the theory of *appreciative systems* and work on soft systems methodology, which can be seen as a way of making practical use of Vickers's concepts. This, it is argued, is more relevant than the current conventional wisdom to managing the problems of the new century.

Checkland, P. B., AMERICAN BEHAVIORAL SCIENTIST **38**(1) pp. 75–91, copyright © 1994 by Sage Publications, Inc. Reprinted by Permission of Sage Publications, Inc.

MANAGEMENT AND ORGANIZATION

In spite of a huge literature—some of it serious, much of it at the level of airport paperbacks—and courses in colleges and universities worldwide, the role of the manager and the nature of the process of managing remain problematic, whether we are concerned with trying to manage global, institutional, or personal affairs. Anyone who has been a professional manager in an organization knows that it is a complex role, one that engages the whole person. It requires not only the ability to analyse problems and work out rational responses but also, if the mysterious quality of leadership is to be provided, the ability to respond to situations on the basis of feelings and emotion.

One of the reasons the manager's role remains obstinately problematic stems from our less than adequate thinking about the context in which managers perform, namely the organization. Some basic systems thinking indicates that if we adopt a limited view of organization then the conceptualization of the manager's role will inevitably also be rather threadbare. Thus a manager at any level occupies a role within a structure of roles that constitutes an organization. The activity undertaken by managers can be seen as a system of activity that serves and supports and makes its contribution to the overall aims of the organization as a purposeful whole. Now, if one system serves another, it is a basic tenet of systems thinking that the system that serves can be conceptualized only after prior conceptualization of the system served (Checkland 1981, p. 237). This is so because the form of a serving system, if it is truly to serve, will be dictated by the nature of the system served: That will dictate the necessary form of any system that aspires to serve and support it.

Now there is a conventional wisdom about the nature of organization that persists in spite of the fact that anyone who has worked within an organization knows that this image conveys only part of the story. The conventional model is that an organization is a social collectivity that arranges itself so that it can pursue declared aims and objectives that individuals could not achieve on their own. Given this view of organization, the manager's role is to help achieve the corporate goals, and it follows that the manager's activity is essentially rational decision making in pursuit of declared aims. This is the conventional wisdom even though intuitively we all have a rich sense that organizations in which we have worked are more than rational goal-seeking machines. The experienced day-to-day reality of organizations is that they have some of the characteristics of the tribe and the family as well as the characteristics necessary if they are to order what they do rationally so as to achieve desired objectives such as, in the case of industrial companies, survival and growth. In spite of this folk knowledge, the orthodoxy has been very strong, and we can see this both in the literature of organization theory and in that of management science.

Organization Theory

This is not the place to discuss the development of organization theory in any detail, but it is useful for present purposes to mark the general shape of this field as it emerges in such wide-ranging studies as Reed's (1985) *Redirections in Organisational Analysis*. The general shape is that of the establishment of an orthodoxy (the systems/contingency model that held sway from the 1930s to the 1960s) and the challenge to that orthodoxy since then, with no single dominant alternative.

Nevertheless, the challenging models do, in general, have in common the fact that they see organizations not as reified objects independent of organizational members, as in the orthodox systems model, but as the continually changing product of a human process in which social reality is socially constructed: the title of Berger and Luckmann's (1966) well-known book—*The Social Construction of Reality*—neatly captures this alternative strand of thinking.

At a broad level of generalization, we can see the two major approaches as reflecting the two main categories of thinking about organizations on which a pioneering sociologist, Ferdinand Tönnies, built his account. In his major work *Gemeinschaft und Gesellschaft* (1887) (translated as *Community and Association* by Loomis, 1955), Tönnies constructed models of two types of society or organization. There is the natural living community into which a person is born, the family or the tribe (Gemeinschaft), and there are the formally created groupings (Gesellschaft) that a person joins in some contractual sense, as when he or she becomes an employee of a company or joins a climbing club.

In general, the orthodox view of organizations emphasizes their Gesellschaft nature, that they are created to do things collectively (*achieve goals* is the usual language) that would be beyond the reach of individuals. The alternatives emphasize rather that all social groupings take on some flavour of Gemeinschaft: being in an organization is something like being part of a family. Intuitively, the lived experience of organizations that we all gradually acquire gives us the folk knowledge that organizations exhibit some of the characteristics of both models.

That the orthodox view has been dominant can be seen by perusing college textbooks, which present students with the conventional wisdom. For example, in Khandwalla's (1979) *The Design of Organizations*, the view of organizations as open systems devoted to achieving corporate objectives is described as 'the most powerful orientation in organization theory today' (p. 251). Much attention is paid to well-known work aimed at correlating an organization's structure with its core tasks carried out in an environment with which it interacts (Lawrence & Lorsch, 1967; Pugh & Hickson, 1976; Woodward, 1965; etc.). Reed's (1985) survey argues that 'systems theorists . . . had dominated organizational analysis since the 1930s' (p. 35) but that by the 1960s there was no common history or intellectual heritage. By the 1970s, a systems-derived approach was 'struggling to retain its grip on organizational studies' (p. 106). This does not mean that the orthodoxy has lost its adherents, however. In the same year that Reed's book was published, Donaldson (1985) brought out his *In Defence of Organization Theory*, the defence being of the 'relatively accepted contingency-systems paradigm' (p. ix).

Both Reed and Donaldson make much reference to a book that marks as much as any other the challenge to the orthodox systems view: Silverman's (1970) *The Theory of Organizations*. Silverman contrasts the systems view from the 1950s and 1960s with what he calls 'the Action frame of reference' in which action results from the meanings that members of organizations attribute to their own and each other's acts. Organizational life becomes a collective process of meaning attribution; attention is displaced away from the apparently impersonal processes by means of which, in the conventional model, a reified organization as an open system responds to a changing environment. Some of Silverman's subheadings convey the nature of his argument: Action not behaviour, Action arises from meanings, Meanings as social facts, Meanings are socially sustained, Meanings are socially changed.

This important work opens the way to various alternatives to the systems orthodoxy. Donaldson's discussion, for example, includes social action theory, the sociology of organizations, and the strategic choice thesis. Just as the orthodoxy draws on a positivist philosophy and a functionalist sociology, the alternatives are underpinned philosophically by phenomenology, and sociologically by an interpretive approach derived from Weber and Schutz.

It has to be said that the orthodox view provides a much clearer model of organization, and hence the manager's role, than is provided by the alternatives. Concentrating on the Gesellschaft aspects of an organization, the conventional view sees it as an open system seeking to achieve corporate objectives in an environment to which it has to adapt. Its tasks are analysed and assigned to groups within a functionalist structure, and the managers' role is essentially that of decision making in pursuit of corporate aims that also provide the standards against which progress will be judged. No similarly clear picture is provided by the alternatives, beyond the notion that organizations are characterized essentially by discourse that establishes the meanings that will underpin action by individuals and groups.

It is not at all surprising that that section of the management literature most concerned with intervening in, in order to influence and shape, real-world situations, namely management science, should itself focus on the orthodox systems model.

Management Science

In examining briefly the state of thinking in management science, it is useful to focus on the work of Herbert Simon. There are two reasons for this. First, it has been a dominating contribution in the field; second, in developing an approach based on the work of Vickers, we find that he explicitly contrasted his approach with that of Simon, drawing attention to the reliance of Simon on a goal-seeking model of human action that he himself was deliberately trying to transcend.

In the period after the Second World War, strenuous efforts were made to apply the lessons from wartime operations research to industrial companies and government agencies. In doing this, a powerful strand of systems thinking was developed— it would now be thought of as 'hard' systems thinking—concerned broadly with engineering a system to achieve its objectives. Systems were here assumed to exist in the world; it was assumed that they could be defined as goal seeking; and ideas of system control were generalized in cybernetics. These ideas mapped the orthodox stance of organization theory discussed in the previous section, and they conceptualized the manager's task as being to solve problems and take decisions in pursuit of declared goals. Indeed, this paradigm is succinctly expressed in Ackoff's (1957) assumption that problems ultimately reduce to the evaluation of the efficiency of alternative means for a designated set of objectives.

This is the field to which Simon has made such a significant and influential contribution, the flavour of which is captured in the title of his 1960 book: *The New Science of Management Decision*.

At a round table devoted to his work, Zannetos (1984) summarized Simon's legacy as 'a theory of problem solving, programs and processes for developing intelligent machines, and approaches to the design of organizational structures for managing complex systems' (p. 75).

Overall, Simon sought a science of administrative behaviour and executive decision making. In an intellectually shrewd move that has no doubt helped to make this body of work so influential, Simon wisely abandoned the notion that managers and administrators seek to *optimize*, replacing it with the idea of *satisficing*: the idea that the search is for solutions that are *good enough* in the perceived circumstances, rather than optimal (March & Simon, 1958). Nevertheless, the flavour of hard systems thinking is retained in the claim that the search is 'motivated by the existence of problems as indicated by gaps between performance and goals' (p. 73).

Similarly in another of Simon's major contributions, the development with Newall of GPS (general problem solver), a heuristic computer program that seeks to simulate human problem solving, the whole work is built on the concept of problem solving as a search for a means to an end that is already declared to be desirable (Newall & Simon, 1972). Simon (1960) stated,

> Problem solving proceeds by erecting goals, detecting differences between present situation and goal, finding in memory or by search tools or processes that are relevant to reducing differences of these particular kinds, and applying these tools or processes. Each problem generates subproblems until we find a subproblem we can solve—for which we have a program stored in memory. We proceed until, by successive solution of such subproblems, we eventually achieve our overall goal—or give up. (p. 27)

This is an especially clear statement of the thinking, derived from the systems theory of the 1950s, that has dominated management science and that underlies organization theory's orthodox model of what an organization is.

It is the argument here that this goal-seeking model, largely adequate though it was in the management science that contributed to post-Second World War industrial development, is not rich enough to support and sustain the management thinking now needed by the crew of Spaceship Earth, that spaceship having become akin to a global village.

An alternative, richer perspective is provided by the systems thinking of the 1970s and 1980s, and in particular by Vickers's development of appreciative systems theory and by an approach to intervention in human affairs that can be seen as making practical use of that theory, namely, soft systems methodology.

These are discussed in the next section, but it may be useful to point out at once that these are developments in what is now known as 'soft' systems thinking, as opposed to the hard systems thinking of the 1950s and 1960s that permeates both orthodox organization theory and Simonian management science. The usual distinction made between the two is that the hard systems thinking tackles well-defined problems (such as optimizing the output of a chemical plant), whereas the soft approach is more suitable for ill-defined, messy, or wicked problems (such as deciding on health care policy in a resource-constrained situation). This is not untrue, but it fails to make an intellectual distinction between the two. The real distinction lies in the attribution of systemicity (having the property of system-like characteristics). Hard systems thinking assumes that the world is a set of systems (i.e. is *systemic*) and that these can be *systematically* engineered to achieve objectives. In the soft tradition, the world is assumed to be problematic, but it is also assumed that *the process of inquiry* into the problematic situations that make up the world can be organized as a system. In other words, assumed systemicity is shifted: from taking the world to be systemic to taking the process of inquiry to be systemic (Checkland, 1983, 1985b).

Thus in the following section both appreciative systems theory and soft systems methodology describe inquiring processes—the former with a view to understanding, the latter with a view to taking action to improve real-world problem situations.

Finally, we may note that soft systems thinking can be seen as representing the introduction of systems thinking into Silverman's action frame of reference, although the organization theory literature is apparently at present innocent of any knowledge of post-1960s developments in systems thinking (Checkland, 1994).

APPRECIATIVE SYSTEMS THEORY

The Nature of an Appreciative System

The task that Vickers set himself in his 'retirement' after 40 years in the world of affairs was to make sense of that experience. In the books and articles that he then wrote he constructed

> an epistemology which will account for what we manifestly do when we sit round board tables or in committee rooms (and equally though less explicitly when we try, personally, for example, to decide whether or not to accept the offer of a new job). (G. Vickers, personal communication, July 1974)

In his thinking as this project developed, Vickers first rejected the ubiquitous goal-seeking model of human activity; then he found systems thinking relevant to his task; but he also rejected the cybernetic model of the steersman (whose course is defined from outside the system), replacing it by his more subtle notion of 'appreciation' (Vickers, 1965, is the basic reference). He expressed his intellectual history in the following terms in a letter to the present writer in 1974:

> It seems to me in retrospect that for the last twenty years I have been contributing to the general debate the following neglected ideas:
>
> (1) In describing human activity, institutional or personal, the goal-seeking paradigm is inadequate. Regulatory activity, in government, management or private life consists in attaining or maintaining desired relationships through time or in changing and eluding undesired ones.
>
> (2) But the cybernetic paradigm is equally inadequate, because the helmsman has a single course given from outside the system, whilst the human regulator, personal or collective, controls a system which generates multiple and mutually inconsistent courses. The function of the regulator is to choose and realise one of many possible mixes, none fully attainable. In doing so it also becomes a major influence in the process of generating courses.
>
> (3) From 1 and 2 flows a body of analysis which examines the 'course-generating' function, distinguishes between 'metabolic' and functional relations, the first being those which serve the stability of the system (e.g. budgeting to preserve solvency and liquidity), the second being those which serve to bring the achievements of the system into line with its multiple and changing standards of success. This leads me to explore the nature and origin of these standards of success and thus to distinguish between norms or standards, usually tacit and known by the mismatch signals which they generate in specific situations, and values, those explicit general concepts of what is humanly good and bad which we invoke in the debate about standards, a debate which changes both. (G. Vickers, personal, communication, 1974)

In developing the theory of appreciative systems and relating it to real-world experience, Vickers never expressed the ideas pictorially, in the form of a model, although this seems a desirable form in which to express a system. (His explanation for this lack was disarming: 'You must remember,' he said, 'that I am the product of an English classical education' [G. Vickers, personal communication, 1979]). What follows is an account of the model of an appreciative system developed by Checkland and Casar (1986) from the whole corpus of Vickers's writings.

From those writings we may highlight some major themes that recur:

- A rich concept of day-to-day experienced life (compare Schutz's [1967] *Lebenswelt*)
- A separation of judgments about what is the case, *reality judgments*, and judgments about what is humanly good or bad, *value judgments*
- An insistence on *relationship maintaining* as a richer concept of human action than the popular but poverty-stricken notion of goal seeking
- A concept of action judgments stemming from reality and value judgments
- A notion that the cycle of judgments and actions is organized as a system

The starting point for the model is the Lebenswelt, the interacting flux of events and ideas unfolding through time. This is Vickers's 'two-stranded rope', the strands inseparable and continuously affecting each other. Appreciation is occasioned by our ability to select, to choose. Appreciation perceives (some of) reality, makes judgments about it, contributes to the ideas stream, and leads to actions that become part of the events stream. Thus the basic form of the model is that shown in Figure A13. There is a recursive loop in which the flux of events and ideas generates appreciation, and appreciation itself contributes to the flux. Appreciation also leads to action that itself contributes to the flux.

It is now necessary to unpack the process of appreciation. From Vickers's writings we take the notion of perceiving reality selectively and making judgments about it. The epistemology of the judgment making will be one of relationship managing rather than goal seeking, the latter being an occasional special case of the former. And both reality and value judgments stem from standards of both fact and value: standards of what *is*, and standards of what is good or bad, acceptable or unacceptable. The very act of using the standards may itself modify them.

These activities lead to a view on how to act to maintain, to modify, or to elude certain forms of relevant relationships. Action follows from this, as in Figure A13.

The model also tries to capture Vickers's most important point and greatest insight, namely, that there is normally no ultimate source for the standards by

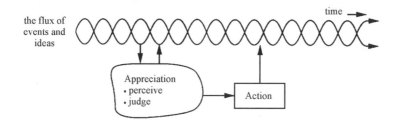

Figure A13   The structure of an appreciative system
SOURCE: Checkland and Casar (1986).

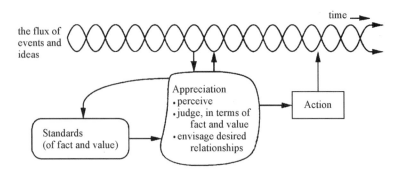

Figure A14   The structure of an appreciative system expanded
SOURCE: Checkland and Casar (1986).

means of which what is noticed is deemed good or bad, important or unimportant, relevant or irrelevant, and so on. The source of the standards is *the previous history of the system itself*. In addition, the present operation of the system may modify its present and future operation through its effect on the standards. These considerations, together with those already discussed, yield Figure A14 as a model of an appreciative system. The most difficult aspect to model is the dynamic one, but it should be clear from Figure A14 that the dynamics of the system will be as shown in Figure A15. The form of the appreciative system remains the same, whereas its contents (its *setting*) continually (but not necessarily continuously) change. An appreciative *system* is a process whose products—cultural manifestations—condition the process itself. But the system is not operationally closed in a conventional sense. It is operationally closed via a structural component (the flux of events and ideas) that ensures that it does not, through its actions, reproduce exactly itself. It reproduces a continually changed self, by a process that Varela (1984) called the 'natural drift' of 'autopoietic systems' (Maturana & Varela, 1980), systems whose component elements create the system itself. Through its (changing) filters the appreciative system is always open to new inputs from the flux of events and ideas, a

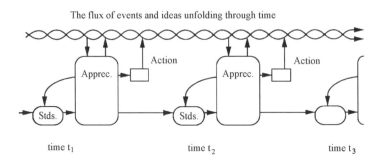

Figure A15   The dynamics of an appreciative system
SOURCE: Checkland and Casar (1986).

characteristic that seems essential if the model is to map our everyday experience of the shifting perceptions, judgments, and structures of the world of culture.

Vickers's claim was that he had constructed an epistemology that can provide convincing accounts of the process by which human beings and human groups deliberate and act. The model in Figures A14 and A15 is a systemic version of the epistemology.

Checkland and Casar (1986) used it to give an account of the learning in a systems study of the Information and Library Services Department of what was then ICI Organics (a manufacturer of fine chemicals within the ICI Group), a study that has been described in detail elsewhere (Checkland, 1985a; Checkland & Scholes, 1990). This study was carried out by a group of managers in the function with some outside help in the use of soft systems methodology (SSM), which was the methodology used. It is a way of making practical use of the notion of an appreciative system, and it will be discussed briefly in the next section. It entails structuring a debate about change by building models of purposeful activity systems and comparing them against perceptions of the real world as a means of examining what the appreciative settings are in the situation in question and how they and the norms or standards are changing. In the study in question, there were three cycles of this learning process.

In the first cycle, the study team's interest and concern were to rethink the role of their function in a changing situation. They perceived many facts relevant to this, which resulted in 26 relevant systems. They selected and judged these facts in terms of a conception of a particular relationship and standards relevant to it: they accepted the relevance of a simple model that took as given that their function was a support to the wealth-generating operations of their company, and they implicitly made use of standards according to which a good version of this relationship would be to make efficient, effective, and timely provision of information to other parts of the company.

These considerations contributed to the ideas stream of the Lebenswelt and led to the action of exploring several perceptions of the relationship between the function and the rest of the company in greater depth. In this second methodological cycle, the focus was still on the relationship between function and company but the appreciative settings began to change. This can be expressed as a change in standards resulting from the first cycle of appreciation. The shift was in the concept of what would constitute a good relationship:

> The focus shifted from ILSD (Information and Library Services Department) as a reactive function responding quickly and competently to user requests and having the expertise to do it, to ILSD as a proactive function, one which could on occasion tell actual and potential users what they *ought* to know. (Checkland 1985a, p. 826)

In the third cycle, the new concept of ILSD was developed and, in the language of Figure A14, several hypothetical forms of relevant relationships were considered. This led to attention being given both to internal relationships within the function (How different would they have to be to sustain a proactive role?) and to the relationship between the function and the company. These considerations led to decisions on actions necessary to broaden the appreciative process. The actions taken were to make both internal (within ILSD) and external presentations of the results of the study. These events entered the company's Lebenswelt and had the effect of starting to bring about the change in the company's appreciative system, as evidenced by the remark made by the research manager at the external presentation, namely, that 'I have known and worked with ILSD for 20 years and I came along this morning out of a sense of duty. To my amazement I find I now have a new perception of ILSD' (Checkland, 1985a, p. 330).

Finally, the company's subsequent allocation of significant new resources to ILSD can be described as illustrating its implicit adoption of new standards with respect to the Information and Library Services function, standards whose change stems from the recent history of the company's appreciative system, involving input of ideas and events from the systems study itself.

The Appreciative Process in Action: Soft Systems Methodology

It is not appropriate here to give a detailed account of SSM, which is described in numerous books and articles since the early 1970s. (The basic books describing its development are Checkland, 1981; Checkland & Scholes, 1990; and Wilson, 1984; a burgeoning secondary literature may be sampled via, for example, Avison & Wood-Harper,1990; Davies & Ledington, 1991; Hicks, 1991; Patching, 1990; and Waring, 1989.)

SSM was not an attempt to operationalize the concept of an appreciative system; rather, after SSM had emerged in an action research programme at Lancaster University, it was discovered that its process mapped to a remarkable degree the ideas Vickers had been developing in his books and articles (Checkland, 1981, chap. 8).

The Lancaster programme began by setting out to explore whether or not, in real-world managerial rather than technical problem situations, it was possible to use the approach of systems engineering. It was found to be too naïve in its questions (What is the system? What are its objectives? etc.) to cope with managerial complexity, which, we could now say, was always characterized by conflicting appreciative settings and norms. Systems engineering as developed for technical (well-defined) problem situations had to be abandoned, and SSM emerged in its place.

The development of SSM has been characterized by four points in time at which what can now be seen, with hindsight, as crucial ideas moved the project forward (Checkland & Haynes, 1994). The first was the realization that all real-world problem situations are characterized by the fact that they reveal human beings seeking or wishing to take *purposeful action*. This led to purposeful action being treated seriously as a systems concept. Ways of building models of human activity systems were developed. Then it was realized that there can never be a single account of purposeful activity, because one observer's terrorism is another's freedom fighting. Models of purposeful activity could only be built on the basis of a declared Weltanschauung. This meant that such models were never models *of* real-world action; they were models *relevant to* discourse and argument about real-world action; they were epistemological devices that could be used in such discourse and debate; they were best thought of as *holons*, using Koestler's (1967) useful neologism, which could structure debate about different ways of seeing the situation. This led to the third crucial idea, that the problem-solving process that was emerging would inevitably consist of a learning cycle in which models of human activity systems could be used to structure a debate about change. The structure was

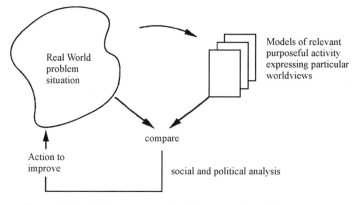

Figure A16   Soft systems methodology as a learning system
SOURCE: Checkland and Scholes (1990).

provided by carrying out an organized comparison between models and perceived real situations in which accommodations between conflicting perspectives could be sought, enabling action to be taken that was both arguably desirable—in terms of the comparisons between models and perceived situation—and culturally feasible for a particular group of people in a particular situation with its own particular history. (The fourth crucial idea, not relevant here, was the realization that models of human activity systems could be used to explore issues concerning what information systems would best be created to support real-world action—which took SSM into the field of information systems and information strategy.)

Given these considerations, SSM emerged as the process summarized in Figure A16. This is a picture of a *learning system* in which the appreciative settings of people in a problem situation—and the standards according to which they make judgments—are teased out and debated. Finally, the influence of Vickers on those who developed SSM means that the action to improve the problem situation is always thought about in terms of managing relationships—of which the simple case of seeking a defined goal is the occasional special case.

CONCLUSION: THE RELEVANCE OF APPRECIATIVE SYSTEMS THEORY AND SSM TO MODERN MANAGEMENT

It is not difficult to envisage the situations in both industry and the public sector in which the thinking about problems and problem solving would be significantly helped by the models underpinned by hard systems thinking, namely the models that see organizations as coordinated functional task systems seeking to achieve declared goals and that see the task of management as decision making in support of goal seeking. These models would be useful in situations in which goals and measures of performance were clear-cut, communications between people were limited and prescribed, and in which the people in question were deferential toward the authority that laid down the goals and the ways in which they were to be achieved. But this image has never accurately described life in most organizations as most people experience it, and it has become less and less true since the end of the Second World War. Since that time the trends have been toward much increased capacity for communication, greater complexity of goals as economic interdependence has increased, much reduced deference toward authority of any kind, and the dismantling of monolithic institutionalized power structures. The dethronement of the mainframe computer by the now ubiquitous personal computer is at once both a metaphor for these changes and one of the catalysts for their occurrence.

In such a situation richer models of organization and management will be helpful, and it has been argued that those based on Vickers's appreciative systems theory and SSM have a role to play here. More important, they do not replace the older models but rather subsume and enhance them. In SSM, focusing on a unitary goal is the occasional special case of debating multiple perceptions and proceeding on the basis of *accommodations* between different interests. For Vickers, managing relationships is the general case of human action, the pursuit of a goal the occasional special case.

Vickers himself has usefully differentiated his stance from that of Simon in remarks that relate to the latter's *Administrative Behaviour* (Simon, 1957):

> The most interesting differences between the classic analyses of this book and my own seem to be the following:
>
> (1) I adopt a more explicitly dynamic conceptual model of an organisation and of the relations, internal and external, of which it consists, a model which applies equally to all its constituent sub-systems and to the larger systems of which it is itself a part.
>
> (2) This model enables me to represent its 'policy makers' as regulators, setting and resetting courses or standards, rather than objectives, and thus in my view to simplify some of the difficulties inherent in descriptions in terms of 'means' and 'ends'.
>
> (3) I lay more emphasis on the necessary mutual inconsistency of the norms seeking realisation in *every* deliberation and at *every* level of organisation and hence on the ubiquitous interaction of priority, value and cost.
>
> (4) In my psychological analysis linking judgments of fact and value by the concept of appreciation, I stress the importance of the underlying appreciative *system* in determining how situations will be seen and valued. I therefore reject 'weighing' (an energy concept) as an adequate description of the way criteria are compared and insist on the reality of a prior and equally important process of 'matching' (an information concept).
>
> (5) I am particularly concerned with the reciprocal process by which the setting of the appreciative system is itself changed by every exercise of appreciative judgment. (Vickers, 1965, p. 22)

As an example of the relevance of SSM to current problems of managing complexity, we offer recent work done within the National Health Service (NHS) in the United Kingdom. (Some of this is described in Checkland & Holwell, 1993; Checkland & Poulter, 1994; and Checkland &. Scholes, 1990, chap. 4.)

In recent years the NHS has been subjected to several waves of government-imposed change. First there was the imposition of a system of accountable management, replacing the previous consensus management of teams of professionals. This had hardly settled down before it was replaced by an internal quasi-market. In this development the old district health authorities (into which the previous change had introduced district general managers) became purchasers of health care for a defined population, whereas hospitals and some general practitioners became providers of health care, the two being linked by contracts (although not legally binding ones) for particular services at a negotiated price. All these changes have entailed a considerable shift in appreciative settings for health professionals, and the NHS has been experiencing a period of considerable turmoil.

In the study described in Checkland and Scholes (1990), the problem was addressed of how a Department of Community Medicine in what was then a district health authority could evaluate its performance. Clearly the evaluation standards would depend completely on this department's image of itself and its role within the district. This is not a casual consideration, because concepts of community medicine range from *providing epidemiological data* to *managing the delivery of health care*. In this work, SSM-type models of purposeful activity relating to concepts of community medicine were built, with participation of members of the department, and eventually an evaluation methodology was developed. This was based on a structured set of questions derived from models that members of the department felt expressed their shared appreciative settings with regard to their image of the role of community medicine, which in their case was a very proactive interventionist one.

More recently, much work has been done in NHS hospitals and purchasing authorities as they assimilate and adapt to the purchaser–provider split (Checkland & Holwell, 1993). The new appreciative settings have been explored with participants via models of notional systems to enact the purchaser and provider roles. These have served to structure coherent debate concerning the requirements of the new roles.

In a recent study in a large teaching hospital, the work was part of a project to re-create an information strategy for the hospital suitable to cope with the new arrangements (Checkland & Poulter, 1994). In this work half a dozen teams of hospital workers representing the different professions were set up; members included clinicians, nurses, professionals from the finance and estates offices, and so on. Over a period of about 6 months, with a plenary meeting of team leaders every month, the teams discussed their activity and its contribution to meeting the requirements of the contracts for providing particular health care services that the hospital would in future negotiate with purchasers. Activity models were built and then used to structure analysis of required information support. This was related to existing information systems, and the information gaps identified helped in the formulation of the new information strategy.

One incident that occurred during this process may be recounted. It illustrates, in microcosm, the change of appreciative settings that can occur in the process of using SSM. It concerns a working group made up of nurses in the teaching hospital, led by a senior nurse. The group was building activity models relevant to providing nursing care, before using them to examine required information support.

Within SSM, when would-be relevant activity models are built, careful concise accounts of them as transformation processes are formulated (so-called Root Definitions). Various questions are asked in clarifying these definitions, one of which is 'If this notional system were to exist, who would be its victims or beneficiaries?' Nurses asking this question naturally wish to answer, 'The patients'. That is what their whole ethos, education, and professionalism tell them. That illustrates why they are in the profession. It was therefore something of a shock to this group—brought to their attention by the structured requirements of the SSM process—to realize in discussion that under the new arrangements the *technically* correct answer is nearer to being 'The hospital contracts manager'. This is because, under the so-called internal market, each contract for a health care service that involves nursing care ought technically to include the cost of providing a certain level (and quality) of nursing care. The nurses' task is then to provide what the contract calls for. Beyond this, of course, there is a theory according to which the interests of patients will, in fact, best be met by the new purchaser–provider contracts.

But it is not easy for nurses to accept this. The senior nurse who described this incident at one of the plenary discussions said that this question, and the issue it exposed, occupied the team for much of one of their meetings. It gave her insight into the NHS changes and helped her to understand her own misgivings about a supposed internal market in health care. Geoffrey Vickers would have appreciated this story.

Given our self-consciousness and the degree of mental autonomy that we seem to possess as human beings, that part of our thinking that is beyond the unreflecting stream-of-consciousness involvement in everyday life can itself be thought about. This can be done by examining the mental models that we use to make sense of our worlds. It is entirely plausible that our perceptions will be coloured by those mental models. And it follows that they need both to be better than primitive and to change as our human and social world changes.

It has been argued here that the models of organization and management that have been useful since the 1950s need to be enriched. It has then been argued that appreciative systems theory and SSM can help to provide such enrichment. They do not replace the earlier functionalist and goal-seeking models: They enclose and enhance them in ways more appropriate to institutional life at the end of the century.

NOTE

This article appeared in a special issue of *American Behavioral Scientist* **38**(1) September/October, 1994 devoted to: Rethinking Public Policy-Making: questioning assumptions, challenging beliefs. Essays in Honour of Sir Geoffrey Vickers on his Centenary. Edited by Margaret Blunden and Malcolm Dando.

The whole issue was republished as a book in 1995: *Re-thinking Public Policy Making* Blunden, M. and Dando, M. (Eds) Sage Publications, London.

The author is grateful to Sage Publications for permission to reprint the article here.

REFERENCES

Ackoff, R. L. (1957). Towards a behavioural theory of communication. In W. Buckley (Ed.), *Modern systems research for the behavioural scientist* (pp. 209–218). Chicago, IL: Aldine.

Avison, D. E., & Wood-Harper, A. T. (1990). *Multiview: An exploration in information systems development*. Oxford: Blackwell.

Berger, P. & Luckmann, T. (1966). *The social construction of reality*. Harmondsworth: Penguin.

Checkland, P. (1981). *Systems thinking, systems practice*. Chichester: Wiley.

Checkland, P. (1983). OR and the systems movement. *Journal of the Operational Research Society*, 34, 661–675.

Checkland, P. (1985a). Achieving desirable and feasible change: An application of soft systems methodology. *Journal of the Operational Research Society*, 36, 821–831.

Checkland, P. (1985b). From optimizing to learning: A development of systems thinking for the 1990s. *Journal of the Operational Research Society*, 36, 757–767.

Checkland, P. (1994). Conventional wisdom and conventional ignorance: The revolution organiza-
tion theory missed. *Organization*, *1*(1), 29–34.
Checkland, P., & Casar, A. (1986). Vickers' concept of an appreciative system: A systemic account.
Journal of Applied Systems Analysis, *13*, 3–17.
Checkland, P., & Haynes, M. G. (1994). Varieties of systems thinking: The case of soft systems
methodology. *System Dynamics Review*, *10*, 189–197.
Checkland, P., & Holwell, S. (1993). Information management and organizational processes: An
approach through soft systems methodology. *Journal of Information Systems*, *3*, 3–16.
Checkland, P., & Poulter, J. (1994). *Application of soft systems methodology to the production of a
hospital information and systems strategy*. United Kingdom: HISS Central Team of the NHS
Management Executive.
Checkland, P., & Scholes, J. (1990). *Soft systems methodology in action*. Chichester: Wiley.
Davies, L., & Ledington, P. (1991). *Information in action: Soft systems methodology*. London:
Macmillan.
Donaldson, L. (1985). *In defence of organization theory*. Cambridge: Cambridge University Press.
Hicks, M. J. (1991). *Problem solving in business and management*. London: Chapman & Hall.
Khandwalla, P. N. (1979). *The design of organizations*. New York: Harcourt Brace.
Koestler, A. (1967). *The ghost in the machine*. London: Hutchinson.
Lawrence, P. R., & Lorsch, J. W. (1967). *Organization and environment*. Cambridge, MA: Harvard
University Press.
March, J. G., & Simon, H. A. (1958). *Organizations*. New York: Wiley.
Maturana, H. R., & Varela, F. J. (1980). *Autopoiesis and cognition*. Dordrecht: Reidel.
Newall, A., & Simon, H. A. (1972). *Human problem solving*. Englewood Cliffs, NJ: Prentice-Hall.
Patching, D. (1990). *Practical soft systems analysis*. London: Pitman.
Pugh, D. S., & Hickson, D. J. (1976). *Organization structure in its context*. Farnborough: Saxon
House.
Reed, M. I. (1985). *Redirections in organisational analysis*. London: Tavistock.
Schutz, A. (1967). *The phenomenology of the social world*. Evanston, IL: Northwestern University.
Silverman, D. (1970). *The theory of organizations*. London: Heinemann.
Simon, H. A. (1957). *Administrative behaviour* (2nd ed.). New York: Macmillan.
Simon, H. A. (1960). *The new science of management decision*. New York: Harper & Row.
Tönnies, F. (1955). *Community and association [Gemeinschaft und Gesellschaft]* (C. P Loomis,
Trans.). London: Routledge & Kegan Paul. (Original work published 1887)
Varela, F. J. (1984). Two principles of self-organization. In H. Ulrich & G. J. Probst (Eds), *Self-
organisation and management of social systems* (pp. 25–32). Berlin: Springer-Verlag.
Vickers, G. (1965). *The art of judgment*. London: Chapman & Hall.
Waring, A. (1989). *Systems methods for managers*. Oxford: Blackwell.
Wilson, B. (1984). *Systems: Concepts, methodologies and applications*. Chichester: Wiley.
Woodward, J. (1965): *Industrial organization: Theory and practice*. London: Oxford University
Press.
Zannetos, Z. S. (1984). Decision sciences and management expectations. In J. P. Brans (Ed.),
Operational research '84 (pp. 69–76). Amsterdam: North Holland.

BIBLIOGRAPHY

Atkinson, C. J. (1984) Metaphor and systemic praxis, Ph.D. Dissertation, Lancaster University,
UK.
Blunden, M. (Ed.) (1984) *The Vickers Papers*, Harper and Row, London.
Bryant, J. (1989) *Problem Management*, John Wiley, Chichester.
Burns, F. (1998) *Information for Health: an Information Strategy for the Modern NHS 1998–2005*,
NHS Executive, Department of Health Publications, Wetherby, UK.
Checkland, P. (1972) Towards a systems-based methodology for real-world problem solving,
Journal of Systems Engineering, *3*(2), 87–116.
Checkland, P. (1981) *Systems Thinking, Systems Practice*, John Wiley, Chichester.
Checkland, P. (1983) OR and the systems movement: mappings and conflicts, *Journal of the
Operational Research Society*, **34**(8), 661–675.
Checkland, P. (1988) The case for 'holon', *Systems Practice*, **1**(3), 235–238.

Checkland, P. (1995) Soft Systems Methodology and its relevance to the development of information systems. In Stowell, F.A. (Ed.), *Information Systems Provision: the Contribution of Soft Systems Methodology*, McGraw-Hill, London.

Checkland, P. (1997) Rhetoric and reality in contracting: research in and on the National Health Service. In Flynn, R. and Williams, G. (Eds), *Contracting for Health*, Oxford University Press, Oxford.

Checkland, P. and Casar, A. (1986) Vickers' concept of an appreciative system, *Journal of Applied Systems Analysis*, **13**, 3–17.

Checkland, P., Forbes, P. and Martin, S. (1990) Techniques in soft systems practice: Part 3: monitoring and control in conceptual models and in evaluation studies, *Journal of Applied Systems Analysis*, **17**, 29–37.

Checkland, P. and Griffin, R. (1970) Management information systems: a systems view, *Journal of Systems Engineering*, **1**(2), 29–42.

Checkland, P. and Holwell, S. (1998) *Information, Systems and Information Systems*, John Wiley, Chichester.

Checkland, P. and Holwell, S. (1998a) Action research: its nature and validity, *Systemic Practice and Action Research*, **11**(1), 9–21.

Checkland, P. and Scholes, J. (1990) *Soft Systems Methodology in Action*, John Wiley, Chichester.

Chorley, R. J. and Kennedy, B. A. (1971) *Physical Geography: a Systems Approach*, Prentice-Hall International, London.

Durkheim, E. (1895) *The Rules of Sociological Method* (Catlin, G. E. G., Ed.) 1964, The Free Press, New York.

Duxbury, J. (1994) The development and testing of a model relevant to the decision making process concerning the provision of health care in Morecambe Bay, M.Sc. Dissertation, Lancaster University.

Flynn, R. and Williams, G. (Eds) (1997) *Contracting for Health*, Oxford University Press, Oxford.

Forbes, P. and Checkland, P. (1987) Monitoring and control in systems models, Internal Discussion Paper 3/87, Department of Systems and Information Management, Lancaster University.

Ham, C. (1992) *Health Policy in Britain* (3rd edn), Macmillan, London.

Ham, C. (1996) The future of the NHS, *British Medical Journal*, **313**, 1277–1278 (23rd November).

HMSO (1997) *The New NHS*, Cm 3807.

Holwell, S. E. (1997) Soft systems methodology and its role in information systems, Ph.D. Dissertation, Lancaster University.

Luckmann, T. (Ed.) (1978) *Phenomenology and Sociology*, Penguin Books, Harmondsworth, UK.

Maturana, H. R. and Varela, F. J. (1980) *Autopoesis and Cognition*, D. Reidel, Dortrecht.

Miller, G. A. (1956) The magical number seven plus or minus two: some limits on our capacity for processing information, *Psychological Review*, **63**(2), 81–96.

Miller, G. A. (1968) *The Psychology of Communication*, Allen Lane, The Penguin Press, London.

Miller, J. G. (1978) *Living Systems*, McGraw-Hill, New York.

Morse, J. M. (Ed.) (1994) *Critical Issues in Qualitative Research Methods*, Sage, Thousand Oaks (Calif.).

Mueller-Vollmer, K. (Ed.) (1986) *The Hermeneutics Reader: Texts of the German Tradition from the Enlightenment to the Present*, Basil Blackwell, Oxford.

Naughton, J. (1977) *The Checkland Methodology: a Reader's Guide* (2nd edn), Open University Systems Group, Milton Keynes.

Optner, S. L. (1965) *Systems Analysis for Business and Industrial Problem-solving*, Prentice-Hall, Englewood Cliffs, NJ.

Rivett, G. (1998) *From Cradle to Grave: Fifty Years of the NHS*, King's Fund Publishing, London.

Simon, H. A. (1960) *The New Science of Management Decision*, Harper and Row, New York.

Simon, H. A. (1977) *The New Science of Management Decision* (revised edn), Prentice-Hall, Englewood Cliffs, NJ.

Stowell, F. A. (1989) Change, organizational power and the metaphor 'commodity', Ph.D. Dissertation, Lancaster University, UK.

Stowell, F. A. (Ed.) (1995) *Information Systems Provision: the Contribution of Soft Systems Methodology*, McGraw-Hill, London.

Tsouvalis, C. N. (1995) Agonistic thinking in problem-solving: the case of soft systems methodology, Ph.D. Dissertation, Lancaster University.

Waring, A. (1989) *Systems Methods for Managers*, Blackwell Scientific Publications, Oxford.

Watson, R. and Smith, R. (1988) Applications of the Lancaster soft systems methodology in Australia, *Journal of Applied Systems Analysis*, **15**, 3–26.

Weber, M. (1904) 'Objectivity' in social science and social policy. In Shils, E. A. and Finch, H. A. (Eds), Max Weber's *Methodology of the Social Sciences*, Free Press, New York.

Webster, C. (1998) *The National Health Service: a Political History*, Oxford University Press, Oxford.

Wilson, A. (1973) *The Concorde Fiasco*, Penguin, Harmondsworth.

Wilson, B. (1984) *Systems: Concepts, Methodologies and Applications*, John Wiley, Chichester, 2nd edn 1990.

Author Index

Subject Index

A65

Soft Systems
Methodology in Action

Soft Systems Methodology in Action

Peter Checkland

Jim Scholes

JOHN WILEY & SONS

Chichester · New York · Brisbane · Toronto · Singapore

Other Wiley Editorial Offices

John Wiley & Sons, Inc., 605 Third Avenue,
New York, NY 10158-0012, USA

Jacaranda Wiley Ltd, 33 Park Road, Milton,
Queensland 4064, Australia

John Wiley & Sons (Canada) Ltd, 22 Worcester Road,
Rexdale, Ontario M9W 1L1, Canada

John Wiley & Sons (SEA) Pte Ltd, 37 Jalan Pemimpin #05-04,
Block B, Union Industrial Building, Singapore 129809

Library of Congress Cataloging-in-Publication Data

Checkland, Peter.
  Soft systems methodology in action/Peter Checkland,
    Jim Scholes.
      p. cm.
  Includes bibliographical references.
  ISBN 0-471-92768-6
  1. System theory. 2. System analysis. I. Scholes, Jim.
  II. Title.
  Q295 C449   1990
  003 dc20   90-12193
                                                   CIP

British Library Cataloguing in Publication Data

Checkland, Peter
  Soft systems methodology in action.
  1. Systems
  I. Title. II. Scholes, Jim
  003

Typeset by Acorn Bookwork, Salisbury, Wiltshire
Printed and bound in Great Britain by
Biddles Ltd, Guildford and King's Lynn

To
Glen, Kris and Kath
PBC

To
Christopher, Cath, Carol and Linda
JS

Contents

Part II Soft Systems Methodology in Action—Learning through Use

Nothing ever goes right. Nothing ever turns out as one has planned. How many times have you heard that said? How many times have you said it? And yet it is never an expression of cynicism, defeatism or nihilism. It is always a confession of faith.

Chester Himes: Introduction to *Pinktoes*

What was normal was not to defy one's upbringing. It was to enact the whole bloody roadshow as scripted by one's aunts and grandmothers.

Barbara Trapido: *Brother of the More Famous Jack*

Preface

It is easy to overestimate the part played by rational thought and reasoned analysis in human life, but most people would agree that they do have some part to play! This suggests that it is worth while trying rationally to find better ways of coping with the complexity of human affairs. This book describes a decade of further development of one well-established way of intervening to improve problem situations, namely, soft systems methodology (SSM), and does so by concentrating on SSM in action in a variety of situations.

SSM was developed in the 1970s. It grew out of the failure of established methods of 'systems engineering' (SE) when faced with messy complex problem situations. SE is concerned with creating systems to meet defined objectives, and it works well in those situations in which there is such general agreement on the objectives to be achieved that the problem can be thought of simply as the selection of efficacious and efficient means to achieve them. A good example would be the USA's programme in the 1960s with its unequivocal objective of 'landing a man on the Moon and returning him safely to Earth' (President Kennedy's words). Not many human situations are as straightforward as this, however, and SSM was developed expressly to cope with the more normal situation in which the people in a problem situation perceive and interpret the world in their own ways and make judgements about it using standards and values which may not be shared by others. Deciding how your company should make use of information technology; launching a new product; choosing your career; running a sports club; setting about improving care of the elderly in a district of the National Health Service; these are all problems of the latter type, for which SSM is appropriate.

A previous book, Checkland's *Systems Thinking, Systems Practice* (Wiley, 1981) described the origins and nature of systems thinking, and showed how it is used in both SE and SSM, giving an account of the decade of involvement in real problems during which SSM was initially established and developed. In 1984, Wilson's *Systems: Concepts, Methodologies, and Applications* (Wiley) described his involvement in the research programme, paying special attention to SSM as a means of creating information systems. Many further experiences since these books were written have extended and enriched SSM, and increased understanding of it. The aim here is to give a mature account of

SSM as it is in the 1990s, after several hundred applications of the approach by a wide range of people and groups in many different countries.

In fact the experiences of the last decade have not simply enriched our view of SSM, they have changed it radically, as Part II indicates. The authors anticipate that there will be an element of surprise, especially in Chapter 2 and in Chapters 6–10, even for knowledgeable readers. And since Paul Forbes's research (1989) suggests that external understanding of SSM is stuck at the point reached in the early 1970s, they also hope that the book will be relevant for some years to come!

Since SSM is intended to offer a very practical use of systems thinking, but one which is soundly based theoretically, the approach adopted is to focus on SSM in action in different kinds of situation and to show how the learning achieved derived from the experiences which generated it.

The book is written in the belief that neither theory nor practice should dominate the other. This is a cogent issue for the systems movement, which is a field very prone to rather vapid theorizing of a broadly holistic kind. Theory which is not tested out in practice is sterile. Equally, practice which is not reflective about the ideas upon which it is based will abandon the chance to learn its way steadily to better ways of taking action. Thus, theory must be tested out in practice; and practice is the best source of theory. In the best possible situation the two create each other in a cyclic process in which neither is dominant but each is the source of the other. This book recounts some experiences of trying to move round that cycle, and it is written out of a number of experiences in organizations of different kinds, in both public and private sectors.

The experiences have constituted an interesting journey of discovery. As a result of it, SSM is no longer perceived as a seven-stage problem-solving methodology, which is probably the form in which most people who know anything at all about it usually think of it. The seven-stage methodology is now seen as one option in a more general approach to trying to tackle coherently the problematical situations in which we find ourselves in professional or private life.

The development of SSM has been a collaborative venture to which many people have contributed. These include not only teachers, postgraduate students and researchers in the (postgraduate) Department of Systems and Information Management at the University of Lancaster, where the work started twenty years ago, but also many people in similar roles in other universities around the world where SSM is taught and used, as well as many managers in organizations public and private who have initiated or carried out or taken part in studies using SSM, and management consultants who have adopted the approach as a way of structuring their work.

Generations of postgraduate students at the Department of Systems and Information Management at the University of Lancaster have been involved in the development of SSM, and we are grateful to them all. Special thanks go to those who worked in systems studies which have enabled this book to be written, even though not all of those studies are described here: Chris Atkinson, Stuart Bamford, Irmela von Bulow, Chris Caiger, Alex Casar, Sheila Challender, Elaine Cole, Dave Culy, Lynda Davies, Andre Dolbec, Jacques Filion, Elizabeth Forbes, Paul Forbes, Ramses Fuenmayor, Bob Galliers, Kek Cheong Goh, Noel Hall, John Hardy, Sue Holwell, Jim Hughes, Ross Inderwick, Paul Jackson, Paul Ledington, Theresa McHugh, Sophia Martin, Carlos Mendizabel, Ana Steenken, Frank Stowell, Costas Tsouvalis, Simon Watson, Julie Weaver, Rebecca Wong.

To our colleagues during the years in question we are very grateful for both their support and their healthy scepticism: Ron Anderton, David Brown, John Denmead, Rod Griffiths, Mike Jackson, Paul Lewis, Neeta Pankhania, Iain Perring, Andrew Taylor, Mike Watson, Brian Wilson, Ian Woodburn.

Many managers in organizations in which we have worked have contributed to the development of our thinking, and we are especially grateful to: Andre van Aken, George Battersby, Bill Beard, Phil Belshaw, Yolanda Buchel, Liz Clarke, Jeremy Cobb, Ninian Eadie, Margaret Exley, Gerard Fairtlough, Mike Fedorski, Alan Fellows, Paul Freeman, Peter Gershon, Raman Hansjee, Cees van der Heiden, Luc Hoebeke, Peter Idenbirg, Jaap Leemhuis, Mike Mockeridge, David Small, Brian Smith, Derek Styles, Jo Teagle, the late Pat Terry, Rob de Vos, John Wales, Gerald Watson.

In the case of Chapters 3 and 4 we are grateful to John Wales and Jeremy Cobb who read early versions of those accounts. Rob de Vos, Jaap Leemhuis and Cees van der Heiden, with whom it was an especial pleasure to work, commented on both pseudonymous and Shell-specific versions of Chapter 9; we are very grateful to the Shell International Petroleum Company Limited for permission to report the work in the Manufacturing Function which is still current, and to Aodh O'Dochartaigh of ICL through whose good offices we were able to include the sequence of studies in that Company, which make a vital contribution to the argument of the book.

Paul Freeman, at the time the Director of The Central Computer and Telecommunication Agency, read and commented on early versions of Chapter 5 and gave his blessing to the version included here; we are especially grateful for that because we were anxious to include an example from work in the Civil Service, which has a different 'feel' to that carried out in industry. Finally, for the production of clean copy from a much worked-over manuscript, our gratitude goes to Gill Williams, ably assisted by Sylvia Johnson. We appreciated both Gill's professional skill and her trenchant good humour.

Peter Checkland   **Jim Scholes**
University of Lancaster   *The Old Vicarage*
Bailrigg   *Vicarage Lane*
Lancaster   *Wraysbury*

Chapter 1

Soft Systems Methodology and Action

Introduction

Consider the scope of the idea of 'managing' anything. The project manager in an engineering company responsible for developing a new product, the doctor running an ear, nose and throat clinic, the single parent with a child of school age, the secretary of a trade union branch, the leader of a guerilla band, all these are 'managers' in the broad sense of the term. To 'manage' anything in everyday life is to try to cope with a flux of interacting events and ideas which unrolls through time. The 'manager' tries to 'improve' situations which are seen as problematical—or at least as less than perfect—and the job is never done (ask the single parent!) because as the situation evolves new aspects calling for attention emerge, and yesterday's 'solutions' may now be seen as today's 'problems'.

Soft systems methodology (SSM) helps such managers, of all kinds and at all levels, to cope with their task. It is an organized way of tackling messy situations in the real world. It is based on systems thinking, which enables it to be highly defined and described, but is flexible in use and broad in scope. The account of it here is based on the last decade of experience, and is complementary to the earlier account of SSM which described its origins and emergence, as well as the systems thinking upon which it is based and the nature of systems thinking in general (Checkland, 1981).

The ideas in this first chapter underpin the whole book. It starts from, but does not dwell on, the basic assumptions behind SSM which make its scope so broad, and describes without going into detail the mature view, after twenty years of research, of the shape and nature of the methodology. Finally an account is given of the shape of the book itself.

Organized Purposeful Action

One of the most obvious characteristics of human beings is their readiness to attribute meaning to what they observe and experience. Indeed, human

1

beings are not simply ready to attribute meanings, they cannot abide meaninglessness. The very existence of the world religions, and the fact that every culture develops its own myths concerning the nature of the world and our place in it, show how important it is to *homo sapiens* to create answers to the most fundamental—ultimately unanswerable—questions. Mankind finds an absence of meaning unendurable. We are a meaning-endowing animal, on both the global long-term and the local short-term level. Members of organizations, for example, tend to see the world in a particular way, to attribute at least partially shared meanings to their world. And that is equally true of corporate members of the Warsaw Pact and individual members of the Batley Ladies Sewing Circle.

Given the creation of an interpreted, not merely an experienced world, we can form *intentions*, we can decide to do one thing rather than another, in the light of how we are interpreting our situation. This seems to be a uniquely human characteristic. The chemist investigating how hydrogen and nitrogen combine to form ammonia never finds it necessary to attribute intentions to the molecules. And if we observe the behaviour of the cuckoo, though we may say casually that 'the cuckoo has the intention of laying her eggs in the nest of another bird', this is an observer's language, not meaningful to the cuckoo. To explain what is observed, we only have to assume that the cuckoo is programmed to a certain behaviour; it has not been found necessary to attribute self-consciousness to cuckoos. But to explain human behaviour, much more erratic than that of cuckoos, we need the additional concept of the 'intentions' which the human beings can formulate, act upon, and change for themselves.

Thus, on this argument, human beings cannot help attributing meaning to their experienced world; and they can then decide to do some things and not do others. They can take *purposeful action* in response to their experience of the world. By purposeful action we mean deliberate, decided, willed action, whether by an individual or by a group.

Now it would seem a good idea if purposeful action deriving from intentions were also based on knowledge rather than consisting merely of random thrashing about—though observation suggests that there may be no shortage of that in human affairs! Where might the knowledge to guide action be found? Probably the most respected source of knowledge is scientific investigation, since it produces 'public knowledge' which can be subject to public refutation (Ziman, 1968; Popper, 1963; Checkland, 1981, Chapter 2). But while the status of scientific knowledge gained in repeatable experiments concerning natural phenomena is unimpeachable in Western culture, the status of knowledge gained in the so-called social or human sciences is much less sure. This is precisely because, as Caws (1988) neatly puts it:

> the causal determinants of the objects of the social sciences always include human intentions, while those of the natural sciences do not (p. 1).

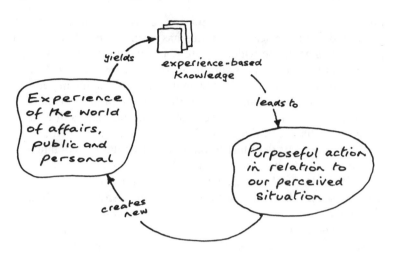

Figure 1.1 The experience–action cycle

In the social sciences repeatable experiments are difficult to achieve and virtually all knowledge gained by social science is heavily meaning-bearing.

If we cannot aspire to a natural-science-like knowledge, perhaps what we seek in human affairs might be described as 'wisdom-based knowledge'. But what one observer sees as wisdom may to another be blinkered prejudice. 'Insight-based knowledge' might be another candidate, but again we have to ask: Insight in relation to whose meaning? The most neutral expression would be 'experience-based knowledge', and this accords at least with the everyday observation that we are all the time taking purposeful action in relation to our experience of the situations we find ourselves in, and the knowledge (shared or individual) which that experience yields.

This, of course, places knowledge acquisition in a cycle, namely that shown in Figure 1.1, since the purposeful action derived from experience-based knowledge will itself result in new experience. This is a cycle whose content will continually change: each time round the cycle the world experienced is a somewhat different place, and hence the cycle embodies fundamentally the possibility of *learning*. If this happens then the purposeful action can be aimed at intended improvements, improvements, that is, in the eyes of those who take the action.

The cycle of Figure 1.1 can itself be seen as the object of concern of so-called 'management science'. This is a curious field, as is indicated by problems surrounding both of the words in its name! Management science clearly ought to be relevant to managing anything, in the same broad sense in which the word is being used here; in fact, management science has devoted itself almost exclusively to the concerns of only one kind of manager: professionals conducting the affairs of private and public enterprises.

Secondly, whether or not management science is a science, or whether it could or should aspire to scientific status has been much debated in the last twenty years, as part of the wider debate about whether natural science and social science are or could be scientific in the same sense (see Checkland, 1983a, for a discussion of the management science debate in Operational Research circles, and Checkland, 1981, for discussion of the broader debate). However, these problems have not been too inhibiting to practitioners in the field. Much work has been done, and one very useful legacy from management science has been its demonstration that a particular kind of language can be very helpful in understanding and articulating the operation of the cycle of Figure 1.1. We refer to the language of systems thinking.

Systems thinking will be discussed briefly in Chapter 2, in which its role in SSM will be made clear. Here it is sufficient to say that in spite of the fact that there are many definitions of the word system in the literature (Jordan, 1965, p. 44–65, for example, offers fifteen) all take as given the notion of a set of elements mutually related such that the set constitutes a whole having properties as an entity. Secondly comes the crucial idea that the whole may be able to survive in a changing environment by taking control action in response to shocks from the environment. We can see at once that the cycle of Figure 1.1 can itself be viewed as 'a system', one which, if self-reflective, could learn, adapt and survive through time. So it is not surprising that systems thinking has had an important role in developing an organized approach to describing and making operational the cycle of Figure 1.1.

For that is where this argument has led us: firstly to the idea that it is probably worth trying to find ways of *formally* operating the learning cycle in which purposeful action is taken in real-world situations in order to bring about what are deemed to be improvements by those carrying out the process; and secondly to the idea that systems thinking may be helpful in this task. SSM is just such a methodology for operating the endless cycle from experience to purposeful action.

Some might deny, either in principle or from despair, that any *formal* account of the necessary process should or could be given. Surely any formal account might either be inhibiting if used prescriptively, cutting off exciting lines of thought, or inadequate if used descriptively, given the glorious richness which human beings can bring to any task? An echo of such concerns surfaced recently in a discussion in the Operational Research Society concerning the extent to which OR embodied a version of the methodology of natural science. Rivett (1989), reiterating thoughts expressed at a conference on 'Systems in OR' a few years earlier (1983), argued that formal accounts of the 'OR process' bore no relation ('complete nonsense' is Rivett's phrase) to the reality, which actually consisted of:

> a complex process which is a mixture, in practice, of fumbling, mind-changing, chaos and political intervention (p. 17).

Critics were quick to respond. Lord (1989) argued that this denied OR's 'aspiration to be a disciplined subject', that is to say, presumably, one in which critical debate can be conducted on explicit and understood premises; while Jones (1989) saw talk of the OR process as a *post hoc* effort to understand what is happening and what developments are taking place through OR work.

The authors here would argue that in SSM a process of tackling real-world problems in all their richness has been developed; that it has benefited considerably from being formally expressed, which enables lessons to be learned and also enables users to know what they are talking about; and that the particular form which SSM takes (helped here by its use of systems thinking) both enables it to be used descriptively to make sense of a complex situation, in Jones's sense, and prescriptively to control Rivett's 'chaos'. This book will seek to illustrate all these themes. Meanwhile we can define its focus using the language of the above argument: the focus is on an organized set of principles (methodology) which guide action in trying to 'manage' (in the broad sense) real-world problem situations; it is systems-thinking-based and is applicable to taking purposeful action to change real situations constructively.

The Basic Shape of SSM

Let any purposeful activity be represented by an arrow (A in Figure 1.2). Such an action, being purposeful, will be an expression of the intention of some person or persons B (also Figure 1.2). Since A is a human action there will be someone (or several people) who take the action: they are C in the figure. The action will have an impact on some person or group, D, and it will be taking place in an environment which may place constraints upon it. These constraints are represented by E. Finally, since human autonomy is rarely total, we can add some person or group F who could stop the action being taken. Of course in real life the same person or persons could be one or more of the elements B, C, D, F, since they are roles, not necessarily individuals or groups. Overall, Figure 1.2 is a simple emblematic model of a purposeful action; it represents one way of thinking about that concept.

Now let A be the purposeful act of you, the reader, reading this book. You are B in the figure, and could now name the other elements. For example, the nature of A could be that you are preparing for an examination in management science, or that you are satisfying a curiosity about SSM, or preparing a critique of it. Given the definitions so far you would also be element D, since reading the book will have a direct effect on you yourself. In the role F might be someone from whom you have borrowed the book, who wants it back, or you might be F yourself if the copy is your own. If someone wants the book back within a week then that might be one of the constraints E. And so on . . . It is clear that you could investigate, or prepare an account of, or intervene to

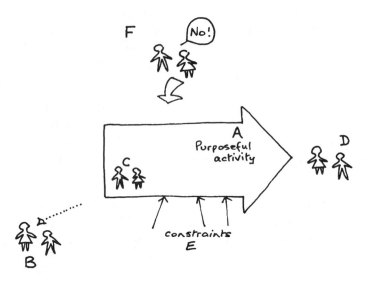

Figure 1.2 An emblematic model of purposeful activity

change your reading of this book by using Figure 1.2 as a tool. This would be an example of using a model, which is itself a pure concept, to investigate or intervene in a part of the real world.

This simple thought experiment in fact illustrates the core nature of SSM. The basic shape of the approach is to formulate some models which it is hoped will be relevant to the real-world situation, and use them by setting them against perceptions of the real world in a process of comparison. That comparison could then initiate debate leading to a decision to take purposeful action to improve the part of real life which is under scrutiny. In SSM the models are not quite like that in Figure 1.2; in fact they are carefully built models of systems to carry out purposeful activity (known as 'human activity systems') and are somewhat more elaborate than the model used in the thought experiment. But the principle is the same: find out about a situation in the real world which has provoked concern; select some relevant human activity systems; make models of them; use the models to question the real-world situation in a comparison phase; and use the debate initiated by the comparison to define purposeful action which would improve the original problem situation. Taking the action would itself change the situation, so that the whole cycle could begin again . . . and is in principle never ending. (Of course your first choice of relevant system might turn out not to be relevant. You will learn your way to true relevance by trying out a number of models.)

The shape of SSM is thus as shown in Figure 1.3, which is a slightly more elaborate version of Figure 1.1. Systems thinking is involved here in two

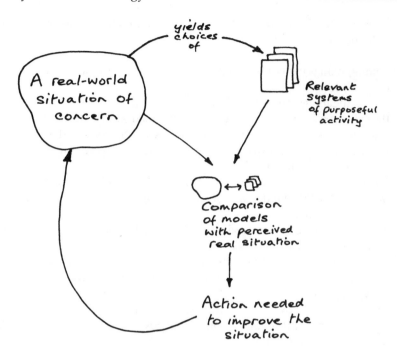

Figure 1.3   The basic shape of SSM

different ways, which make SSM doubly systemic. Firstly, the process of Figure 1.3 can itself be viewed as the operation of a cyclic learning system (a point to be discussed later in Chapters 3 and 10, in the guise of formulating ' a system to use SSM'); secondly, within the process of Figure 1.3, systems models are used to initiate and orchestrate the debate about purposeful change. The first use of systems ideas, that the whole enquiring process can be articulated as a system, is more fundamental than the second, the fact that within SSM the models used to set up a comparison/debate happen to be systems models.

We have now set out the basic ideas which underlie this book. The basic concept of Figure 1.1 has been elaborated in Figure 1.3, and the process of the latter figure has been naively illustrated using the simple model of Figure 1.2. In later chapters the process of Figure 1.3 will be expressed in more sophisticated terms, and the sequence of systems studies described in Chapters 3–9 illustrate it in action, showing it in very different contexts and making the point that the formal expression of SSM does not mean that it has to be used rigidly. It is there to help in the face of real-life's richness, not constrain. Chapter 10 then summarizes the learning from the last decade of the use of, and research into, SSM which the chosen studies illustrate. It remains to set out the structure of the book. Meanwhile the argument of this introduction can be condensed in the following way:

(1) Human beings cannot help but attribute meaning to their perceptions of the world.
(2) Those meanings constitute interpretations of the world which can be thought of as deriving from experience-based knowledge of the world.
(3) The interpretations can inform intentions which can translate into purposeful action to improve situations perceived as lying somewhere on a scale from 'less than perfect' to 'disastrous'.
(4) Purposeful action when taken changes the world as experienced (as indeed does the mental act of interpreting it) so that 1, 2, 3, above constitute a cycle.
(5) The cycle can be expressed and operated by making use of systems thinking as an epistemology.
(6) SSM does that in a coherent process which is itself an enquiring or learning system (and within the process uses models of purposeful activity systems).
(7) SSM seeks to provide help in articulating and operating the learning cycle from meanings to intentions to purposeful action without imposing the rigidity of a technique.

The Structure of this Book

Although any of the systems studies described in Chapters 3 to 9 can be read on its own and will make sense to anyone with a basic knowledge of SSM, the intention is that the book as a whole conveys a sense of the overall learning achieved during the second decade of the development of the methodology. This chapter has laid foundations. Chapter 2 gives an account of SSM in its 'late-80s' form. It stands in line with three other accounts. The first is the initial coherent account (Checkland, 1972) which was seminal but now reads as too close to the functionalism of the 'systems engineering' which was its parent. The second and third are those in *Systems Thinking, Systems Practice* (1981) and Wilson's *Systems: Concepts, Methodologies, and Applications* (1984).

Chapters 3, 4 and 5 then illustrate SSM in use in three different contexts: industry, the National Health Service and the Civil Service. The authors worked together in the study in Whitehall (in the Central Computer and Telecommunication Agency) and then carried out a sequence of studies in ICL. This sequence is described in Chapters 6, 7 and 8. The experiences in ICL changed and enriched the authors' perception of SSM: in particular the work showed that SSM is not necessarily a methodology for carrying out a special highlighted study, but can be applied to any situation in which purposeful action to bring about improvement is sought—such as in the day-to-day work of managers of any kind at any level.

Chapter 9 describes another industrial study, one which illustrates a sophisticated flexible use of SSM in its modern form, although some would see this as a 'post-modern' form of SSM!

Finally Chapter 10 summarizes the learning from the last decade, a decade in which every use of a basically established SSM has also been a piece of research on its development.

For ease of reference the shape of the book is summarized in Table 1.1

Within the accounts of uses of SSM, in Chapters 3 to 9, *inset tinted*

Table 1.1   Soft systems methodology in action

Section	Focus	Comment
Intro Chapter 1	this book	This concerns the context of this book and its shape.
Chapter 2	*methodology* of SSM	Here is an updating of SSM *as a methodology*, bringing up to date what was in *Systems Thinking Systems Practice* (1981)
Chapter 3 Chapter 4 Chapter 5	SSM in three different *application areas*	These accounts look at SSM in action in *Industry, NHS* and *Civil Service*. They enrich the account of SSM in Chapter 2, exemplifying its development to the position in Chapter 2.
Chapter 6 Chapter 7 Chapter 8	*using SSM* in a sequence of studies in one company	These accounts focus not on methodology or application areas but on *learning from use*; they happen to be based in the computer industry but that is incidental. The studies led the authors to a new view of SSM and its use. This material sheds light on what has gone before in Chapter 2–8.
Chapter 9	*current late 80s SSM in action*	Current work is described which exemplifies what has been learnt in the last decade.
Chapter 10	*more general lessons* and *future directions*	This discussion gathers the learning, reviews it, discusses the changes of perspective, and points to future directions of research and use.

paragraphs indicate a *meta-level commentary* on the content of the work, usually focused on aspects of methodology.

In 1988 Wang and Smith suggested that SSM could be combined with IDEFO, a procedure for modelling functionality developed as part of the US Air Force programme on computer-aided manufacturing. They wrote of SSM that it is

> . . . very sophisticated and mature: it requires highly experienced analysts and is dependent on a high intellectual input. Its sophistication tends to obscure the rigour. (Wang and Smith, 1988, p. 17).

It is the authors' intention to dispel the kind of fears that such a statement may evoke. They hope to establish absolutely that SSM does not require 'highly experienced analysts'; and while any use of SSM will probably benefit from 'high intellectual input' it is not dependent upon it. SSM can work with whatever 'intellectual input' is available! After reading *SSM in Action* interested persons should be well able to use the approach in a way comfortable for them in relation to situations about which they are concerned.

Part I

Soft Systems Methodology and its Application in Different Contexts

Chapter 2

The Developed Form of Soft Systems Methodology

Introduction

In January 1986 a world-wide television audience watched aghast as the Challenger space shuttle, moments after lift-off from the Kennedy Space Center at Cape Canaveral, exploded in an orange fireball, killing all seven members of its crew. This included, for political reasons, school teacher Christa McAuliffe as well as the professional astronauts. The Mission Control spokesman uttered the words which instantly became one of the most famous understatements in world history: 'Obviously a major malfunction'.

What kind of a failure was the Challenger disaster? Obviously in one sense it was a failure of engineering. Inadequately designed joints of the solid rocket booster, sealed with O-rings, failed on take off, allowing combustion gases at 5000 °C to blast and breach the external fuel tanks containing tons of liquid hydrogen. This initiated the rupture which then led to the ignition of hundreds of tons of liquid oxygen propellant. This was a clear failure of technology. But wait: NASA was under pressure to get Challenger into orbit before the President made his 'state of the Union' television broadcast. The launch was being made in freezing conditions, conditions colder than those of any previous launch. At these temperatures the O-rings could not seat properly, being less plastic than under normal conditions. This initiated the catastrophe. And NASA's Associate Administrator for Space Flight, in charge of the launch, later testified to the Presidential Commission on the Challenger accident that he had no knowledge of the concerns about the low-temperature performance of the O-rings which the relevant engineers had expressed the night before the decision to launch was taken in the morning. The Commission later saw this as a prime example of NASA's 'flawed' decision-making process (McConnell, 1988, p. 235).

This suggests that it would be equally plausible to see Challenger as a managerial or organizational failure rather than simply a failure of engineering design. The Commission took this view, describing the Challenger tragedy as 'an accident *rooted in history*' (authors' italics). In McConnell's carefully researched account of the disaster (1988) he makes the telling point that within NASA the emphasis had inevitably shifted over the years from scientific and technological considerations to managerial, commercial and political ones. It became a question, not of: Is it technically safe to launch? but of Why shouldn't we launch? with consequent changes in organizational culture, as expressed in such things as what constitutes an issue, what counts as a defensible position, a problem, or a solution.

This crucial distinction between the world of the engineer and that of the manager, so dramatically illustrated by Challenger, is important to an understanding of the so-called 'management science' which tries to bring rational discourse into problems in human affairs.

Initially, as the phrase 'management science' indicates, the focus was on resolute attempts to treat the phenomena of management in a scientific manner. In the case of Operational Research (OR), for example, the pioneers were agreeably surprised when they found that they could bring their scientific and technological thinking to bear on at least some aspects of human affairs. Blackett's celebrated note on the methodology of OR, written in the early 1940s, records that:

> . . . many more useful quantitative predictions can be made than is often thought possible. This arises to a considerable extent from the relative stability over quite long periods of time of many factors involved in operations. This stability appears rather unexpected in view of the large number of chance events and individual personalities and abilities that are involved in even a small operation. But these differences in general average out for a large number of operations, and the aggregate results are often found to remain comparatively constant.
>
> (Blackett, 1962, p. 178)

It is those 'aggregate results' which are condensed in the algorithms which apply to the queuing situation, or to decisions on depot location, bidding or equipment replacement, etc. What Blackett and his colleagues had discovered in early OR was that *the logic of situations which recur* can be treated by applied mathematics, even though any particular situation, rooted in particular history, will probably be dominated by Blackett's 'chance events' and 'individual personalities and abilities'. Unfortunately for early management science, in the real life of managers it is the details which make a particular situation unique (such as the weather on the day of the Challenger launch and the communications between engineers and managers) which usually dominate a problem situation, rather than the fact that the *form* of the situation may be one of a general class. So the early approach to management science, though useful to a degree, has had limited impact.

Historically, management science in the last twenty years has tried to expand its concerns to cover more than the logic of situations, and significant progress has been made within such fields as industrial engineering, operational research, systems analysis and systems engineering. Again using OR as an example, we find in the late 1980s that many people now assume that the phrase 'soft OR' is meaningful. It is taken to cover such approaches as soft systems methodology (SSM) and cognitive mapping (Eden, Jones and Sims, 1983). The phrase is now accepted as meaningful even though there is no shared and agreed definition of it, and some well-established figures still argue that 'this . . . is not OR' (Machol, 1980). Pruzan (1988) lists a number of the shifts entailed in a move from classic to soft OR (though he himself does not use that phrase): from optimization to learning; from prescription to insight; from 'the plan' to the 'planning process'; from reductionism to holism, etc.

Management science has thus moved to try to increase its relevance to the world of management, though this has been at the expense of some of its aspirations to be scientific, at least in the same sense that natural science and engineering are scientific.

The development of SSM during its own history has shown a shift of just this kind, in its case from the world of engineering thinking to the world of management thinking (using the word 'management' in the broad sense discussed in the previous chapter). Its parent was systems engineering, and it moved experientially from an approach aimed at optimizing a system to an approach based on articulating and enacting a systemic process of learning.

This chapter takes as given the fundamental difference between the thinking of science/engineering and that needed to learn one's way to action to improve human problem situations. (Further discussion of these issues is found in Checkland, 1980, 1981, 1983a, 1985a). Here we present an account of the developed form of SSM in the late 1980s, with only brief allusion to its early history (covered in detail in *Systems Thinking, Systems Practice*). The purpose of the present account is simply to provide the intellectual base which underlies the illustrations of SSM in action in Chapters 3–9. Together with those accounts it then leads to the discussion of 'the system to use SSM' developed in Chapter 10.

THE EMERGENCE OF SSM

Context

When Gwilym Jenkins moved to the new University of Lancaster in the mid-1960s he was somewhat disenchanted with what he felt to be the lack of contact with the real world in the older university from which he had moved. He established a postgraduate Department of Systems Engineering at Lancaster, and the initial funding for the Department was a gift from ICI, the UK's largest company. ICI was realizing at that time that the operation of a petrochemical complex, in which the output from some plants is the input to others, entails engineering the whole as a single complex system. This was different from building and operating a plant to make a single product, and it seemed to some forward-looking managers in ICI's then Agricultural and Heavy Organic Chemicals Divisions that it would be useful if this kind of problem were being addressed by some university-based group: hence the gift to Lancaster.

Jenkins, however, always interpreted the word 'engineering' in the broad sense that you can 'engineer' a meeting, or the release of hostages, as well as a nitric acid plant. He started with work on 'hard' systems engineering (engineering a well-defined system to achieve its objectives) but soon recruited Checkland from industry (from ICI, as it happens) to research the application of the methodology of (hard) systems engineering to less-well-defined problems of the kind with which managers of all kinds and at all levels try to cope. He had already established the Department on the basis that the only way to develop this new subject was by interaction with real problem situations in an 'action research' mode. Such work cannot test hypotheses in the classic manner of scientists in laboratories. Action research requires involvement in a problem situation and a readiness to use *the experience itself* as a research object about which lessons can be learned by conscious reflection. In order to make this possible it is absolutely essential to declare in advance an intellectual framework which will be used in attempts to make sense of both the situation and the researcher's involvement in it. It is with reference to the declared framework that 'lessons' can be defined. The action researcher thus has two hopes: that the framework will yield insights concerning the perceived problems which will lead to practical help in the situation; and that experiences of using the framework will enable it to be gradually improved. (For some discussions of action research—which on the whole unfortunately neglect the crucial importance of declaring the intellectual framework—see Susman and Evered, 1978; Hult and Lennung, 1980; Warmington, 1980; Gilmore, Krantz and Ramirez, 1985.)

The Lancaster researchers started their action research programme by taking hard systems engineering as a declared framework and trying to use it in unsuitable situations, unsuitable, that is, in the sense that they were very messy problem situations in which no clear problem definition existed. 'See what you can do to help us', said the Marketing Director of a textile business sliding to extinction; 'Make me plan', said the owner of a small carpet-making business; 'Try to improve the project', said a Director of British Aircraft Corporation, where the world's first supersonic passenger aircraft, Concorde, was taking longer to develop and was costing very much more than the British taxpayer had been led to believe it would. These were the kinds of situation in

which systems engineering used in an action research mode {
emerged as an alternative.

Systems Engineering and its Breakdown

What we think of as engineering begins when a need is established; and the engineer's task is to provide something which meets the need, whether in the form of a physical object or a procedure or both. The best engineer is the one who provides with a minimum of resources a solution which both works and is aesthetically pleasing. The history of the engineers' contribution to human life is largely a history of achieving more with less, as experience is gained. Buckminster Fuller points this out in his last major work *Critical Path* (1983) and indeed we note that where the dome of St Peter's weighs 30 000 tons, Fuller's first geodesic dome, of the same size, weighed 30 tons. This is all very inspiring, and the engineering world attracts many people who are motivated by the achievement of demonstrable results and have a feel for elegant solutions.

Note, however, the status of the engineer's contribution. It is a 'how-oriented' activity; it answers the question *How* can this need be met? *What* the need is has already been defined. This fundamental point is especially important if the engineer is a systems engineer. The approach then boils down to expressing the need to be met in the form of a named *system with defined objectives* (say, a system to build a supersonic aircraft meeting a defined specification within a stated time and to a stated budget; or a system to supply a small town with pure water at a certain rate for a given cost). If the system and its objectives are defined, then the process is to develop and test models of alternative systems and to select between them using carefully defined criteria which can be related to the objectives. Nowadays the testing of alternatives will often involve running simulation models using computers, and many techniques are available to help the systems engineer select the best alternative. But note that the whole of this approach is predicated on the fact that the need and hence the relevant need-meeting system, can be taken as given. Systems engineering looks at 'how to do it' when 'what to do' is already defined.

This was found to be the Achilles' heel of systems engineering, however, when it was applied, in the Lancaster research programme, to ill-defined problem situations. Problem situations, for managers, often consist of no more than a feeling of unease, a feeling that something should be looked at, both from the point of view of whether it is the thing to do and in terms of how to do it. Being a manager, in the broad sense argued in the first chapter, is to be a decider of what to do as well as how to do it. This means that naming a system to meet a need and defining its objectives precisely—the starting point of systems engineering—is the occasional special case. The normal situation is to speculate about both 'whats' and 'hows', and to come to the business of engineering systems to achieve stated objectives only late on in problem handling.

What was found to be needed was a broad approach to examining problem situations in a way which would lead to decisions on action at the level of both 'what' and 'how'. The solution, already outlined in Figure 1.3 in the previous chapter, was *a system of enquiry*. In it, a number of notional systems of purposeful activity which might be 'relevant' to the problem situation are defined, modelled and compared with the perceived problem situation in order to articulate a debate about change, a debate which takes in both 'whats' and 'hows'. To explain that process in detail it is first necessary to examine briefly the nature of the core ideas upon which it is built, namely those of systems thinking.

THE NATURE OF SYSTEMS THINKING

The failure of systems engineering to cope with anything other than well-structured problem situations led to the basic rethink of the fundamentals of systems thinking which is described in the first four chapters of *Systems Thinking, Systems Practice*. Eventually the rethink provided a theoretical base for the new methodological moves which the limitations of systems engineering made necessary. Here the discussion of systems thinking as such will be parsimonious. Only enough will be covered to make sense of the discussion of the process of SSM which follows.

Systematic and Systemic

Most literate people, asked to name an adjective from the noun 'system', would offer 'systematic'. But there is another adjective from 'system' which is more important than 'systematic' if the nature of systems thinking is to be understood: the adjective 'systemic'. For most people who know it, the word would have biological or medical connotations. A condition is 'systemic' if it pervades the body as a whole, and it is noticeable that the recently published *Oxford Dictionary of Current English* gives only a medically oriented definition, namely 'of the bodily system as a whole'. But that is unnecessarily limiting, and a better definition would be: 'of or concerning a system as a whole'.

The adjective 'systemic' implies that we have a clear concept of what we mean by the notion of 'system'. There *is* such a notion, and systems thinking is simply consciously organized thought which makes use of that concept.

The concept itself starts with the most basic core idea of systems thinking, namely that a complex whole may have properties which refer to the whole and are meaningless in terms of the parts which make up the whole. These are the so-called 'emergent properties'. The vehicular potential of a bicycle is an emergent property of the combined parts of a bicycle when they are assembled in a particular way to make the structured whole. But the best example of an emergent property known to the authors was supplied spontaneously by a Lancaster Master's student, Pat Lazenby. In an animated discussion with class colleagues of this idea of emergence, and its well-known everyday expression in the popular phrase 'The whole is more than the sum of its parts', Pat turned to a fellow student alongside her and said 'You're certainly more than the sum of your parts, you're an idiot!' Pat is here asserting that if we take this man to be a single whole entity, then at that level of reality it becomes meaningful to apply the label 'idiot' to him as a whole. Clearly there would be literally no meaning in applying that label to his ankle or his elbow.

The concept of emergent properties itself implies a view of reality as existing in layers in a hierarchy (there being no connotations of authoritarianism in this technical use of the word). In the biological hierarchy, for example, from atoms to molecules to cells to organs to organism, an observer can describe emergent properties at each layer. In fact it is the ability to name emergent properties which defines the existence of a layer in hierarchy theory.

To complete the idea of 'a system' we need to add to emergence and hierarchy two further concepts which bring in the idea of *survival*. The hierarchically organized whole, having emergent properties, may in principle be able to survive in a changing environment if it has processes of *communication* and *control* which would enable it to adapt in response to shocks from the environment.

Although there is no common account of the concept 'system' in the literature, Atkinson and Checkland (1988), examining many accounts of basic systems ideas, find that all authors draw on the same cluster of ideas and that the image underlying all accounts can be expressed in the two pairs of ideas: emergence and hierarchy, communication and control, as suggested by Checkland in 1981. These ideas together generate the image or metaphor of the adaptive whole which may be able to survive in a changing environment. To make mental use of that image is to do systems thinking.

Systems Thinking: Thinking with Holons

When the Spanish conquistadores arrived in what is now Mexico, the indigenous people, unfamiliar with horse riding and seeing riders dismount from horses, thought that creatures had arrived on their shores who could divide themselves in two at will. This story provides a good illustration of the way in which we have in our heads stocks of ideas by means of which we interpret the world outside ourselves.

Philosophically the story supports the view of Kant, that we *structure* the world by means of already-present, innate ideas, rather than the view of Locke that our minds are blank screens upon which the world writes its impressions. But it seems clear that the supposedly 'innate' ideas may have two sources. They may indeed be part of the genetic inheritance of mankind, truly innate; or they may be built up as a result of our experience of the world. In the late 1980s in Europe, for example, the concept 'lager lout' has been established as meaningful. Once it exists, with its strong connotations of certain kinds of socially undesirable behaviour on the part of young males with time on their hands, it becomes a structuring device which will colour future interpretations of observed behaviour. A decade ago neither the phrase nor the connotations it trails after it, existed, so that the interpretations now based upon it could not have been made in quite that way.

What is being argued is that we perceive the world through the filter of—or using the framework of—the ideas internal to us; but that the source of many (most?) of those ideas is the perceived world outside. Thus the world is continually interpreted using ideas whose source is ultimately the perceived world itself, in a process of mutual creation like that shown in Figure 2.1, which can be seen as another version of Figure 1.1. As human beings we enact this process every day, usually unconsciously. But if we now add the thought that thanks to our human nature we are able consciously to think about our own mental processes, we arrive at Figure 2.2 in which the ideas (x in the illustration) are used in some methodology, M, to interpret perceived reality (Checkland, 1984). The figure is perfectly general, and has many specific manifestations. For example, the recently developed ideas of plate tectonics now structure the methodology through which geologists perceive the basic structure of the Earth's crust.

Systems thinking is another way in which Figure 2.2 is made manifest; the idea 'system' is explicitly used, and there are formally defined methodologies such as systems engineering or systems analysis. Systems thinking is in fact a very general manifestation of Figure 2.2, with multiple applications in different areas of our perceptions.

'System' itself is simply one of the x's in Figure 2.2. The idea of the adaptive whole is a very general idea, but it has also turned out to be a very powerful one, suggesting that the world is characterized by complex structure and much connectivity. This at least matches our intuitive feeling that the world is probably more like a dense privet hedge than a glass of water or a handful of marbles.

In fact the process of using 'system' consciously to interpret the world has been rather too successful. It has been so successful that the system idea, though abstract, has long been used as a *label* for parts of the world. In the third book of Newton's *Principia* (published in 1687) he writes that 'I now demonstrate the frame of the system of the world', and in everyday language

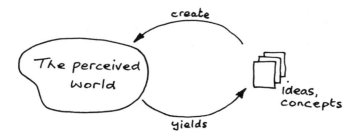

Figure 2.1 The world interpreted by ideas whose source is the world itself

we follow Newton when we casually refer to 'the education *system*', 'the legal *system*' or 'health care *systems*'. . . . We could indeed examine the problems of coherently providing health care, or education, or the application of the law, by making use of the idea of a system, that is to say a whole with emergent properties, a layered structure and processes which enable it to adapt in response to environmental pressures. But it is too easy casually to say 'education system' *as if* the arrangements for providing education automatically meet the requirements of the notion 'system'. Most people engaged in education will probably deny that the arrangements they encounter actually map the system concept! The error here is to confuse a *possibly plausible description* of perceived reality, with perceived reality itself.

The source of this confusion, which leads to untold misunderstandings of systems thinking and its nature, is the mistake made by the pioneers of systems thinking when the idea 'system', long used in everyday language, was being elaborated as a more tightly defined technical term.

The concept 'system' began to be elaborated when Ludwig von Bertalanffy, an organismic biologist interested in the organism as a whole rather than any

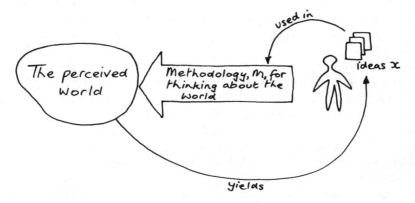

Figure 2.2 Figure 2.1 expanded

of its constituent parts, suggested that the ideas he and his colleagues had developed concerning organisms could be applied to wholes of any kind (Gray and Rizzo, 1973). Unfortunately he made a bad mistake in using the word 'system' for the name of the abstract notion of a whole he was developing. In *General System Theory*, Bertalanffy (1968) clearly regards 'system' as an abstract concept, but unfortunately he immediately starts using the word as a label for parts of the world. Now, going back to the idea of 'an education system', it is perfectly legitimate for an investigator to say 'I will treat education provision *as if it were* a system', but that is very different from declaring that it *is* a system. This may seem a pedantic point, but it is an error which has dogged systems thinking and causes much confusion in the systems literature. Choosing to think about the world as if it were a system can be helpful. But this is a very different stance from arguing that the world *is* a system, a position which pretends to knowledge no human being can have. What Bertalanffy should have done in the late 1940s was to make up a new word for the abstract concept of a whole which might then be used to understand or create real-world systems, with the latter word conceded to everyday language.

In fact a number of words have been suggested as an alternative to 'system' for the name of the concept of a whole; adopting any of them would correct Bertalanffy's error. Proposals include 'org' (Gerard, 1964), 'integron' (Jacob, 1974) and 'holon' (Koestler, 1967, 1978). Only the latter has been taken up to any significant extent, but it would clarify the whole field of systems thinking if it became more popular—and especially if the field became known as 'holonic thinking' or 'thinking with holons' (Checkland, 1988b)! If the word 'holon' were adopted for the abstract idea of a whole having emergent properties, a layered structure and processes of communication and control which in principle enable it to survive in a changing environment, then Figure 2.2 would be readily understood. It would make it clear that systems thinkers are people who formulate some holons (x) relevant to aspects of perceived reality which they are interested in, and then use the holons in a methodology, M, to find out about, or gain insight into, or engineer, some of the world outside themselves.

The adoption of the word 'holon' would also make clearer the two complementary schools of thought within systems thinking, usually labelled 'hard' and 'soft'. Probably most people aware of the distinction imagine that it marks the difference between the kinds of problem tackled. Hard systems engineers tackle rather well-defined problems, while soft systems methodologists address messy, ill-structured, problem situations. This is true, but it is not the fundamental difference between the complementary schools. Referring again to Figure 2.2, hard systems thinking assumes that *the perceived world* contains holons; soft systems thinking takes the stance that the methodology, M, *the process of enquiry*, can itself be created as a holon. In the case

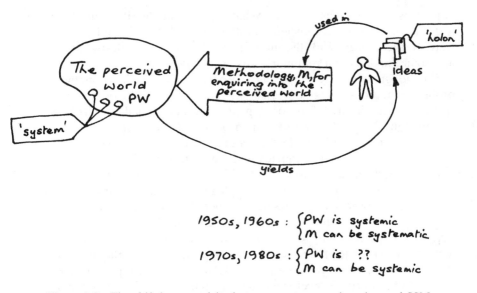

Figure 2.3 The shift in systemicity between systems engineering and SSM

of SSM we have a cyclic methodology which is itself a systemic (we would better say, holonic) process, one which within its procedures happens to make use of models of holons.

In summary, then: we engage with the world by making use of concepts whose source is our experience of the world; this process of engagement, usually unconscious as we live everyday life, can be made explicit; one way of doing so is embodied in so-called 'systems thinking', based on the idea of making use of the concept of 'a whole'. In systems thinking, accounts of wholes are formulated as holons, and these can be set against the perceived world, in order to learn about it. Within the systems movement two schools are complementary: that which takes the world to be holonic ('hard systems thinking') and that which creates the process of enquiry as a holon. SSM is such a holon, a cyclic process of enquiry which happens also within its processes to make use of holons. In everyday language, we say that SSM is systemic in two senses. It is a systemic process of enquiry, one which happens to make use of 'systems models'. Figure 2.3 provides a summary reminder of this argument.

Purposeful Holons: Human Activity Systems

When the Lancaster researchers tried to apply the methodology of systems engineering to ill-defined problem situations, they quickly found themselves in difficulties because the questions: 'What is the system?' and 'What are its objectives?' were not answerable. What in fact made the situations ill-defined was that objectives were unclear and that both what to do and how to do it were problematical. With the intended methodology manifestly inappropriate, some fundamental rethinking was necessary.

The thought which led eventually to progress was the realization that all the problem situations being addressed, whether in the public or in the private sector, whether in small firms or in giant corporations, had one characteristic in common. They all featured human beings in social roles trying to take purposeful action. The textile firm going out of business, the small carpet manufacturer wanting his company to do more than live from day to day, the managers and engineers concerned with the Concorde project, these and many similar situations all revealed people immersed in complex action which they were trying to make purposeful rather than instinctive or merely random. They were involved in purposeful action in the world, and they also hoped to act purposefully to make a situation seen as problematical somehow 'better'. Why not then take a set of activities connected together in such a way that the connected set makes a purposeful whole, to be a particular kind of holon? Why nor try to develop the idea of a purposeful whole to put alongside the ideas of *natural* wholes developed by ecologists and physical geographers, and the ideas of *designed* wholes developed by engineers and others? That was the thought which launched SSM.

In order to label such a holon, the phrase 'human activity system' was borrowed from a text on industrial engineering in which that phrase was used but not developed (Blair and Whitston, 1971). The phrase is useful because it enables us to distinguish between what gets executed in the (always abstract) holon, namely 'activity' and what characterizes the real world, namely 'action'. Those who write about 'human activity systems' as if they exist in the world, rather than being holons which can be compared with the world, are failing to grasp the essence of soft systems thinking, namely that it provides a coherent intellectual framework, namely a structured set of '*xs*' in Figure 2.2, as *an epistemology* which can be used to try to understand and intervene usefully in the rich and surprising flux of everyday situations.

Holons which are 'human activity systems' were defined in such a way that they meet the characteristics of a whole as developed in systems thinking. The emergent property of a defined human activity system is the ability, in principle, to pursue the purpose of the whole. Any such whole may itself be thought of as containing similar smaller wholes as constituent parts, and may itself be part of a larger whole. For example, a holon relevant to provision of a district nursing service might contain a number of purposeful sub-holons including, among others, ones relevant to nurse recruitment and to service delivery. Similarly, the parent holon might itself be thought of as part of a wider holon which mapped total health care provision for a defined geographical area. Finally, to meet the holistic requirement, a purposeful holon must have within it activities and structure concerned with communication and control so that the holon could in principle (were it to exist) adapt and survive in a changing environment. This concept was elaborated in the development of SSM in ways which will be described later in this chapter. Meanwhile, that development required facing up to one other crucial characteristic of a notional 'human activity system'.

This characteristic is simply that the description of any purposeful holon must be from some declared perspective or worldview. This stems from the special ability of human beings to interpret what they perceive. Moreover, the interpretation may, in principle, be unique to a particular observer. This means that multiple perspectives are always available. (The letter from the tax collector may seem to be unequivocally a message concerning your financial affairs, but no one can stop you perceiving it as a bookmark!) This is nowhere more apparent than in the case of purposeful activity. Officially a university may at core be concerned with teaching and research, but for would-be holiday makers and hard-pressed university finance officers it may be appropriate to regard as relevant the view that it is a source of cheap and cheerful hotel accommodation—as, in real life, the growth in the campus holiday trade in recent years demonstrates.

What this means is that whenever we develop human-activity-system holons relevant to real purposeful action, it is important to envisage a number of different worldviews and to develop a handful of holons. Figure 2.4 gives a few obvious possibilities for newspaper publishing, guerillas and professional football.

Finally we may summarize this condensed account of the systems thinking needed to understand SSM in the following bald terms:

(1) Systems thinking takes seriously the idea of a whole entity which may exhibit properties as a single whole ('emergent properties'), properties which have no meaning in terms of the parts of the whole.

(2) To do systems thinking is to set some constructed abstract wholes (often called 'systems models') against the perceived real world in order to learn about it. The purpose of doing this may range from engineering (in the broad sense of the word) some part of the world perceived as a system, to seeking insight or illumination.

(3) Within systems thinking there are two complementary traditions. The 'hard' tradition takes the world to be systemic; the 'soft' tradition creates the process of enquiry as a system.

(4) SSM is a systemic process of enquiry which also happens to make use of systems models. It thus subsumes the hard approach, which is a special case of it, one arising when there is local agreement on some system to be engineered.

(5) To make the above clear it would be better to use the word 'holon' for

the constructed abstract wholes, conceding the word 'system' to every-day language and not trying to use it as a technical term.

(6) SSM uses a particular kind of holon, namely a so-called 'human activity system'. This is a set of activities so connected as to make a purposeful whole, constructed to meet the requirement of the core system image (emergent properties, layered structure, processes of communication and control).

Figure 2.4 Ideas for holons relevant to some familiar concepts

(7) In examining real-world situations characterized by purposeful action, there will never be only one relevant holon, given the human ability to interpret the world in different ways. It is necessary to create several models of human activity systems and to debate and so learn their relevance to real life.

THE ENQUIRING PROCESS WHICH IS SSM

The Process Overall

The brief excursion into basic systems thinking in the previous section has provided enough ideas to make detailed sense of the enquiring process which is SSM. First we shall expand Figure 1.3 into a form which reflects current practice; then its two streams of enquiry, logic-based and cultural, will be described.

The usual general description of SSM is that shown in Figure 2.5, in which it is presented as a seven-stage process (Checkland, 1975). In the late 1980s the 1975 version seems rather bald, and in any case gives too much an impression that SSM is a seven-stage process to be followed in sequence, a

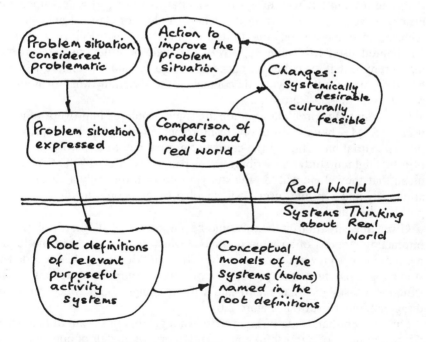

Figure 2.5 The conventional seven-stage model of SSM

point which will be discussed in the final chapter. The version of SSM to be used here is that shown in Figure 2.6 (Checkland, 1988a). That figure describes a process which has recently been succinctly summarized by von Bulow (1989):

> SSM is a methodology that aims to bring about improvement in areas of social concern by activating in the people involved in the situation a learning cycle which is ideally never-ending. The learning takes place through the iterative process of using systems concepts to reflect upon and debate perceptions of the real world, taking action in the real world, and again reflecting on the happenings using systems concepts. The reflection and debate is structured by a number of systemic models. These are conceived as holistic ideal types of certain aspects of the problem situation rather than as accounts of it. It is taken as given that no objective and complete account of a problem situation can be provided.

Referring to Figure 2.6, we have a situation in everyday life which is regarded by at least one person as problematical. There is a feeling that this situation should be managed (in the broad sense discussed in Chapter 1) in order to bring about 'improvement'. The whats and hows of the improvement will all need attention, as will consideration of through whose eyes 'improvement' is to be judged. The situation itself, being part of human affairs, will be a product of a particular *history*, a history of which there will always be more than one account. It will always be essential to learn and reflect upon this history if we are to learn from the relative failure of classical management science, since that is surely due to its attempt to be ahistorical. In so doing it has limited itself to dealing only with the logic of situations. We are not indifferent to that logic, but are concerned to go beyond it to enable action to be taken in the full idiosyncratic context of the situation, which will always reveal some unique features.

Facing up to the problem situation are some 'would-be improvers' of it, the users of SSM. They consist of one or more persons motivated to improve the problem situation, either in the course of normal day-to-day work or as part of a highlighted study. Describing these people as 'users of SSM' does not mean that they alone will 'do the study': SSM is intrinsically a collaborative approach, and sensible 'users' will involve other people in the process of problem handling.

Given the situation and the would-be improvers of it, there follow two interacting streams of structured enquiry which together lead to the implementation of changes to improve the situation. Both may be regarded as stemming from both the perception of various purposeful actions in the situation ('tasks' in Figure 2.6) and various things about which there are disagreements ('issues' in Figure 2.6).

On the right-hand side of Figure 2.6 is a logic-driven stream of enquiry in which a number of purposeful holons in the form of models of human activity

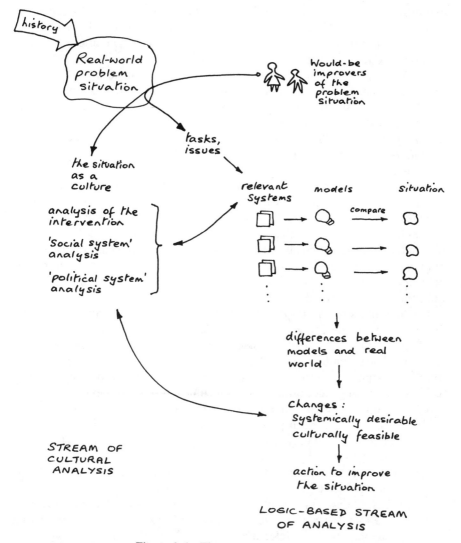

Figure 2.6 The process of SSM

systems are named, modelled and used to illuminate the problem situation. This is done by *comparing* the models with perceptions of the part of the real world being examined. These comparisons serve to structure a debate about change.

What is looked for in the debate is the emergence of some changes which could be implemented in the real world and which would represent an *accommodation* between different interests. It is wrong to see SSM simply as consensus-seeking. That is the occasional special case within the general case of seeking accommodations in which the conflicts endemic in human affairs are still there, but are subsumed in an accommodation which different parties are prepared to 'go along with'. [In macro-politics the anomalous post-Second World War status of Berlin provides a good example of an accommodation, in this case concerning the multiple occupation of that city by East and West as represented by Warsaw Pact and NATO countries. In this example the accommodation was tested to the limit by the Russian blockade of West Berlin in the late 1940s, which provoked the airlift to break the blockade; the accommodation over Berlin held, just; there was never East–West consensus (see Tusa and Tusa, 1988)].

What has been described of SSM so far is a stream of thinking and debate which is essentially logic-driven. It uses the purposeful holons as logical machines which can be used to question the real world. Some successful studies have done no more than this. But in general, though logic has a part to play in human affairs, we need also to pay attention to the cultural aspects of human situations, the aspects which make them specifically human. In recent years a second stream has been developed in SSM which interacts with the first, the logic-driven stream.

The 'cultural stream', on the left-hand side of Figure 2.6, consists of three examinations of the problem situation. The first examines the intervention itself, since this will inevitably itself effect some change in the problem situation. The second examines the situation as a 'social system', the third as a 'political system'. In both cases the phrases within inverted commas are used as in everyday language, rather than as technical terms. And in the case of all three 'cultural' enquiries, general models are used which relate respectively to problem solving, the social process and the power-based aspects of human affairs. (A more detailed account will be given below, and the following chapters will provide illustration.)

It is clear that the logic-driven stream and the cultural stream will interact, each informing the other. Which selected 'relevant' human activity systems are actually found to be relevant to people in the problem situation will tell us something about the culture we are immersed in. And knowledge of that culture will help both in selection of potentially relevant systems and in delineation of changes which are culturally feasible. Here we need to remember that what in the end turns out to be feasible will itself be affected by the learning generated by the project itself: human situations are never static.

Changes implemented as a result of the use of SSM of course change the problem situation as originally perceived, and in the new situation the cycle of learning stimulated by the methodology can begin again. . . . It is in principle never ending, and ending a systems study is an arbitrary act.

Overall, the aim of SSM is to take seriously the subjectivity which is the crucial characteristic of human affairs and to treat this subjectivity, if not exactly scientifically, at least in a way characterized by intellectual rigour.

The Stream of Logic-based Enquiry

If a user of SSM is adopting the approach in the day-to-day situation, then it is likely that he or she will feel that they know a lot about the situation and can get straight on the logic-based stream of thinking. In other cases, and especially in highlighted studies tackled by a team constituted for the purpose, it will be necessary to do some organized finding out. Approaches to this will be discussed as part of the stream of cultural enquiry. First we consider the logic-based thinking in which relevant systems are chosen, named, modelled and compared with perceptions of the real-world situation.

Selecting Relevant Systems

No human activity system is *intrinsically* relevant to any problem situation, the choice is always subjective. We have to make some choices, see where the logical implications of those choices takes us, and so learn our way to truly 'relevant systems'. In the early years of SSM development much energy was wasted in trying at the start to make the best possible choices. (This at least was better than the very earliest attempts to name *the* relevant system, in the singular!) Users of SSM have to accept this initial dousing in subjectivity, and though this is never a problem for those whose inclinations are towards the arts and humanities, it can be difficult for numerate scientists and engineers whose training has not always prepared them for the mixed drama, tragedy and farce of the social process.

Two kinds of choice of relevant systems can be made (Checkland and Wilson, 1980). In many cases there will be visible in the real world some organized purposeful action which could be reflected in the choice of a notional human activity system whose boundary would coincide with the real-world manifestation. This is the kind of choice made axiomatically in hard systems thinking, and it is often the only kind of choice with which unreflecting hard systems thinkers are comfortable. In SSM this kind of choice is referred to as a 'primary-task system'.

Consider the charity organization Oxfam, subject of an early study using SSM, aimed at improving the operation of its management committee. Oxfam can be observed providing relief, providing aid, running retail shops and begging. It would be possible to name relevant systems based on these actions ('a system to provide relief', etc.), and we might in the real world anticipate finding functional divisions of Oxfam which map these choices. Or an overall relevant system with the four named systems as subsystems could map the organization boundary of Oxfam as a whole. This would be a primary task system for Oxfam with each subsystem being itself a choice of the same kind. But within Oxfam, as in any organization undertaking a portfolio of different tasks, there will always be debate about its core purposes and about the fraction of resources which should be devoted to each. From this consideration we could make the second kind of choice of relevant system. We could name as relevant such conceptualizations as 'a system to resolve disagreements on resource use' or 'a system to define information flows to and from the management committee'. Here we would not necessarily expect to find institutionalized versions of such systems in the real world. In SSM these are called 'issue-based relevant systems'; in general their boundaries would not map on to real-world organization boundaries.

Experientially it has been found important to make choices of both 'primary task' and 'issue-based' systems if the thinking in the study is to be of the mind-opening variety. In Chapter 5 is described a study in the UK Government's Central Computer and Telecommunication Agency. One relevant system was based on the Agency mission statement, which yields a primary task system whose boundary would map the organizational boundary of the Agency as a whole. One issue-based choice was of 'a system to reconcile the conflicting demands made by the Agency, and by its 'customer' Government Departments, on the Agency's Departmental Liaison Officers'.

The distinction between primary task and issue-based relevant systems is not sharp or absolute, rather these are the ends of a spectrum. At the extremes, primary task systems map on to institutionalized arrangements; issue-based systems, on the other hand, are relevant to mental processes which are not embodied in formalized real-world arrangements.

Working with both kinds of relevant system frees the thinking, but perhaps because of this an initial tentativeness about choice-making is often observed. There is a fear perhaps that the initial choices made will inevitably have a blinkering effect on subsequent thinking, causing hesitation or even freezing. Helpful at such times (and not to be neglected in general) can be a conscious lifting of the thinking to the level of metaphor. For example, the problem situation addressed will always contain many relationships between parties A and B. The authors have often found it useful to think of a number of metaphors for an A–B relationship and reflect these on to the real situation to stimulate thought. Is the relationship between A and B like that between: policeman/robber, parasite/host, husband/wife, mistress/lover, master/slave, judge/accused, equal partners, brother/sister, mother/child, organism/virus, etc., etc.? This can get the thoughts moving.

Davies and Ledington (1987) report the interesting case of an engineer who pleaded that he was incapable of developing metaphors but in the end gained much by thinking of a project he was engaged in as similar to a car. Was it an old banger, a sporty model, a family saloon etc.? This led to consideration of roles related to the metaphor (mechanic, car salesman etc.) and to speculation about such relevant systems as ' a system to sell an old banger as a classical model', a thought which could then be rephrased in the language of the real situation. Richie (1987) gives another example of the value of temporarily shifting to the level of abstraction which metaphors represent.

Davies and Ledington argue that this kind of abstracting has been found useful in using SSM in studies characterized by:

> conservative thinking, premature judgement of solutions and politically difficult situations (p. 184).

This describes most problem situations in our experience!

Naming Relevant Systems

In the development of SSM it was quickly found necessary to pay close attention to the formulation of the names of relevant systems. These had to be written in such a way that they made it possible to build a model of the system named. The names themselves became known as 'root definitions' since they express the core or essence of the perception to be modelled.

A root definition expresses the core purpose of purposeful activity system. That core purpose is always expressed as a transformation process in which some entity, the 'input', is changed, or transformed, into some new form of that same entity, the 'output'. Figure 2.7 sets out a prescription for formulating a transformation process, and gives an example in which a public library is conceptualized in several different ways according to different worldviews, in the same way that was demonstrated earlier in Figure 2.4. This simple notion is astonishingly misunderstood not only in the everyday world and among careless users of SSM, but also, alas, in the systems literature.

It is worth quoting some inadequate expressions of the input–output idea. This is done not to pillory their authors but because learning from mistakes is a good way to learn. In an expression of a health care system Passos (1976) gives 'knowledge', 'facilities and equipment', and 'manpower' as inputs and 'population served' and 'health care needs met' as outputs! This is a jumble, but various defensible answers could be constructed from these ideas. Thus if the input were 'health care needs' then the given output of 'health care needs met' would be acceptable. If 'facilities and equipment' were chosen as input then 'used facilities and equipment' would be technically correct as an output. If the output is to be something like 'population served' then we would have to express this as 'population with health care needs' transformed into 'population with health care needs attended to'. Or perhaps Passos means 'need to define population served' transformed into that need met. In another health care example Ryan (1973) considers a nursing care system. She has 'nursing assessment' transformed into 'nursing orders', which could be a legitimate account of a transformation carried out mentally by people directing nurses. Unfortunately she also has 'observe, infer, hypothesize' transformed into 'prescribe, schedule, inform'! The error here is to name the input and output as verbs instead of entities. Actions do not get transformed into

Input ⟶ (Transformation process) ⟶ Output

(some entity)

(that entity in a transformed state)

Example: a public library

a local population ⟶ that population better informed

books ⟶ 'dog-eared' books

local need for information and entertainment from books, records, etc. ⟶ that need met

local provision of education ⟶ that provision enhanced

books and other materials on the shelves ⟶ books and other materials out in the local community

etc., etc.

Figure 2.7 The idea of 'transformation process'

anything; they may *lead to* conclusions or other actions, but 'lead to' is a different concept from 'are transformed into': a causal sequence is not the same as a transformation. It is vitally important always to express inputs and outputs as entities: the concept of 'transforming' demands it. Finally, Easton (1961) provides a bad example at an abstract level. He suggests that for a political system the inputs are 'demands' and 'support' and that the outputs are 'decisions and actions'! To be technically correct in systems terms, his

Formulate root definitions by considering the
elements C A T W O E :

C 'customers' : the victims or beneficiaries of T

A 'actors' : those who would do T

T 'transformation: the conversion of input to output
   process'

W 'Weltanschauung': the worldview which makes this
   T meaningful in context

O 'owner(s)' : those who could stop T

E 'environmental: elements outside the system which
   constraints' it takes as given

Figure 2.8 The CATWOE mnemonic

concept would have to be expressed as 'need for decisions on demands having a degree of support' into 'that need met by a decision-taking process'.

An input–output transformation is, on its own, too bald to be modelled richly, and root definitions came to be written as sentences elaborating the core transformation. Smyth and Checkland (1976) researched historical root definitions and suggested that well-formulated root definitions should be prepared by consciously considering the elements shown in Figure 2.8. The elements make the word CATWOE, and much experience has shown this to be a most useful mnemonic. (If the elements seem familiar it is because the emblematic model of purposeful activity shown in Figure 1.2 is in fact an undeclared representation of CATWOE.)

The core of CATWOE is the pairing of transformation process T and the W, the *Weltanschauung* or worldview which makes it meaningful. For any relevant purposeful activity there will always be a number of different transformations by means of which it can be expressed, these deriving from different interpretations of its purpose. Consider how different are the examples for a public library given in Figure 2.7: they derive from different *Weltanschauungen*.

The other elements in CATWOE add the ideas that someone must undertake the purposeful activity, someone could stop it, someone will be its victim or beneficiary, and that this system will take some environmental constraints as given. A root definition formulated with attention to these elements will be rich enough to be modellable. Each one does not have to be explicit in the definition, but if they are to be omitted that should be a conscious act.

The structure of CATWOE implies that the simplest version of a root definition would be 'a system to do X' where X is a particular transformation process. This leaves the system itself to select a means of doing X (there may be several available); it would freely choose a 'how' for the 'what' defined by X. Or it may be felt useful to constrain the system to a particular 'how', so that the next most complicated form of root definition will be 'a system to do X by Y'. Now, the existence of O in CATWOE implies the concern of someone (or some group) who could stop the activity of the system if it were not meeting their aspirations. This implies that a 'full' root definition's core transformation would be '*a system to do X by Y in order to achieve Z*', where the T will be the means Y, Z is related to the owners' longer term aims, and there must be an arguable connection which makes Y an appropriate means for doing X. In general it is useful to write root definitions with the XYZ formula in mind.

Later chapters contain many examples of root definitions and CATWOE used in recent studies, and in the next section the modelling of a simple root definition will be illustrated in detail.

Modelling Relevant Systems

Root definitions and CATWOE are the source of the purposeful holons known as 'human activity systems'. The modelling language is based upon verbs, and the modelling process consists of assembling and structuring the minimum necessary activities to carry out the transformation process in the light of the definitions of the CATWOE elements. The structuring is based upon logical contingency: 'convert raw material', for example, is contingent upon 'obtain raw material', and this dependent relationship will be shown by linking the activities with an arrow from 'obtain raw material' to 'convert raw material'.

Consider the following simple root definition:

> A householder-owned and manned system to paint a garden fence, by conventional hand painting, in keeping with the overall decoration scheme of the property, in order to enhance the visual appearance of the property.

Do-it-yourself enthusiasts might not feel it necessary to conceptualize such a system before getting on with the job, but it will serve to illustrate the process of model building without any problems arising from the sophistication of the task named! This definition follows the schema: do X by Y in order to achieve Z, with X as 'paint the fence', Y as 'conventional hand painting', and Z as 'enhance the visual appearance' of the property. CATWOE is as set out in

Root Definition :

> X | A householder- owned and manned system to paint
> a garden fence, by conventional hand painting, in
> keeping with the overall decoration scheme of
> y | the property) in order to enhance the visual
> appearance of the property

C - householder

A - householder

T - unpainted fence ⟶ painted fence meeting
criterion in the definition

W - amateur painting can enhance the appearance

O - householder

E - hand painting

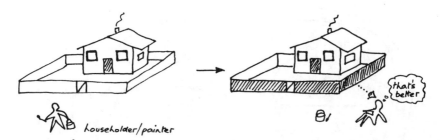

householder/painter

Figure 2.9 A root definition, CATWOE and pictorial representation of a fence-painting system

Figure 2.9 and presents no problems. Now we have to assemble the minimum necessary activities to meet the requirements of the root definition and CATWOE. It is usually useful to bridge the gap from definition to model via an informal pictorial representation of the concept of the definition. Although this is hardly necessary here, it is done in Figure 2.9. (Chapter 9 contains a number of serious examples of this procedure.) Clearly the main activity in the model will be 'paint the fence', and this will be surrounded by other activities which fit with CATWOE. In general, we aim to express the main operations to bring about the transformation (in the light of CATWOE) in a handful of activities. The guideline is: aim for 7 ± 2, this coming from Miller's celebrated paper in cognitive psychology in which he suggests that the human brain may have a capacity which can cope with about this number of concepts simultaneously (Miller, 1968). If this seems sparse, there is no problem: each activity in the model can itself become a source of a root definition to be expanded at the next resolution level.

The core activity 'paint the fence' will be contingent upon obtaining the necessary materials, and this will be contingent upon deciding the colour in the light of the overall decoration scheme of the property and taking a decision on the scope or extent of the task, since this is an *amateur* effort. These considerations yield the operational subsystem shown in Figure 2.10. Because this is a *system* (of course, we really mean a holon) we need, as always with such constructions, to add also the processes of monitoring and control which embody the guarantee that the entity could in principle survive in a changing environment. [For example, notionally, if the paint were to run out near the end (an amateur may not calculate the amount needed correctly) we might modify the definition of the scope of the task and treat it as finished—as long as the result could count as 'a painted fence', and as long as it meets the system owner's aspirations.] Figure 2.10 shows a general form for

Figure 2.10 A first model from the root definition in Figure 2.9

a monitoring and control subsystem, but this aspect is so important in the use of SSM that it is worthy saying more about it.

Logical analysis of the notion of a transformation shows that any conversion of input to output would be judged successful or unsuccessful on three different counts (Forbes and Checkland, 1987). A first dimension checks whether the means chosen actually works in producing the output. A second then considers whether the transformation is being carried out with a minimum use of resources. Finally, a transformation which works and uses minimum resources might still be regarded as unsuccessful if it were not achieving the longer term aim, the aim expressed by Z in: do X by Y in order to achieve Z. The three criteria need three names. In SSM we now use the '3 Es':

efficacy (for 'does the means work?')
efficiency (for 'amount of output divided by amount of resources used')
effectiveness (for 'is T meeting the longer term aim?')

and in general a model builder ought to decide what the criteria would be for the efficacy, efficiency and effectiveness of the system modelled. This adds a useful richness to the later comparison between the model and perceptions of the real world.

[Here we may remark that our definitions are clearer than those of the '3 Es' which public servants refer to in reviewing whether a public service gives value for money. They ignore efficacy and define *economical* as meaning acquiring appropriate resources at lowest cost. This then overlaps with their definition of *efficient* as meaning producing a maximum output for a given use of resources. Their *effectiveness* definition matches ours. (See Goddard, 1989.)]

For the model being built here, the criteria for SSMs '3 Es' would be those given in Figure 2.11. It is clear, however, that effectiveness is at a different level from efficacy and efficiency, and it is often useful to indicate this in the final model. This is done for the fence-painting model in Figure 2.11, which shows a complete and defensible conceptual model from the root definition shown in Figure 2.9.

In this particular model it is likely that the question of what is contingent upon what causes no problems. However, in teaching model building of this kind to undergraduates, Woodburn has found that the abstract notion of 'contingent upon' or 'logically dependent upon' can cause difficulties. He has found it useful in deciding which way to draw the arrows in models of human activity systems to get the modellers to ask whether a particular activity yields an output (matter, energy or information, etc.) which is a significant input to any other activity: if so then the latter is contingent upon the former (Woodburn, 1985).

Figure 2.11 The final model from the root definition of Figure 2.9

In the case of the model in Figure 2.11 there is no need for discussion of this kind, and in general it is always wise to draw a version of a model in which the arrows show simply contingent dependencies, even if it is also felt useful in the particular instance to prepare other versions in which material or abstract flows between activities are indicated.

The value of sets of models of the type shown for the fence-painting system, based on different worldviews and different Ts, is now well established in SSM. Only occasionally have other types of systemic model been used. Apart from using such ready-made models as the cybernetic 'Viable System Model' (Espejo and Harnden, 1989) one possibility is to make more complicated models. These may involve more than one T in various relationships, such as a parasite/host relation, or a syndicate of equals, or an imperial T dominating subject Ts. Atkinson and Checkland (1988) show how this can be done, and describe a study in which possible future changes in an inner-city health centre's management were conceptualized as a move from state A to an improved state B, where A and B were themselves thought of in terms of a number of more complicated (multi-T) models including a syndicalist system, a contradictive system and an imperialistic system.

Such use of the more complex models is really a resort to the conscious use of metaphor more formal than the informal use in naming relevant systems described above. The positive aspect of the use of more complex models is that it might enrich the debate when models are compared with the real world. The negative aspect is that the increased complexity of the models might lead to our slipping into thinking in terms of models *of* parts of the real world, rather than models *relevant to* debate about change in the real world. There is of course nothing to stop our building models on the basis of a single T and then using *the debate itself* to tease out the possible forms of relationships between them.

Once a model of a purposeful holon exists in a form like that in Figure 2.11, then it may be used to structure enquiry into the problem situation. However, before using the model as a tool in this way, most modellers will probably be asking themselves if their intellectual construct is adequate, or 'valid'. Since the model does not purport to be a description *of* part of the real world, merely a holon relevant to debating perceptions of the real world, adequacy or validity cannot be checked against the world. Such models are not, in fact, 'valid' or 'invalid', only technically defensible or indefensible. Whether or not they can be defended depends upon each phrase in the root definition being linked to particular activities and connections in the model; and each aspect of the model should be capable of being shown to stem from words in the definition. It is a worthwhile discipline to check that this is so, since credibility (and participants' confidence in the process) can be diminished if some smart person in the situation points out a basic logical flaw in the model. Whisper it abroad, though, that it is not unknown for useful progress to be made in real situations using models which might not meet the stringent requirements of the present authors!

For some years use has been made of a general model of purposeful activity known as the Formal System Model (Checkland, 1981, pp. 173–177) against which conceptual models of activity systems could be checked. It was expressed as a set of entities (boundary, subsystems, resources, etc.) and later expressed by Atkinson and Checkland (1988) as an activity system. However, its use has declined in the last decade, CATWOE has virtually eliminated it, and in any case its language tended to blur the distinction between the real-world language of problem situations and the language of systemic thinking *about* the real world, a distinction it is vital to maintain if the status of holons

Table 2.1 The core method of the SSM logic-driven stream of thinking and some variants

Core method within SSM	Possible elaborations
Name relevant systems, both 'primary task' and 'issue-based'	Use metaphors to examine relationships in the situation, or other aspects of the situation. (Davies and Ledington, 1987)
Formulate root definitions meeting the CATWOE requirements; think of the schema: a system to do X by Y in order to achieve Z	(See (b) below)
Build models based on one T, '7 ± 2' activities in an operational system, and a monitoring and control system using criteria for efficacy, efficiency and effectiveness	(a) Use more criteria than the '3 Es' (e.g. add Ethicality, Elegance) (b) Use more complex model structures entailing several Ts in various relationships (e.g. parasite/host or syndicate) (Atkinson and Checkland, 1988)
Make the links in the model indicators of which activities are contingent upon which other activities.	Develop flow versions of the model (abstract or concrete flows), or use this to decide on dependencies. (Woodburn, 1985)

as epistemology not ontology is to be maintained. von Bulow (1989) usefully points out some of the confusions surrounding the use of the Formal System Model, and it can probably now be cheerfully dropped.

Finally, concerning model building, we may remark that the '3 Es' for judging the in-principle performance of a human activity system cover only the most basic idea of transformation. They can be supplemented with other considerations of a broader nature if it seems appropriate in a particular field. For example, considerations of ethicality and elegance would bring in ethics and aesthetics, incidentally making the '3 Es' into the '5 Es'! Atkinson (1989) has suggested the use of Seedhouse's ethical grid (1988) as a means of thinking about the ethical dimension of studies using SSM.

This discussion has covered the core method of conducting the first part of the logic-driven stream of SSM, and it has also indicated some experientially derived variants. To avoid confusion these are disentangled in Table 2.1.

Comparing Models with Perceived Reality

Models are only a means to an end, which is to have a well-structured and coherent debate about a problematical situation in order to decide how to improve it. That debate is structured by using the models based on a range of worldviews to question perceptions of the situation.

Checkland (1981) describes four ways of doing the comparison (informal discussion; formal questioning; scenario writing based on 'operating' the models; and trying to model the real world in the same structure as the conceptual models). Of these the second has emerged as by far the most common. The models are used as a source of questions to ask of the real world; answering those questions initiates debate, which may be conducted in any way which seems appropriate to the particular situation. It may be carried out by a group of people gathered in one place at one time to have discussion, or carried out in one-to-one interviews or dialogues spread over a period of time. It is impossible to generalize. In the study related to decision support systems described in Chapter 6, the debate took place at one meeting between Scholes and the initiator of the study; in the reorganization of the Shell Group's Manufacturing Function (Chapter 9) all 600 members of the Department, as well as many managers from Shell Operating Companies, were given the opportunity to make their contribution to the debate over a period of months.

What we can say by way of generalization is that this mode of comparison by model-defined questions can usefully be got under way by filling in a matrix like that shown in Figure 2.12. The right-hand column is then a summarising source of ideas for change in the situation or new ideas for relevant root definitions.

The second most common way of setting models against perceived reality is one which is less abstract than the matrix approach. It consists of notionally operating a model, doing its activities either mentally or on paper, in order to

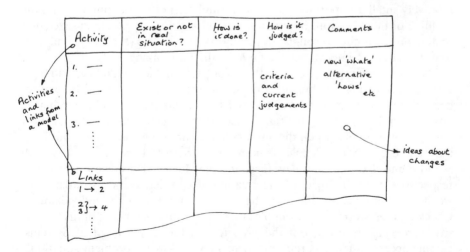

Figure 2.12 A matrix for comparing a conceptual model with a real-world situation

write a scenario which can then be compared with some real-world happenings. For example Dave Thomas, working with Checkland on the study on the Concorde project, took a model developed from the idea of 'a system to create the aircraft to meet a particular technical specification within a certain time at a certain cost' and operationalized it on paper, with respect to the air intake structure, a part of the Concorde engines. The resulting narrative was then revealingly compared with the actual history of the creation of the air intake structure for the first two pre-production aircraft. (None of the design and manufacturing documents concerning the structure in question was found to have any mention of cost!)

No matter how the models are used for a comparison with the real world, the aim is not to 'improve the models'—as management science enthusiasts sometimes tend to think—it is to find an accommodation between different interests in the situation, an accommodation which can be argued to constitute an improvement of the initial problem situation. But getting to that accommodation, and to the motivation to action which is an equal concern, requires cultural knowledge. That will have been being gained in parallel with logic-driven modelling work just described. It is to this overlapping stream of cultural enquiry that we now turn.

The Stream of Cultural Enquiry

One of the reasons why Figure 2.6 is a better representation of SSM in action than Figure 2.5 is that, in the latter, finding out about the culture in which the work is being done—which will be crucial to its success—is apparently lost in 'Stages 1 and 2' of the approach. This can give the impression both that the finding out can be done once and for all at the start of a study, and that it is of relatively smaller importance than the formulation of root definitions and the building of models. Both impressions are wrong. Figure 2.6 emphasizes that finding out continues throughout a study, right up to its (arbitrary) end, and that it is equal in importance to the logic-driven thinking.

That this is the case stems from the acceptance that although facts and logic have a part to play in human affairs, the *feel* of them, their felt texture, derives equally (or more) from the myths and meanings which human beings attribute to their professional (and personal) entanglements with their fellow beings. For example, when the authors made career moves, Checkland from industry to university, Scholes from the Civil Service to industry, they both found themselves learning new facts and new logics. These were concerned respectively with the logics of the creation of knowledge and the creation of wealth. But more compelling was the need to learn new myths and meanings. Checkland discovered, for example, that some senior members of his university were very unsure about the *propriety* of the Department of Systems' gaining access to its research object, namely real-world problem situations, via the business operations of a university-owned limited liability company.

(The myths underlying this attitude changed quickly and dramatically in later years under the stimulus of cuts in Government grants to universities!) Scholes, moving to ICL from Whitehall, found himself learning the myths which drive a reward-seeking culture after many years in what, at the time he experienced it, was the punishment-avoiding culture of the Civil Service. If we are going to intervene in human affairs and grapple with their full complexity, we had better have available some ways of enquiring into the 'systems' of myths and meanings which constitute what we mean by 'a culture' (Schweder and Le Vine, 1984).

Rich Pictures

A characteristic of fluent users of SSM is that they will be observed throughout the work drawing pictures and diagrams as well as taking notes and writing prose. The reason for this is that human affairs reveal a rich moving pageant of relationships, and pictures are a better means for recording relationships and connections than is linear prose. The significant number of figures in this book is witness to that fact.

Representing root definitions pictorially is one example of the use of pictures in SSM (see Chapter 9 especially) but the best known is the policy of representing the problem situation itself in the form of so-called 'rich pictures'. There is no formal technique or classic form for this, and skill in drawing is by no means essential (though it's not a hindrance!) in the production of pictures which are found to be very helpful. Figure 2.13 shows a rather formal rich picture 'drawn' on a computer screen by Stuart Bamford in a current study. It shows a situation in a large engineering-based company in which the study is concerned with problems of tendering for contracts and then organizing to meet their requirements profitably and on time. The picture, though properly idiosyncratic, shows the preoccupations of compilers of rich pictures: expressing relationships and value judgements; finding symbols to convey the correct 'feel' of the situation; indicating that the many relevant relationships preclude instant solutions. Figure 2.14 shows similar concerns in a less formal picture.

More generally, if we take the figures in Chapter 4 as an example of diagrams in systems work: Figure 4.1 is a rich picture of a conventional kind; Figures 4.2, 4.4, 4.7, 4.9, 4.11 and 4.12 show diagrams of various kinds in the service of systems thinking. Always, though, they express in a condensed way relationships which would require much prose to expound.

Analysis of the Intervention

Early in the development of SSM it was found useful to think of an intervention in a problem situation as itself being problematical (Checkland, 1981, pp. 238–240). It was found useful to think of the intervention structurally as

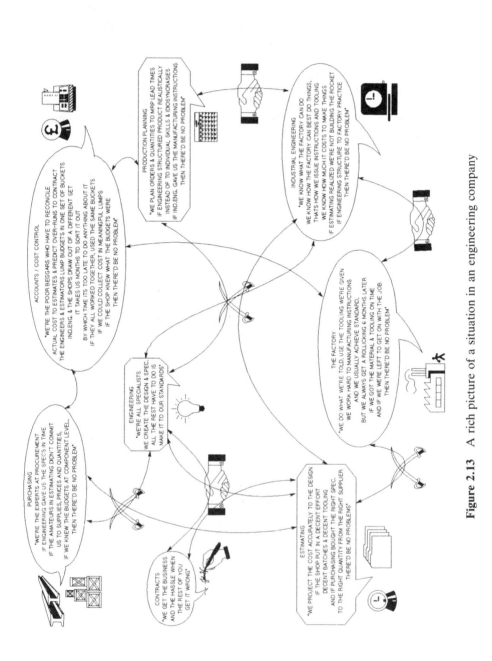

Figure 2.13 A rich picture of a situation in an engineering company

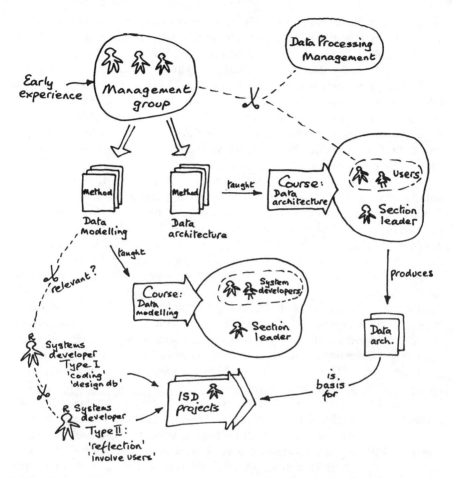

Figure 2.14 A rich picture of a problem situation in the DP department of an insurance company. (After Mathiassen and Nielsen, 1989, reproduced by permission)

entailing three roles. The role 'client' is the person or persons who caused the study to take place. There will always be a real-world answer to the question: Who is in the role client? And it is a question worth asking because it is wise to keep in mind (but not be dominated by) the client's reasons for causing the intervention to be made. In the role 'would-be problem solver' (and it could be whoever is also 'client') will be whoever wishes to do something about the situation in question, and the intervention had better be defined in terms of their perceptions, knowledge and readiness to make resources available. Finally, and of crucial importance, is the role 'problem owner'. No one is intrinsically a problem owner. The 'problem solver' must decide *who to take possible 'problem owners' to be*. There will always be many possibilities. The list should include, but never be limited to, whoever is in the roles 'client' and 'problem solver', and this list is the best source of choices of relevant systems in the logic-driven stream of enquiry. It will be noted that making the 'problem solver' one possible 'problem owner' often means that the first relevant system looked at is 'a system to do the study'. The first model built is often a model of the structured set of activities which the problem solver(s) hope to turn into real-world action in doing the study. This is what Rodriguez-Ulloa (1988) means when he argues for seeing 'the problem solving system' as part of the problem content.

This role analysis, now known as 'Analysis One' in SSM, is always relatively easy to do and is very productive, especially through the list of possible problem owners. For example, in a study of policing vice in the West End of London based on SSM, some immediately obvious possible 'problem owners' would be: Parliament, the courts, the general public, the police, organizers of the vice trade, providers of its products, purchasers of its products, its potential recruits, their families, social service departments, local hospitals, local residents, respectable West End traders, etc. How to use models deriving via relevant systems from these choices of problem owner would depend upon who was undertaking the study and who caused it to occur: the client. Flood and Gaisford (1989) describe just such a study which they claim used SSM, but their account does not distinguish clearly between 'client' and 'problem owner'. This illustrates a problem for SSM, one which is endemic in all applied social science. Because the technical terms of any explicit approach will use words which have meanings in everyday language (in SSM technical terms include 'client', 'problem solver', 'problem owner', 'relevant system', 'comparison', etc.) it is important in using the approach to use its language with unusually careful rigour. Life is easier for natural scientists. They can talk of 'spin Hamiltonians' and 'diffusion coefficients' and know that they will not be misunderstood.

'Social System' Analysis

Rich pictures will continue to be drawn and amended throughout any use of SSM, and new occupants of the roles 'problem solver' and 'problem owner' in Analysis One may emerge in the course of a study, so these are not once-and-for-all analyses. Neither is a study of the problem situation as a 'social system', using that phrase in its everyday language sense. What the phrase conveys is not easy to pin down, though we all have an intuitive feel for it. The social science literature does not easily yield a usable model, and it has been found necessary to develop one experientially for use in SSM's 'Analysis Two' (Checkland, 1986).

The model which has been used in the last five years derives ultimately

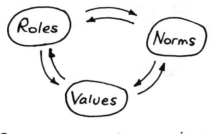

Each element defines, and is defined by
the others

Figure 2.15   The model used in Analysis Two

from the work of Vickers (1965) but is drastically simpler than that of the model of what Vickers calls 'an appreciative system' which Checkland and Casar (1986) derived from all his writings. To be usable 'on the hoof' throughout a study, a model has to be very simple indeed: the problem is to find a simple model which is not too simplistic.

The model in question assumes a 'social system' to be a continually changing interaction between three elements: roles, norms and values. Each continually defines, redefines and is itself defined by the other two, as shown in Figure 2.15. By 'role' is meant a social position recognized as significant by people in the problem situation. Such a position may be institutionally defined ('classroom teacher', 'team captain', 'shop steward') or may be defined behaviourally ('licensed jester', 'nutter', 'solid citizen'). A role is characterized by expected behaviours in it, or *norms*. Finally, actual performance in a role will be judged according to local standards, or *values*. These are beliefs about what is humanly 'good' or 'bad' performance by role holders.

This model has been found to be widely applicable and very useful as long as it is accepted that the account of the 'social system' it leads to is never either complete or static. For example, in a highly sophisticated but functionally organized engineering company, in difficulties over project management, the important recognized roles were essentially *professional* ones. To get on in the company you needed to be, for example, a highly skilled hydraulic engineer, *in the eyes of the peer group*. The values were ones from the world of professional engineering, and technical excellence was prized above all other qualities; it was the kind of company which was ready to make technically excellent products which no one wanted to buy. Part of its project management problem was analysed to be due to the fact that its inappropriate 'Rolls Royce' values were only reluctantly shifting to more realistic commercial ones.

Finally, it needs to be said that the nature of the social system is not likely to emerge in response to direct questions. Direct questions will probably receive as responses the official myths of the situation. Instead, the SSM user needs, mentally or actually, to open a file labelled 'Analysis Two'. Subsequent to every conversation, interview, or perusal of documents, etc., the exchange experienced needs to be reviewed for what the analyst can infer with regard to roles, norms and values. Appendix 1 to Chapter 4 gives an 'Analysis Two' carried out in a District Health Authority.

'Political System' Analysis

'Analysis Three' in the stream of cultural analysis accepts that any human situation will have a political dimension, and needs to explore it (Checkland, 1986). As in the case of Analyses One and Two, this is done via a general model, in this case of 'a political system'. (Again, this phrase is used in an everyday-language sense.)

This is not the place for a deep discussion of the nature of politics, but, for the practical purposes of Analysis Three, politics is taken to be a process by which differing interests reach accommodation—a view which may be supported with reference to the literature of political science. (See, for example, Lasswell and Kaplan, 1950; Miller, 1962; Crick, 1962; Dahl, 1970; Blondel, 1978). The view derives ultimately from Aristotle, whose *Politics* argues that what he means by politics are the processes by which order is maintained in a *polis* (for Aristotle, the Greek city-state) which is an aggregate of members who will have different interests. Accommodating those interests is the business of politics, and the concept will apply to a company or work group or a sports club as well as to a city or a nation state. Finally, the accommodations which are generated, modified or dissolved by politics will ultimately rest on dispositions of power. So politics is taken to be power-related activity concerned with managing relations between different interests. As such it is endemic in human affairs, and there will be few purposeful acts which do not have a political dimension. In the late 1980s, for example, a long-running public debate concerns whether or not there should be sporting links with South Africa. In this debate in the UK, those who support such links constantly cry 'Keep politics out of sport' as if the two were separable. They make a category mistake. If, say, the England rugby team were to go to play the Springboks, this will be perceived as, and hence will be, a political act in support of apartheid. There is an unavoidable political dimension to sport, just as there is a political dimension to theatre, to management, to sexual relations, to health care provision or, indeed, to any human affairs which entail taking deliberate action.

In Analysis Three, political analysis is made practical by asking *how power is expressed* in the situation studied. Following a suggestion developed by

Stowell in action research using SSM in a medium-sized manufacturing company (Stowell, 1989), we ask: What are the 'commodities' (meaning the embodiments) through which power is expressed in this situation? How are these commodities obtained, used, protected, preserved, passed on, relinquished? Through what mechanisms?

In the last five years very many 'commodities' of power have been observed in different situations. Examples include: formal (role-based) authority, intellectual authority, personal charisma, external reputation, commanding access (or lack of access) to important information, membership or non-membership of various committees or less formal groups, the authority to write the minutes of meetings, etc.

In one recent study in an organization which was rethinking its future after the death of its charismatic founder, a standard half-joking classification of people in the company was into 'KTs' and 'NKTs'. This referred to those who 'knew Tom' and those who had 'never known Tom'. The fact of having been in the company in the days of the charismatic Tom was used by the KTs as a commodity of power in this organization: the organization joke was very serious, as they usually are.

Answering the power-oriented questions in Analysis Three enriches the cultural appreciation built up in Analyses One and Two, and all three complement the work on selecting, naming and modelling relevant human activity systems going on simultaneously in the logic-driven stream of thinking. Working with SSM, in the 'ideal type' methodology being described here, entails carrying out simultaneously the two streams of thinking and action set out in Figure 2.6. They complement each other and should unfold through time interactively. It is especially important never to regard Analyses One, Two and Three as finished; and delicate judgements are usually required concerning the public visibility of Analysis Three.

This sensitivity of Analysis Three to public exposure does not imply that SSM is just the methodology Machiavelli needed for giving advice to his Prince! The sensitivity stems from the fact that politics is ultimately concerned with power and its disposition, issues not usually faced overtly in human dialogue. There is a natural reluctance to be blunt about the crudities of power, and there is a sense in which the real politics of a situation, not publicly acknowledged, will always retreat to a tacit level beyond whatever is the explicit level of analysis. Given the concept of politics outlined above, if the results coming from Analysis Three are all bluntly made public, then those results can themselves easily become a potent commodity of power in the 'real' politics of the situation. There is potentially an infinite regress here in which the politics of the situation forever escapes open analysis. So it behoves users of SSM to be circumspect about the use of the cultural enquiry, and especially Analysis Three.

Making Desirable and Feasible Changes

Whether SSM is being used by an individual to help tackle his or her everyday work, or whether it is the adopted methodology in a highlighted study, its aim will be to *do* something about a situation regarded as in some way unsatisfactory. The two streams of thinking and action in SSM converge on a structured debate concerned with defining changes which would help remove the dissatisfaction. But beyond definition of the changes, the SSM user looks for implementation of them.

This implementation is, of course, itself 'a problem situation', and it is not unusual to use SSM to tackle it. We may conceptualize and model systems to implement the changes, and do that according to several relevant *Weltanschauungen*. Finally we may pinpoint 'a system to make the changes' whose activities can then become real-world action. We can set about doing the activities of that final model, in the real-world situation.

The changes themselves are usually described as 'systemically desirable' and 'culturally feasible' (Checkland, 1981), and it is worth dwelling briefly on these phrases because if they are understood, SSM is understood.

The models of purposeful activity systems built within SSM are selected as being hopefully relevant to the problem situation. They do not purport to be models of the situation. It is because of this that any changes coming out of the debate initiated by comparing the models with the real situation are (only) arguably desirable, not mandatory. They are systemically desirable if these 'relevant systems' are in fact perceived to be truly relevant.

Implementation of changes will take place in a human culture, and will modify that culture, at least a little, and possibly a great deal. But the changes will be implemented only if they are perceived as meaningful within that culture, within its worldview. A Western observer's proposal to a remote culture that they abandon tribal rain dances on the grounds that they manifestly do not work in controlling the weather, will be ignored in a culture for which the dances are meaningful. [Indeed, Western thinkers have had to invent a distinction between 'manifest' and 'latent' functions of such rituals in order to cope with such observations within a Western intellectual framework (Merton, 1957).] What is perceived as 'meaningful' by a particular culture might range from a tiny incremental change to a major revolutionary one—it is not the amount of change which determines its feasibility, but whether or not it is seen to be *meaningful*. Hence the changes introduced by SSM have to be culturally feasible in the sense that they have to be regarded as meaningful within the culture in question.

Thus the two criteria for the changes sought by SSM are 'systemically desirable' and 'culturally feasible'. Understanding that this is so provides a test of whether or not the distinction between hard and soft systems thinking is understood. Someone intellectually locked within the 'hard' paradigm, believing the world to be systemic, will imagine that changes have to be systemically feasible and culturally desirable!

SSM IN THE CREATION OF INFORMATION SYSTEMS

In recent years there has emerged a particular area of application for SSM to which it is well suited; we refer to its use in the creation of information systems. Information systems are here described as being 'created', rather than simply 'designed', because the connotations of the 'design' activity are that what is required has been specified, and design is concerned with the question of how to realize the specification. Creating something implies a broader perspective.

SSM has a major contribution to make in tackling a crucial question which is prior to this, a question neglected in much of the literature of information systems, namely: which of the huge number of information systems that we could put together, should we? Then, beyond that issue, SSM's activity models can form a cogent basis for information flow models upon which the information system design process itself can be based (Checkland and Griffin, 1970; Wilson, 1984).

This section reviews this field briefly in order to relate SSM to it. An appendix to the book as a whole sets out a view of the field from a systems-thinking perspective. Here will be set out only the main arguments concerning the nature of 'information systems' as a field of practical endeavour and related theory. The purpose is to establish how SSM can make a contribution to a ubiquitous problem.

Information Systems in Theory and Practice

Information Systems (IS) as an area of theory and practice has been dominated by a particular means, the computer, and by the anthropomorphic language about computers which the earliest pioneers of computers, alas, encouraged. Since major figures like von Neumann and Turing themselves used the high-flying rhetoric of metaphors based on human mental processes, speaking of 'memory' (rather than storage) and casually applying the word 'intelligent' to machines, it is not surprising that the humbler workers in the vineyard have failed to remind themselves that what computers can do is manipulate tokens for data, rather than 'process information'.

In addition, the history of the development of computer technology has meant that most organizations' purchase of their first computer was a major capital investment. It was therefore treated as a capital project, and not surprisingly the IS field adopted without question the thinking developed by engineers to cope with such developments. A 'project life cycle' model was adopted, and this has survived longer than it should have done, given the rapid development of computer technology. That development has seen the technology handed over to users, and made its acquisition no longer a major capital expenditure. In fact the thinking underlying the provision of IS has been dominated by hard systems thinking, as Miles (1985) has argued; see also Veryard (1986) and Miles (1986).

However, the thinking which starts from a means (the computer) rather than ends (an organization's conceptualization of its world) and which adheres to a now less-appropriate model (that of engineering projects) has not been totally dominant. An important minority strand in the IS debate has come from an alternative perspective, from those whose fundamental response to 'information' is to see it as symbol rather than signal. Such a school of thought is humanistic, and treats IS as a cultural rather than a technical phenomenon. (For the flavour of this tradition, see, for example: Boland and Hirschheim, 1987; Davies and Wood-Harper, 1989; Galliers, 1987; Hirscheim, 1985; Land and Hirscheim, 1983; Lyytinen, 1988; Mumford, 1983; Mumford et al., 1985; Schafer et al., 1988; Symons and Walsham, 1988.) The use of SSM in IS work is a contribution within this school; and below we mount the argument which relates SSM to the creation of information systems.

From Worldviews and Meanings to Data Manipulations

Let us accept that there are myriad facts about the world which can be stated neutrally. Today is a Thursday; the authors were brought up in Birmingham and Blackpool, not Brisbane and Buffalo; Philips manufactures light bulbs in Trois Rivières, Quebec, etc. These are the givens of the world, the data which it yields to our inspection. But the data do not keep their virginity. Usually, in everyday life, we accept them in a context of meaning: we interpret them, and in principle may do so autonomously. The position of the hands of the clock are an item of data but we may read them not as that but as an indication that we have time for another cup of tea before catching the train. The sales figures for the new product in its first month are an item of data, but will not be seen neutrally by the Sales Manager, the Sales Representative, the Production Planner or the person responsible for the advertising campaign. We turn such data into information by interpreting them in a particular context at a particular time. Data with attributed meaning in context we may define as 'information'.

Now consider the nature of a so-called 'information system'. Why do we need such things? It seems most appropriate to assume that the purpose of creating an organized IS is to serve real-world action. Organized provision of information in organizations is always linkable in principle to action: to

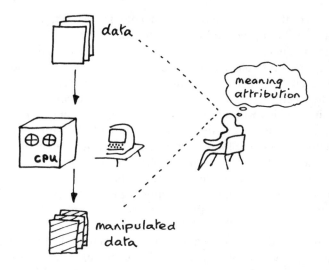

Figure 2.16 An 'information system', in the true sense, entails data manipulation and meaning attribution

deciding to do things, doing them, observing and recording the results—and then if necessary modifying the deciding, doing and recording.

From these considerations (that 'information' is data to which meaning has been attributed in a particular context, and that information systems serve action) two consequences flow.

Firstly, the boundary of an IS, if we are using that phrase seriously, will always have to include the attribution of meaning, which is a uniquely human act. An information system, in the exact sense of the phrase, will consist of both data manipulation, which machines can do, and the transformation of the data into information by attribution of meaning. Of course the designers of the data manipulating machine will have in mind a particular set of meaning attributions and will hope that the manipulated data will always be interpreted as particular information—but they cannot guarantee that, since users are ultimately autonomous. Figure 2.16 illustrates this view of the nature of an information system.

Secondly, designing an IS will require explicit attention to the purposeful action which the IS serves, and hence to the meanings which make those particular actions meaningful and relevant to particular groups of actors in a particular situation.

Thus if we wish to create an 'information system' in the exact sense of the phrase, we must first understand how the people in the situation conceptualize their world. We must find out the meanings they attribute to their perceptions of the world and hence understand which action in the world they regard as sensible purposeful action, and why.

Having obtained that understanding we shall be in a position to build some of the purposeful holons known as 'human activity systems' and use them to stimulate debate aimed at defining some human activity systems widely regarded by people within the situation as *truly* relevant to what they see as the required real-world action. (Achieving this agreement may be easy or difficult, depending upon the characteristics of the situation. In a small manufacturing company there is unlikely to be disagreement on the relevance of 'a system to obtain continuous supplies of raw materials of an appropriate quality at an acceptable price'. In the systems study in Shell, described in Chapter 9, a process which took about eighteen calendar months resulted ultimately in agreement on an entirely new concept of the manufacturing function.)

Once an agreed 'truly relevant' system has emerged, it may be treated in the following way. Ask of each activity in the model: What information would have to be available to enable someone to do this activity? From what source would it be obtained, in what form, with what frequency? Similarly, ask: What information would be generated by doing this activity? To whom should it go, in what form, with what frequency? In this way an activity model may be converted into an information-flow model.

Given an information-flow model which is agreed to be a necessary feature of the situation studied (deriving from the way people in the situation see their world as meaningful and define necessary purposeful action) we may now ask: What data structures could embody the information categories which characterize these information flows? This then takes us into the design of a suitable data manipulation system, something generally known, alas, in everyday language, as 'an information system'. Taking a deep breath, we may say that this data manipulation system will yield the information categories and information flows required by the structured set of activities which are regarded as truly relevant to the real-world action which is itself relevant according to the meanings which people in the situation attribute to their world as a result of their worldviews.

Figure 2.17 illustrates this sequence, which connects a worldview to necessary data structures. SSM provides a way of traversing the sequence in a lucid manner.

Recent practice suggests that it is in fact difficult resolutely to pursue the logical sequence of Figure 2.17 in the step from information categories to data structures. At this point it is seemingly necessary to allow what are perceived as meaningful data structures in the real situation to enter into the design discussion. What is required to move from step 6 to step 7 in Figure 2.17 is a *reconciliation* of common real-world data structures with those which in principle would provide the information categories required by the logical analysis (Miles, 1987).

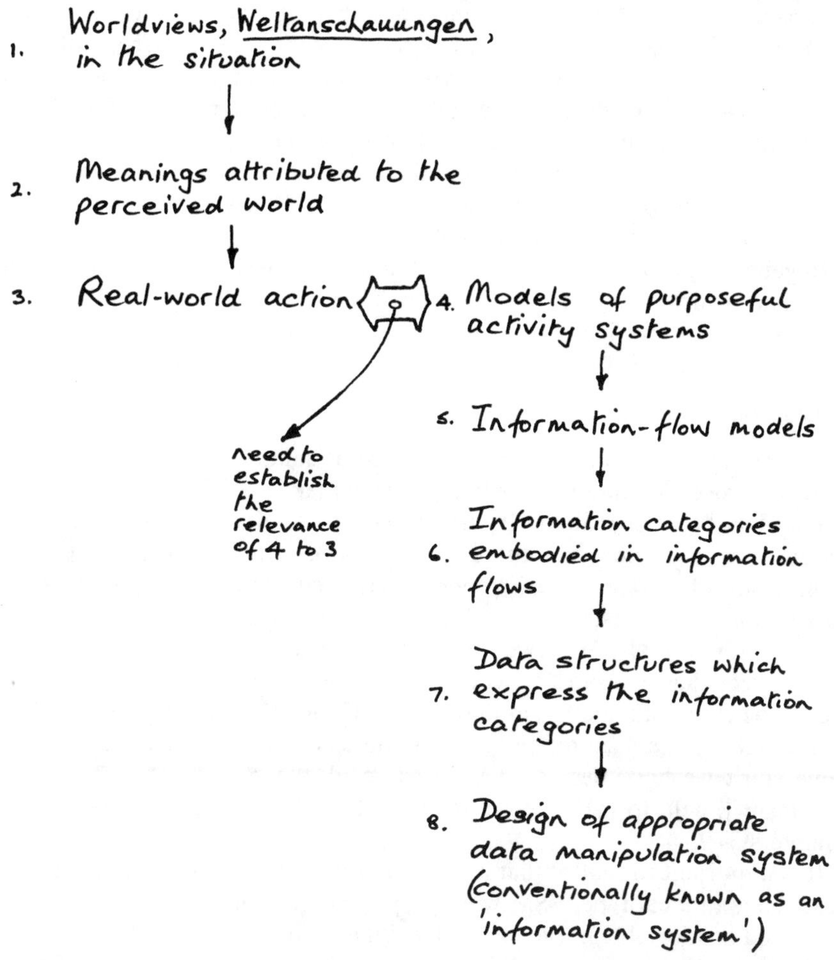

1. Worldviews, Weltanschauungen, in the situation

   ↓

2. Meanings attributed to the perceived world

   ↓

3. Real-world action ⟷ 4. Models of purposeful activity systems

   ↓

   need to establish the relevance of 4 to 3

   5. Information-flow models

   ↓

   Information categories
   6. embodied in information flows

   ↓

   Data structures which
   7. express the information categories

   ↓

   8. Design of appropriate data manipulation system (conventionally known as an 'information system')

Figure 2.17   The links from worldviews to the manipulation of appropriate data

Much work is currently underway in this area. In particular SSM could enrich those poverty-stricken stages of systems analysis and design methodologies in which information requirements analysis is assumed to be straightforward, or organizations are naively documented as a set of unproblematical entities and functions (see Maddison, 1983). But the detailed linking of SSM to detailed design of computerized data manipulation systems has not yet been accomplished.

One suggestion in the area of applying SSM to the problem of information provision is the development by the UK Government's Central Computer and Telecommunication Agency of the so-called 'Compact' approach to 'the analysis, design and introduction of small single-site office systems' (CCTA,

Compact Manual, 1989). This attempts to link a rather mechanical version of SSM to the Government's 'standard methodology for the analysis of information systems', SSADM. Compact uses SSM for carrying out 'business analysis'. (One point in it will seem strange, if not hilarious, to readers of this book. Evidently considering the notion of *Weltanschauung* or worldview too obscure for the Government service, the consultants who recommended the approach changed CATWOE's W into V, for 'viewpoint', and rewrote CATWOE as the more vapid VOCATE!)

In general, this is a richly developing area of application for SSM, and the argument which yields Figure 2.17 provides an intellectual skeleton for conceptualizing this field.

Conclusion

We have now filled out the account of SSM as summarized in Figure 2.6. A systemic process of enquiry which happens to make use of models of systemic holons has been described. Its use in the particular field of information systems has also been signposted. What has been described is SSM in the form of an 'ideal type'. Max Weber (1904) used the phrase to describe the pure intellectual constructions which can then be used analytically in the investigation of real-world examples. These will always be found to be richer and messier than the pure concept. Weber's account of 'bureaucracy' as an ideal type is not intended as as description of something in the world; it is an idea which can be used to study real-world examples of bureaucracy. In the same spirit we have described SSM as an ideal type. The account in this chapter will help to make sense of the real SSM-based studies described in Chapters 3–9.

It is important to realize that any account of methodology is in fact always describing an ideal type. SSM will always emerge in use in a form which its users find comfortable in the particular situation they are in (Atkinson, 1984, 1987). SSM, in the accounts which follow, will always be there, but never, in all aspects, precisely in the form described in the above account. That *mouldability* by a *particular user* in a *particular situation* is the point of methodology. That is why a methodology is so much more powerful than mere method or technique.

Chapter 3

An Application of Soft Systems Methodology in Industry

Introduction

This and the following two chapters describe SSM in action in three very different areas: industry, the National Health Service, and the Civil Service. The example from Whitehall selected itself for inclusion, since it was the first occasion on which the authors worked together, and subsequently led to the sequence of studies in ICL described in Chapters 6, 7 and 8, studies which changed our view of the methodology. In the case of the industrial and health care examples, studies have been selected which contributed to the overall learning which this book describes, since the book as a whole is intended to be more than simply a string of beads with each representing one more example of SSM in action.

The industrial example described in this chapter is selected because the experience drew attention to new concerns—namely the nature of a 'system to use SSM'—which the Lancaster developers of the methodology had not previously regarded as of special significance. Another important feature of this study is that it makes an interesting comparison with the 'late 1980s' study described in Chapter 9. There is almost a decade of experience between these two industrial studies, and because of this, and the fact that they have some similarities, they make a particular contribution to the argument developed in the final chapter (Chapter 10).

As has been described in detail in *Systems Thinking, Systems Practice* (Checkland, 1981) and briefly alluded to in the previous chapter, SSM grew out of unsuccessful attempts to apply the methodology of systems engineering to ill-structured management problems. Two features of the situation in which SSM was developed helped to dictate the way in which the developers thought about it. Since one methodology (systems engineering) was failing and was being replaced by another, the replacement was naturally thought of as a series of methodological steps or stages which could be applied whenever the new methodology (SSM) was used instead of the old; Checkland and his colleagues were evolving, they thought, a different version of the (very different) stages of systems engineering. Secondly, because the developers happened to be based in a postgraduate department in a university, the development necessarily took place in external situations in which outsiders were making studies of problem situations owned by others. Not surprisingly, SSM came to be thought of as a sequence of stages for making some kind of formal systems study in problem situations; and both the developers and the managers in whose situations they worked accepted rather easily the implicit model which perceives SSM as a consultant's tool. It can be that, and is now often used in that way (see, for example, Watson and Smith's (1988) account of the use of SSM in Australia) but that is not an intrinsic feature of SSM. It was clear to those developing SSM in the form described in the previous chapter, that SSM was not necessarily the professional expertise of 'soft systems engineers'. Rather it was simply an organized approach to tackling real-world problems which could in principle be used by anyone. Furthermore, the uses felt to be *best* were invariably those in which there was collaboration between people in the problem situation and the outsiders who came along armed with SSM. The methodology was thus thought of, and this was not regarded as in any way problematical, as a sequence of stages by means of which a joint insider–outsider problem-solving team tackled a messy problem situation.

The study to be described in this chapter did not quite fit that pattern, and as a result was a source of useful learning. The study was carried out under a constraint imposed by its client, Phil Belshaw of the Organics Division of ICI, and this constraint usefully drew attention to *the process of using* the stages of SSM, rather than allowing all attention to be directed to the stages themselves, as usually happened. Belshaw's constraint was simple and explicit: that we should not do the study! He wanted it done by three managers in the department of which he was head. We were only to provide enabling help, even though the three managers themselves initially knew nothing of SSM! The effect of this was to make us more than usually aware of the process of using SSM, and it can now be seen as the start of experiences which led eventually to the kind of study to be described in Chapter 9, in which very many managers take part in an extended study in which no explicit reference is made to SSM unless the participants themselves specifically request it.

The Nature of the Study

Phil Belshaw had led a varied career in ICI which included both line and staff responsibilities—from managing production plants to working directly for the Organics Division Board in carrying out special studies concerning strategic issues. Near the end of his career he was asked to head the Information and

Library Services Department (ILSD) with a view not simply to managing it but rethinking its role and refurbishing it. He decided that a systems study of ILSD using SSM would be a way of doing this. But he understood the importance of both participation in, and psychological ownership of, such studies, and he decided to make time available for three of his managers to make a fundamental systems study of ILSD, working on it together at least one day a week. Checkland and his colleague Iain Perring were asked to provide some help, since the three managers concerned had no knowledge of SSM. The three managers were: the manager of ILSD under Belshaw, Derek Styles; a technical expert knowledgeable about modern information technology, Mike Fedorski; and Jo Teagle, a professional librarian who ran the Division library and its services.

In a four-page letter at the start of the study setting out his aims for it, Belshaw indicated that he looked for 'a fundamental study not constrained by the present situation'. At that time in ICI (the late 1970s) there was much emphasis on 'doing more with fewer resources', and a target reduction of staff numbers in ILSD had been named by the Board, from 32 to 25, but the question was not to be 'how to run ILSD with 25 people', it was to be 'what business are we [ILSD] in and what capability is required for the 1980s?' Another objective would be a process one: 'to upgrade the problem solving capability of ILSD', and Belshaw added: 'I wish the task force to carry out the study, rather than, say, bringing in the Department of Systems at Lancaster University to carry it out on our behalf.' There followed a plan for a two-day event to get the study going, after which the team would work together on the study one day a week, calling upon Checkland and Perring for help when necessary and reviewing the work with the outsiders every six weeks or so.

The account of the study which follows will complement rather than duplicate an earlier, pseudonymous, account (Checkland, 1985b, 1989). Different aspects will be emphasized here, with a greater emphasis on the process of using the methodology, and the experience will be used to establish the rudiments of 'a system to use SSM' which will be further developed in the final chapter. The point is made in the earlier account that no description of any study can approach the felt richness of the experience itself: every account will be partial, from a particular point of view. And every study, if well documented at the time, can be mined repeatedly for insights relevant to new concerns in an ongoing programme of research.

The First Methodological Cycle: Relevant Systems through Modelling

Four complete or partially complete cycles of SSM can be discerned in the study during its formal existence. They will be described in sequence to build up the four-column matrix of the history of the work which is shown in Table 3.1.

Table 3.1 The methodological history of the study

	Cycle 1 (hours)	Cycle 2 (days spread over weeks)	Cycle 3 (days spread over weeks)	Cycle 4 (days spread over weeks)
Stage 1 (Finding out)	Discussions with the team. Written accounts of structures, processes	Further informal finding out done by alert involvement in the problem situation	—	—
Stage 2 (Expressing the problem situation)	Definition of 26 problem situations/themes, all considered relevant	Frequently this involved asking questions arising from Stages 3–7	—	—
Stage 3 (Formulating root definitions)	The EROS model: a general model applicable to all 26 problem situations	Root definitions 1–6	More detailed RDs formulated	—
Stage 4 (Building conceptual models)	The EROS model	Build 6 models	More detailed models built (e.g. Figure 3.5 expands two activities from Figure 3.3)	—
Stage 5 (Comparing models and perceived real world)	Classification of the 26 problem statements in terms of the model	Table comparisons compiled	Table comparisons compiled	Discussion at presentations
Stage 6 (Debating, defining changes)	Team select those thought most significant (6 of them)	An assembly of issues from the comparisons	Examination of desirable/feasible 'hows' (e.g. Figure 3.7). Message of study worked out	Discussion of report message with a wider audience. Proposal from Belshaw to Board
Stage 7 (Taking action).	Move to 2nd cycle	Decision to expand some models	Report written, presentations made	Study proposals accepted. Investment in the new ILSD

At the start of the study, having explained the rudiments of SSM (putting the emphasis on models of purposeful activity systems as devices for exploring reality rather than as descriptions of it), Checkland and Perring urged that it would be a good idea formally to record a 'finding out' phase (Stages 1 and 2 of SSM) even though Fedorski, Styles and Teagle were steeped in the problem situation, it being the context of their professional lives. They were understandably reluctant to spend much time on this, since they felt they 'knew all about' the situation. Nevertheless, for the benefit of the outsiders, they did prepare material recording the structures and processes of ILSD which provided a rich picture of their problem situation and enabled the relation between structure and process to be discussed. (The project file contains 22 pages of this material covering ILSD's various roles: provider of general library and information services; guardian of a remarkable collection of specimens of every organic compound synthesized in the research laboratories; keeper of a secure collection of company reports with security classifications defining allowed classes of reader; and provider of access to computerized information systems enabling ICI scientists to consult various relevant databases around the world.) Discussion of this material yielded an image of ILSD as a group whose professional esteem derived essentially from its ability always to react quickly and efficiently to requests from users. This was informally organized in ILSD's carefully nurtured and maintained network of relationships with certain influential users of their services who were seen as 'gatekeepers' for their part of ICI Organics Division. Some trouble was taken to identify and maintain links with such people.

Once the picture of the situation was established and shared, attention could turn to 'problem solving'. The ICI team of three were very ready to perceive their problem situation as complex, with no simple unitary definition of 'the problem'. Possible 'problem owners' included Phil Belshaw, the team itself, ICI Organics, ILSD as a whole, or users of ILSD services. Encouraged to name 'problem themes', Styles, Fedorski and Teagle had no difficulty in naming 26, all felt to be important! The themes related to ILSD's services, technological developments, relations with users and with the Division as a whole, as well as to the immediate need to respond to the imposed manpower constraints. The team were reluctant to rank the 26 expressions of concern from 'most important' to 'least important', and the first cycle of the methodology was devoted to coping with the 26-strong list. This first cycle created a debate which enabled a manageable handful of relevant systems to be expressed as root definitions and conceptual models.

All 26 expressions of concern at least had in common that they were related to ILSD as a service function within a wealth-generating Division of ICI, this being an acknowledged 'given' of the study. Following from this, the first model produced represented a notional system within which the 26 concerns (and there could have been more) arose. The root definition of this system is

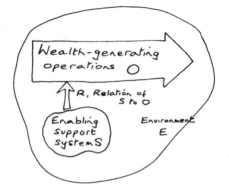

Figure 3.1 An emblematic version of ILSD's role in ICI Organics (as one of the systems S)

shown emblematically in Figure 3.1. Figure 3.2 shows the root definition (and its CATWOE) in conventional form, as well as the conceptual model to which it leads.

Setting the real-world expressions of concern against this simple model enabled the problem themes to be grouped and classified according to which element in the model they addressed. The discussion this provoked enabled ideas for a handful of relevant systems to be agreed. They covered:

(1)   an environment-appreciating system (reflecting ILSD's felt need to understand its context better);
(2)   a relationship (O–S) establishing and maintaining system;
(3)   a system to manage information as a resource (reflecting the client's wish to rethink ILSD radically); and
(4)   an 'aiding the business' system.

This first methodological cycle enabled the ILSD team to sharpen their ideas on problematical issues and to do so holistically, in a way which did not arbitrarily ignore aspects felt to be important. It also introduced the idea of making models of purposeful activity systems and using them to structure debate, and did this by means of a very simple model, that of Figures 3.1 and 3.2. The role of the outsiders had been to furnish this process, rather than attempt to provide the study's content.

More generally, this first cycle of SSM illustrates what has emerged as one of the two common ways of

Root Definition

> An ICI-owned and staffed system to operate wealth-generating operations supported by enabling support systems which tailor their support through development of particular relationships with the main operations

C ICI

A ICI people

T Need for supported wealth generation
   ⟶ Need met via a structure of main operations and enabling support

W A belief that this structure will generate wealth

O ICI

E Structure of main operations plus support ; ICI ethos

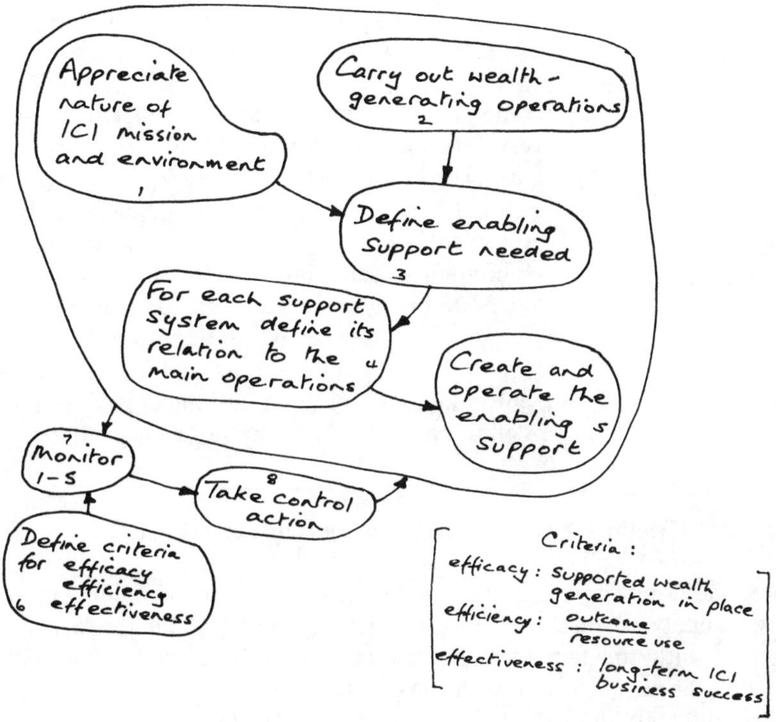

Figure 3.2 The formal root definition and model of the EROS concept

embarking upon a study using SSM. Many people, including usually the authors, now carry out the 'finding out' phase by making use of such guidelines as 'structure/process/climate', assembling rich pictures and doing Analyses One, Two, Three. The strategy is to allow that process to throw up problem themes, or suggest relevant systems. This is, however, too uncertain a process for some. A common alternative is quickly to take the organization (or part of the organization) in which the study is being done to be 'a relevant system', formulate a primary task root definition of it, build a model and then use the comparison of this with the real world as a major means of finding out. The model will generate questions to ask of the situation, and that questioning initiates and, in part, constitutes the finding out. What has been described in the ILSD study is a version of this, done for pragmatic reasons, with Figure 3.2 representing a high-level very abstract primary task model of ICI Organics with ILSD as one of its support functions.

Of these two approaches the first (using Analyses One, Two and Three, etc.) is intellectually the richest, but is less structured and calls for more confidence. The approach through quick primary task modelling has the advantage of providing a highly structured entry, which reassures the nervous, but can be inhibiting. It carries the danger of cutting off radical lines of thought. It can tend to point only in the direction of improving the efficiency of existing operations. Probably the best guideline is always to entertain both primary task and issue-based root definitions as the 'finding out' phase is done.

The Second Methodological Cycle: Forming New Perspectives

Aided, no doubt, by the extremely simple nature of the modelling in the first cycle, namely the use merely of the EROS model of Figures 3.1 and 3.2, together with the fact that that model had been demonstrably useful in enabling the team to cope with the 26 ideas for 'relevant systems', the team had no difficulty in moving into a second cycle of the methodology in which the now handful of relevant systems were formally expressed as root definitions and modelled. Checkland and Perring were required at this stage only to provide some technical help with the modelling.

Of the four main ideas for root definitions and models listed above, the third (managing information as a resource) was seen to be the central one to which the others were adjuncts. This was the idea focused on first, and a number of root definitions and models were prepared, from the points of view of the ICI Organics Division, ILSD itself, and users of its services. Figure 3.3, for example, gives the root definition, CATWOE and model of a system which provides a comprehensive information service, this image being seen from the point of view of the professional providers.

Many such models were built over the next few months during which the team met weekly to progress their project. As root definitions were formulated and models were built, comparisons of these images with the real-world day-to-day life of ILSD were made by the team.

> Although the content and sequence of such work can be reconstructed from the project files—and can for any use of SSM which is adequately documented— what is probably more important (but usually escapes the files) is the *change in perceptions* which take place in the heads of users of SSM as the methodology is used. The cycle from perceptions to relevant systems, to models, to new perceptions, is an organized way of thinking one's way to clearer, or new, perceptions. In terms of ultimately taking purposeful action to improve a problematical situation, it is the original perceptions and the new perceptions which are crucial, not the models. And it is the difference between the two sets of perceptions which stimulates the debate about change. The importance of the methodological skeleton is that it makes the thinking process coherent and capable of being *shared*. And if the new thinking is not seen as useful, but leads into a mental cul-de-sac, then the explicit record of the finding out, root definitions, models and comparisons enables an orderly retreat to be made to new directions of thought through new choices of relevant systems and new root definitions.

In this work in ICI Organics Division, it seemed to the outsiders, who were visiting the three doing the study only intermittently, that the team was markedly changing its concept of ILSD during this phase of the work. They were becoming more comfortable with a vision of ILSD as a proactive

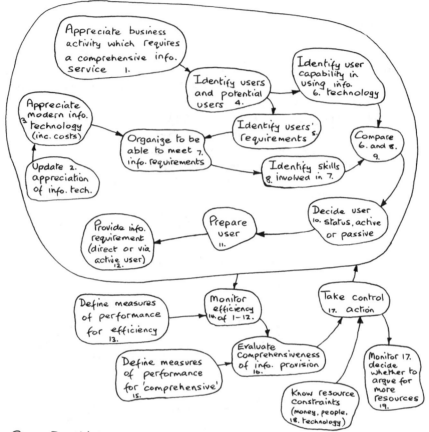

Root Definition

A system, organized by ILSD, which provides Comprehensive information to active or passive users employing technical and other skills, assisted by modern technology, so that the service is regarded as comprehensive.

C User
A ILSD, some users
T user ⟶ user helped by information provision
W modern technology and the local culture make this feasible, useful
O (implied) The Company using ILSD as agent
E existing structure, modern technology, company resources

Figure 3.3 Root definition and model for an information provision system from the providers' point of view

support function rather than as a reactive service function. The difference was that between on the one hand reacting to requests for information efficiently and effectively, and on the other *supporting the business* through information provision, and doing this proactively.

Indicative of the changing perceptions was the feeling that a more detailed level of comparison was needed for some models and parts of models which included activities which might be thought of as 'providing support' rather than 'responding to requests'. This led to a third cycle of methodology in which some of the activities in existing models were expanded as root definitions and models.

The Third Methodological Cycle: From 'Whats' to 'Hows'

Further model building was felt to be necessary in order to question the current procedures in more detail, so that the comparison-stage questioning could address questions of *how* rather than *what*. Figure 3.4 shows a root definition, CATWOE and concept of a system in which ILSD and users jointly identify the level and nature of the information requirements of those who should be recipients of ILSD's support. This concept is an expansion of Activities 4 and 5 in the model of Figure 3.3. Figure 3.5 shows the model from the new, more detailed, root definition.

At this stage comparisons between models and real world were being done either by the team of Fedorski, Styles and Teagle on the basis of their collective knowledge of ILSD and ICI Organics or by individual team members in discussions with other colleagues. The comparisons were usually recorded in the form of a simple three-column table which for each activity in a model indicated whether or not it existed, and if so in what form; how it was currently perceived and judged; and what changes in it might be contemplated as both desirable and feasible.

In carrying out the comparisons during this third methodological cycle there was the usual shift from comparison at the level of 'what exists' to that of 'how activities are (or could be) carried out'—the latter leading to detailed examination of how things could be improved. The models of the kind shown in Figure 3.5, an expansion of two activities in the model of Figure 3.3, reflect this shift.

By this time the ICI team were using SSM like seasoned veterans, this illustrating the relative ease with which methodological skills can be acquired when users are immersed in a difficult real situation which they see as calling for urgent action. The project file contains, for example, a note written on

Root Definition

> A system owned by ILSD which, together with users, identifies those scientific, technical staff in Research, Technical Service, Development Production and business functions who require specific information to do their jobs effectively, and 'key' users in particular; and which identifies the level and nature of these requirements, ie. breadth and depth of subject matter, detail and precision of presented output.

This expands activities 4 and 5 of Figure 3.3.

C : ILSD and users

A : ILSD and users

T : potential users and 'key' users ⟶ users and users
        identified together with info. requirements

W : this kind of collaborative approach is essential
    for effective and efficient information system development

O : ILSD

E : ICI Organics organization structure

Concept :

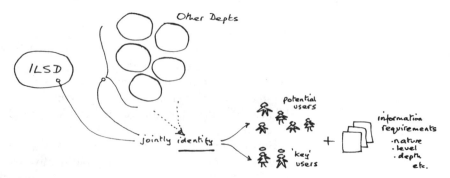

Figure 3.4 Root definition, CATWOE and concept of a system based on activities 4 and 5 of Figure 3.3

Figure 3.5 The conceptual model from the root definition in Figure 3.4

behalf of the team by Mike Fedorski which illustrates the sophistication of their methodological discussions at this period. It is a record of a team session which raises points for discussion with Checkland and Perring. Some extracts give its flavour:

(1) Discussion of RD6 (that in Figure 3.3 here) and our commitment to it suggested we should stay with it for the time being and model some sub-systems. RD6 gives rise to sub-systems RD6.1 . . . . RD6.x. When we come to model sub-systems we must ensure that they are compatible with the model

of RD6—i.e. avoid overlap between sub-systems or the introduction of new features not present in the parent. At the comparison we have two alternatives: compare RD6 with 'what is' in the real world, using RD6.x to help us; or compare the models from RD6.x, etc., with the real world. Either method should lead to a list of differences leading to possible changes. Will this bring us to a 'how' stage? We can break the differences into two lists:

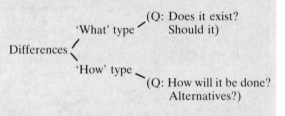

This should lead to possible changes and an action plan.

(2) Possible techniques for looking at RD6's components. 1st method (the one used so far): make a new RD and model it. 2nd method (simpler?): take RD6's activities as black boxes, list their inputs and outputs only, then fill in the required activities for the sub-system. For example if we had done this for activities 4 and 5 of RD6 (Figure 3.3 here) we would have had

appreciation   identified users
of . . .   ⟶ (4 & 5) ⟶
                         identified requirements

The first method introduced richness. Advantage of the second is that it is quicker. We shall use this second method to model activities 2 and 3, activities 6, 7, 8 and 9 (taken together) and activities 10, 11, 12 (also together).

(3) PC suggested we expand activites 13, 15, 16 of RD6. We need the users thoughts/feelings about 'comprehensiveness'.

These extracts underline the fact that by now the ICI team were themselves confidently doing the study, with the outsiders providing advice on the process of SSM rather than substantive contributions.

The many detailed comparisons carried out by the team both served to build the new concept for a re-aligned and reorganized, more proactive

ILSD, and confirmed the team's commitment to it, even though it would entail many specific changes. The team were rather meticulous in deriving these changes explicitly from the comparison debates, which were carefully documented. Figure 3.6, as an example, shows a comparison for a single (but important) activity in a model which had ILSD as actors, and contained the activity 'Appreciate ICI Organics Business'. It is examined in terms of both commercial and technical appreciations. For each the team recorded an 'ideal' version of the activity, the present version in real life, the differences between the two and the changes implied.

This careful procedure allowed the possible changes to be examined for desirability and feasibility, and enabled possible actions to be recorded. Figure 3.7 shows extracts from a 21-page report prepared by the team based on the comparisons deriving from the model in Figure 3.3 and from its modelled subsystems (of which Figure 3.5 is an example).

Detailed work of this kind enabled the team to build up a very rich account of a refurbished ILSD and to examine logically and in detail required staff numbers and roles. It now became clear that the team was coming to the end of the work which they themselves could do. Reference to a wider audience was now necessary; in technical terms the team needed to involve a new audience in SSM's Stage 6 if major change were to be achieved.

The Fourth Methodological Cycle: Widening the Debate, Taking Action

Expanding the range of people concerned with the ILSD study took several forms. The team took their colleagues in the Department through the whole study so that the members of ILSD could make their contributions and also assess how the new concept would impinge upon them. The outsiders were not present at this discussion, and the obvious joke was exploited: 'We're being Lancastrated!' Secondly, with Phil Belshaw's help, a presentation to senior managers in Organics Division was arranged, since changes in their interactions with ILSD were now required; and thirdly an internal report on the study was published by the team.

Both the presentation and the report required prior agreement by the team on the summary message of the study. This was argued out in a half-day discussion between the team and Checkland and Perring. Here the outsiders were able to make substantive contributions, on the principle that the spectators may see the game as a whole more easily than the players. The summary statement produced at the discussion was as follows:

(1) We have made a fundamental appraisal of the role of ILSD.
(2) The method was to examine several 'systems' relevant to ILSD and examine the forms they might take, depending upon how the fundamental purpose of ILSD is defined. Possible purposes examined ranged from

Part of a comparison table having the form :

Activity :
Ideal :
Present :
Differences:
Changes:

Activity : Appreciate ICI Organics Business

(a) Commercially

Ideal: — know about present and projected future business of the Division, e.g. what product areas are running down/closing what product areas are we moving into

Present: — knowledge of current manufactures (from Production Dept.)
— new fields—learn from requests for information coming in from Research Dept and New Projects Section (informal and irregular).
— closing down/changing emphasis. Get some information at a late stage and for individual projects from the Sales Circulars.

Differences: — Quality, quantity and effectiveness of information

Changes: — (cannot know about everything. Concentrate on cost–benefit: avoiding information provision no longer required.)
— structural: Obtain a presence on appropriate committees to obtain most effective feedback
— procedural: Obtain and monitor all reports from appropriate meetings. Develop better antennae in useful departments
— attitudinal: Division management to appreciate need of information providers to be kept up to date
— amount: there is a large gap between concept and reality here at present

(b) Technically

Ideal: — know about present and possible future developments of interest to the Division

Present: — current technical documentation, e.g. period reports, programmes for experimental work, annual plans, consultants' reports, etc.

Differences: — no formal system although informal system is pretty good. Not involved at early-enough stage (so that we could organize full information flow in these areas).

Changes: — structural: obtain involvement in relevant early planning meetings
— procedural: improve antennae
— attitudinal: as for commercial
— amount: gap seen as being much smaller than for the commercial area.

Figure 3.6 An example of a comparison stage done by the ICI team

      'respond quickly to requests for information' to 'manage information as a
      major resource'.

(3) A particularly fruitful definition of ILSD purpose has been found to be
      both not too far from reality and suggestive of useful changes. That
      definition is:
          A system organized by ILSD to provide a continuing comprehensive
          information service to users, employing technical and other skills avail-
          able within the system, and employing modern technology so that the
          service is effective in furthering ICI Organics' business interests.

(4) A system which would meet this definition looks like this . . .

(5) We have compared this (and other) models with present operations and
      structures in order that the identification of differences can stimulate
      discussion about desirable and feasible change.

(6) The comparison led to identification of four major differences with major
      implications:
        (i)   the modelled ILSD plays a more central role in the Company;
        (ii) a relationship with users is implied which requires 'active' users
            (trained by ILSD, perhaps, to help themselves) and a richer user–
            ILSD dialogue;
        (iii) ILSD would have to develop its training function;
        (iv) a more organized monitoring and control of ILSD activity is implied.
            (Organizational implications are still being worked out.)

(7) These implications imply some changes which ILSD can make, but some
      which require the active collaboration of others.

(8) Hence we require endorsement at this stage.

This message was the basis both of the report on the study written by the
team and of the subsequent presentations to senior managers. The report
'The Role of ILSD in the Organics Division Information Network: a fundamental study using systems-based methodology' discussed the changing context of ILSD work (namely, that of the 'information explosion' since the
1960s and the availability of new technical means to cope with it) and argued
that 'information is a prime resource of increasing importance and centrality
to any function'. It described how ILSD had taken a thorough look at itself.
The final expanded version of the model of which Figure 3.3 is the parent was
included as an illustration, and the comparison between it and existing
activity was discussed in terms of issues well known to members of ICI
Organics. The conclusion argued for seeing ILSD as the Division's required
information support function, and indicated the changes necessary to act out
this role, which would involve not only ILSD itself but also users and the
computer experts of Management Services Department. This is why the
progress of the work now required endorsement by the Division.

Appendices to the report include a summary of the range of alternative
root definitions which were used in the study, an example of a detailed
comparison, and a short piece written by Checkland: 'A More Radical View'.
This latter contribution summarized thinking during the project about the
organizational implications of treating information as a resource. It is inclu-

An Example of Comparison Stage detailed procedure

Assessment of Desirability/Feasibility of 'Hows'
via the format : 'How'; Desirable?; Feasible?; Action?

'HOW'	DESIRABLE?	FEASIBLE?	POSSIBLE ACTION
'Identify/obtain those reports that we do not see at present.'	Suspend judgement for the time being	Yes	Identify *all* reports and assess for 'desirability' in order to maximize on usefulness, then make a selection.
'Form a direct link with Management Conference.'	Yes—gives early warning of new events	Head of Patents Dept. (PLB) attends this Conference. Therefore we have a ready-made link; but there are problems of security/confiden-tiality.	Discover from PLB what types of issues are discussed. If security okayed, ask him to give us a post-Conference run-down.
'Link into the Division Business Investment Con-ference (DBIC).'	Yes.	Yes—we have an ILSD staffer who attends this Conference—but also 'Organics News' gives summary of main items.	Requires only slight effort to get our representative to report back.
'Get invited to RD Section Meetings.'	Yes. Emphasize continuous ongoing two-way process rather than the one-off type of meeting used at present by us to give talks to RD on our services	How many such meetings are there? Could be too many for us to attend them all. Acceptability to section managers is a major problem.	Consult with Research Manager, get approval and make arrangements with Section Managers.
'TILO (Technical Information Liaison Officer) concept.'	Uncertain. This is one of the possible ways of getting technical information.	More suitable to those units where there is only a narrow band of specialism. We already have this kind of specialism in our unit.	Needs clarification—action will have to wait until we have a firmer job description for Information Officers.

Figure 3.7 Desirability and feasibility of 'hows' as examined by the ICI team

ded here as an appendix to this chapter. Written in 1978, it makes slightly quaint reading twelve years later; but its core message is still relevant and has not yet been acted upon coherently in most organizations. The message is that both technological and social considerations now lead to the view that organizations can be usefully treated as a net of semi-autonomous work groups linked by an information network.

After publication of the draft report internally, and the presentation by the team to their ILSD colleagues, came the presentation to a group of senior managers assembled by Phil Belshaw to hear the message that *they* now needed to think about ILSD in a different way and to rework their links with that group. For the team the most significant moment came when a senior manager from Research Department said, unprompted, during the discussion: 'I have had a big surprise. I have known and worked with ILSD for 20 years, and I came along this morning out of a sense of duty. To my amazement I find I now have a new perception of ILSD.' On the other hand there was little response to the 'More Radical View' which Checkland presented. Perhaps it served a purpose in making the audience comfortable with the more limited changes actually being proposed!

Subsequently, with ILSD having reached the limit of its political power with the completion of the work described in the team's report, Belshaw, in the months following the presentation, prepared the case to the Board for the management changes and the capital investment in new technology required by the new ILSD. This was described as 'bearing little resemblance to the present ILSD. It will require fewer people, will have a simple unified structure, and will make an enhanced contribution to the Division's business.' In an early draft Belshaw included a paragraph which was omitted from the final version, on the sound grounds that it was of more interest to those doing the study than to the Board, but which is relevant to this book:

> The approach used . . . has general application and is based on the branch of systems thinking evolved in the Department of Systems at Lancaster University this past decade. The approach . . . bears directly on the problem of 'doing more with less' . . . It involves getting back to the purpose or *raison d'etre* of a function, and leads to the development of radically different ideas about resources and their organization.

In the final version, the note on the approach taken in the study argued that the problems addressed

> need to be worked first at the higher levels of WHY and WHAT, rather than at the structural level, which is about HOW.

A prior requirement was 'a transformation in the level of thinking' in the organization which the use of SSM could itself promote.

The main body of the note to the Board argued the case for the significant investment needed to transform ILSD itself. An appendix on the approach used in the study focused on: 'Thinking at the right level; Thinking in the right context; and Thinking in the right time perspective', making the point that 'the planning process is much more important than the plan'.

The reorganization of ILSD was accepted, and the Division signalled its acceptance of the study's findings by making two kinds of investment in the new information unit. Money was provided for the new information technology needed to foster a proactive ILSD, and, with Belshaw's retirement imminent, a new Head of Department, John Wales, was appointed from the Company's cadre of young managers of high potential. In the local culture this was a significant signal.

Methodologically, this partial fourth cycle of SSM drew senior managers into the study at the point of establishing 'desirable and feasible' changes sufficiently strongly that action ensued. At the end of the third cycle, Stage 7, for the team of Fedorski, Styles and Teagle, consisted of making some internal changes and seeking endorsement of larger changes with Phil Belshaw's help. Now the fourth cycle was 'owned' by the Division. The whole methodological history of the study is shown in the matrix of Table 3.1. At any moment during the study the team would have been able to define exactly what they were doing ('Further model building in the third cycle, . . .' etc.) but the methodology quickly became satisfyingly transparent, so that attention could focus on the substantive content, not the methodological framework. This was perhaps helped by the fact that the outsiders were there to keep an eye on the process, a point which will be taken up in the Conclusion below.

The study also illustrates the fact that to 'end' a systems study is an arbitrary act: the flux of events and ideas continues to unfold and the need to cope with it continues. . . . This study put new items on the agenda of debate in this part of ICI; it contributed to changing ICI Organics' readiness to see the world in a particular way, and this led to structural, procedural and attitudinal change. But this itself creates a new 'problem situation' (in this case, installing the new, refurbished ILSD) and the relevance of SSM is no less than it was to initiating the rethink which produced this outcome.

As a final methodological point it is worth recording an understandable, but foolish, criticism which has been made of this study. In discussing an account of this study which he had just listened to, a professional librarian offered the comment that in his view the study had not achieved much since it was clear to forward-thinking people in his profession that all libraries and information units *ought* to provide proactive support rather than a passive service. He was missing the point. What was significant here was not the shift of concept, it was the shift of concept for a particular group, in a particular situation, *allied to a generated motivation to make changes*. The crucial outcome was *action*. There is never any shortage of talk, ideas and rhetoric. Words are cheap. What is in short supply in organizations is an organized sharing of perceptions sufficiently intense that concerted action gets taken corporately. Enacting the process of SSM can help with that.

Conclusion: Towards 'a System to Use SSM'

The study described above illustrates in its several cycles the stages of the overall cycle of SSM. This experience made us realize with greater clarity than hitherto, that most of our effort in our systems studies was normally on the stages of the methodology, on carrying them out and moving round the mosaic of stages flexibly. Wider considerations were thought of, and dealt with, informally. In the ILSD study, however, we accepted from the start Phil Belshaw's constraint that Lancaster people should not do the study; their role was to help or *enable* Fedorski, Styles and Teagle to do it. This was a somewhat different role to that of enacting the stages of SSM participatively with people in the problem situation. It made us more conscious of the considerations outside the stages of SSM itself. It raised thoughts of the nature of 'a system to use SSM'.

Given Belshaw's constraint, much attention at the start was on the nature of the study itself: who would do it, how they would do it, how would SSM be used, etc.? These were all considerations which could not necessarily be decided once and for all; they might need redefinition as learning accrued. Therefore we could think of our situation as in Figure 3.8, with the study being defined by these initial considerations and itself modifying them.

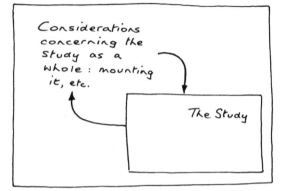

Figure 3.8   The study and the meta-study

Secondly, the study consisted of carrying out the various stages of SSM *in a particular situation* which had to be understood. That situation would affect the enactment of the stages and itself be affected by them. As an example of this, the appreciation of the situation in which the study was being carried out led to calling a halt to further cycles of SSM by the team; producing the report and giving the presentation to senior managers at the end of those cycles itself

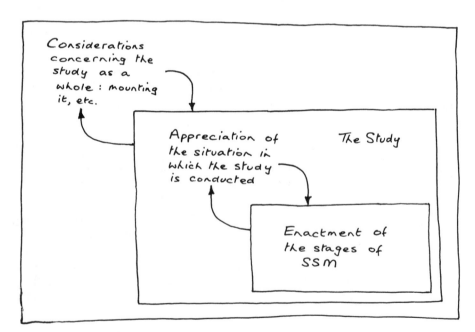

Figure 3.9   The rudiments of a system to use SSM

changed our appreciation of the situation in which the study was being made. This kind of consideration suggests expansion of Figure 3.8 into the form shown in Figure 3.9—which should be thought of as overlapping rather than as representing three stages of work—hence the two-way arrows. [Interacting activity throughout all three levels of this rudimentary model is possible at any time. This arises because although the cyclic model of SSM's stages happens to be, technically, an organizationally closed system, it consists of elements (the stages of SSM) some of which (Stages 1, 2, 5, 6, 7) are themselves open with respect to the environment of a study and to yet wider considerations. This means that SSM is always open to the real world in which it is being used. That is why it can evolve (Varela, 1984).]

Previous experience had directed our attention mainly to the stages of SSM, with only informal consideration of the other two levels. But with hindsight we can map many other experiences on to this epistemology. In Chapter 10, drawing both on this rudimentary thinking and on the more sophisiticated study described in Chapter 9, the model of Figure 3.9 will be refined into a form which might be tentatively thought of as a *prescriptive* account of how to use SSM.

The ILSD Project: A More Radical View

This appendix was included as an appendix to the team's report on the ILSD study. It summarizes the content of part of the presentation made to senior managers, and goes speculatively beyond the recommendations of the study itself.

As has been described in the main body of the report, the project methodology entailed defining a number of possible views of the role played by information, and by an information function, in the Division. Possibilities examined in the course of the work ranged from the view that ILSD should be a unit which simply responds quickly and efficiently to requests made to it, to the view that a whole Division of ICI might be looked at from an information-processing point of view. The main work of the project concentrated upon studying the implications of viewing ILSD as a support unit to the Division business activity. The notes which follow are part of the outcome of examining the more radical view that information is a prime Divisional resource. They take the form of notes on a presentation given by Professor Checkland of the University of Lancaster as part of the project presentation made on 26 September 1978.

(1) An industrial firm may be regarded as a *purposeful enterprise* which acquires and manages various *resources* in carrying out its *tasks*.

(2) Traditionally the resources needed can be expressed as: *money* and *men*. The former subsumes *materials* and *machines* which can both be expressed in financial terms. Management literature of the 1950s and 1960s frequently spoke of resources as 'the four Ms'.

(3) This traditional view takes as 'given' the communications, the information flows which will characterize all of the enterprise's tasks. It assumes that the provision of information is not problematic.

(4) A more sophisticated view of the set of *contracts* (financial and psychological) entailed in the creation and operation of an industrial enter-

prise suggests that *information* is itself a prime organizational resource which has to be managed rather than taken as 'given'.

(5) Information is obtained as a result of attributing meaning to data (the latter being 'recordings' related to events, objects or people). A view which considers information as a prime resource therefore has to explore the meanings which people in the enterprise will attribute to the happenings they observe. This will entail initially an examination of the fundamental *purposes* embodied in the organization's tasks, sub-tasks and their inter-relation, and then a study of attitudes towards those purposes.

(6) If our industrial enterprise is a manufacturing company we may view it as a mechanism for generating wealth by transforming raw material into saleable products, this being the physical embodiment of a parallel (abstract) transformation of a defined market need into a satisfied need.

There will be a set of organizational tasks directly concerned with these transformations and a further set of tasks which provide essential support. These support tasks will be focused respectively on *knowledge* ('R & D', 'Process Development', 'Corporate Planning', etc.), on *money* ('Accounts Department', etc.) and on *people* ('Personnel Department', 'OD unit', etc.). In addition, if we are taking an information-processing view of an organization, to these we must add a support task focused on *information*. (In real life examples there are frequently *data* processing functions but these are not usually part of an information function.)

(7) Now, if we envisage an enterprise which takes seriously the proposition that information is a prime resource, *and organizes itself according to that view*, that enterprise will have to take into account two recent real-world revolutions.

Firstly, there has been a technical revolution in the world of computers and data processing. Over the decade 1965–1975 there has been an exponential drop in costs:

	Cost of processing 1 000 000 multiplications	Cost of disc storage of 1000 characters	Cost of main computer storage for 1 000 000 characters
1965	14p	25p	20p
1975	1p	3p	2p

allied to physical miniaturization. *Cheap distributed* computing is now technically possible. Achieving the techical revolution requires three things: *networks, hardware* and *protocols*. Provision of networks and hardware is a *technical* problem which is virtually solved. Provision of protocols (rules and methods for access and usage) is an *intellectual* problem which is not yet solved.

Secondly, there has been a social revolution in advanced western societies which is still underway. This may summarized in the words 'democratization' and 'participation'. The dominant philosophy of this revolution is that of egalitarianism and this is manifest in many different aspects of life in our society, in work, in play, in politics, in education, even in art, where the cry is for the subsidy of 'community arts' rather than the individual with talent. Ten years ago, to describe something as 'uneconomic' was to rule out any further discussion of the matter; the 'environmental' movement has stopped that. Nowadays, it is the description 'elitist' which is assumed to end all argument. (This is so even in universities, which by their very nature are embodiments of one kind of elitism!)

(8) When we bring together the technical revolution and the social revolution and ask how they impinge upon a model of the organization which grants a prime importance to information (gathering it, processing it, disseminating it) we obtain a picture of the following kind:

- organizations are likely to move towards a mode of operation in which autonomous groups, where possible small enough to be self-organizing, are linked in an information network; out of the question will be the management style, obsolescent for some years now, in which the manager's power is based on the withholding of information;
- the need for the high-level management of information (in defining protocols, etc.) will lead to an information function having the kind of status now accorded to other support functions focused on knowledge, money or people;
- the information function of the future will fill an enabling role, defining the rationality of the information net and the rules governing access to it, and beyond that making sure that resources are available which will allow the net to adapt to changing circumstances.

(9) I shall be surprised if ICI takes the lead in appointing Directors of Information as peers of Directors of Production, Finance, Personnel and R & D; but equally I shall be surprised if there is not, in 5–10

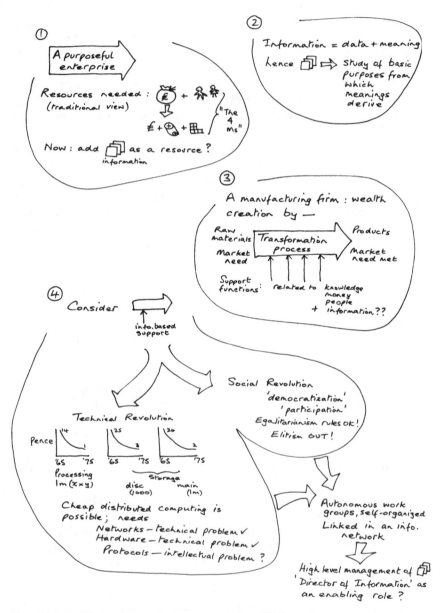

Figure 3A.1 Illustration used at presentation to ICI Organics management, September 1978

years' time a Director of Information in every Division. And the Directors of Information of the Divisions will, as a group, be much concerned with—ironically—both the *greater fragmentation* of ICI into smaller groups and its greater cohesion through a computer network.

(10) There is one aspect in which ICI is well fitted to face the information revolution to come. Information networks and their protocols will increase the influence of rationality in the unfolding flux of events. And ICI is well stocked with minds trained in rational thinking. ICI is already strong on rationality, and that will help it to cope with the coming revolution.

Acknowledgement

I am grateful to my colleague Iain Perring of ISCOL for his work in this area which has materially helped in the preparation of this note.

P. B. Checkland

Chapter 4

An Application of Soft Systems Methodology in the National Health Service

Introduction

In developing SSM the members of the Lancaster group always took the view that the work called neither for theoretical work on its own, nor practical work for its own sake, but rather a cycling between the two as theory informed practice which was itself the source of theory. Work in real problem situations would lead not only to practical outcomes but also to reflections which would modify the approach in further practice. This cycling of practice and theory, out of which SSM emerged, took place in many different kinds of problem situation, but the main experiences which led to the basic formulation of SSM all took place in work in industrial companies. Only in recent years have experiences outside industry contributed significantly to SSM development.

There are several reasons for the dominating role of industrial settings in SSM's development. The original group of people developing the approach were in a Department of Systems Engineering, and even though they were interpreting the word 'engineering' very liberally, systems engineering, as normally thought of, is very much an activity developed in and characteristic of industry (Hall, 1962; Chestnut, 1967). So in the early years of work it seemed entirely natural to look for industrial applications of the ideas. Secondly, it was industrial companies which were most receptive to the ideas and hence were most ready to host a systems study of some part of their activity. These are both very practical reasons for the early concentration on working in industry, but there is an intellectual reason at least as cogent. An industrial company, for all its complexity, is much simpler than, say, a local authority, a department of the Civil Service, or a district health authority. It is simpler in the sense that its ultimate constraints and measures of performance will be clearer than those of a public authority. The logic of the industrial situation is such that unsubsidized companies simply have to produce their goods or services at a lower cost than their customers are prepared to pay for them, or they will not survive. In addition, in an industrial company trying to survive and/or grow under these terms, there will have to be some power structure in place which can make certain things happen and stop others. It is easier in circumstances of this kind to define and carry out studies and to know at the end whether or not something useful has been accomplished. This makes industry an intrinsically easier environment in which to develop an approach to tackling real-world problems.

The Lancaster archives reveal some studies carried out in local authorities during the early years, but rereading them more than a decade later suggests that in these early projects the intellectual energy went into trying to grasp the greater complexity of the area of application rather than into developing the use of the systems ideas. This is not to deny that work outside industry contributed to the continuing development of SSM; and some significant experiences *were* gained from which the whole Lancaster group learned. One such study, carried out by Galliers, Whittaker, Clegg and Mouthon in the London Borough of Camden, usefully illustrates the way in which the public sector calls for less arrow-like studies than those carried out in, say, a company manufacturing gumboots, or a fast-food chain.

The Camden study (Galliers *et al.*, 1981) began with a request from the Social Services Department to ascertain what resources would be required to enable a substantial number of mentally handicapped people in the Borough to be placed in 'open' (as opposed to 'sheltered') employment. The thinking behind the request can itself be seen as part of the general move towards providing care in the community rather than in institutions, a shift of thinking which was taking place during the 1960s and 1970s. In Camden an internal report in 1979 had suggested that care for the mentally handicapped was relatively weak compared with that provided for other groups, and it was suggested that some of the people attending the local adult training centre would benefit from the opportunity to work in a normal ('open') environment rather than in the 'sheltered' environment of the training centre.

The initial finding-out phase revealed a very complex situation with many statutory and voluntary agencies involved, patchy inter-agency communication, many different values and attitudes regarding service for the mentally handicapped, a lack of information about that group in Camden, and no commonly held policy with respect to the employment of mentally handicapped people. The project brief was soon renegotiated to cover *establishing a policy* with respect to the open employment of the mentally handicapped, and a complicated project ensued. A pilot job placement scheme was initiated, not viewed necessarily as the precursor or a major initiative but rather treated as a source of experiential learning for those trying to finalize a policy. It

provided experience of job placement and illustrated the inter-agency relationships which needed to be handled with care. It proceeded in parallel with
the work on policy formulation, and that work included such unusual actions
(unusual, that is, for systems studies at that time) as using a questionnaire and
organizing a one-day workshop with the explicit aim of teasing out and
debating conflicting attitudes and *Weltanschauungen* towards placing the
mentally handicapped in employment.

The account of the project by Galliers's team (1981) describes a number of
eventual outcomes which include the adoption by the Council of a coordinated policy, the setting up of regular meetings to improve cooperation and
communication between the Social Services Department, the Manpower
Services Commission and the Employment Unit of Camden Chief Executive's
Department, and the reorganization of a specialist group in the Training
Centre to ease the transition from sheltered to open employment. The team
remarked that:

> In situations such as this, we found that the question of resources is inextricably
> bound up with the inter-related problems of attitudes, policies, information and
> communication

and that

> . . . the wide range of values and attitudes towards the employment of mentally
> handicapped people held by the various 'actors' in the problem situation had to
> be taken into account.
>
> (Galliers *et al.*, 1981, p. 113)

They could have added that the complexity is increased by the fact that these
'actors' are themselves in many cases *autonomous professionals* exercising
their professional judgement. What is more, they are distributed among a net
of autonomous and semi-autonomous groupings all having a concern for the
problem situation but not being part of any unified power structure such as
exists, and can ultimately be invoked, in an industrial company.

With hindsight (though its significance was not particularly noted at the
time) we can see that the Camden study and some of the other involvements
in the public sector at that time, when compared with the industrial studies,
revealed a more sophisticated attention being paid to *the process of using
SSM*. This was the result of the greater complexity of public-sector problem
situations. It suggests that with the basic shape and content of SSM established by carrying out projects in industry, studies in the public sector ought
to provide both a more difficult test for the approach and a very suitable arena
for its further development. In recent years a specific effort has been made to
use SSM within the National Health Service, and this has been fruitful. For

example the development of Analyses One, Two and Three (Chapter 2) has been much stimulated by the complexities of public sector studies.

The remainder of this chapter describes a study carried out in the East Berkshire District Health Authority. First, however, it is useful to make a brief general examination of both the NHS as a whole and the specialty of community medicine.

Context—General: the NHS

Established in 1948 as a tax-funded service freely available to all, the UK's National Health Service (NHS) is the country's largest employer, with almost a million staff. Although health spending has declined since the early 1970s as a fraction of national income (from around 5% to about 1½%; Huhne, 1989), the budget in the late 1980s is around £20 000 million (Bowden and Gumpert, 1988), so that the NHS is a major national enterprise whose efficacy, efficiency and effectiveness will attract much attention.

In the 1970s and 1980s the NHS saw major attempts to bring about improvements in service provision through structural change on a large scale. In the 1980s this has settled down to a hierarchical structure descending, in principle, from the NHS Board through around 20 Regional Health Authorities to the level at which health care is delivered to the population, that of the 192 District Health Authorities (DHAs). (East Berkshire, the DHA in which the study to be described below took place, serves a population of about ⅓ million people and spends about £44 million annually.) Whether structural change ever brings about, as a natural consequence, desired procedural and attitudinal changes is always debatable, and the most recent moves to improve the NHS's performance have focused mainly on managerial, i.e. process, rather than structural changes.

Prior to 1983 the management of a DHA was spoken of as being the 'consensus management' of a team containing an administrator, a treasurer, and representatives of the main professional groups: hospital clinicians, community physicians, nurses, etc. In October 1983 the report of an NHS management inquiry team under Sir Roy Griffiths, a manager with experience in the retail industry, was accepted by the Secretary of State. The Griffiths Report (1983) recommended that the NHS should change its mode of management by appointing general managers at all tiers in the organizational hierarchy 'charged with the general management function and overall responsibility for management's performance' (p. 5). The report contains the striking sentence:

> In short if Florence Nightingale were carrying her lamp through the corridors of the NHS today she would almost certainly be searching for the people in charge. (p. 12).

Griffiths aimed to make clear who were the people in charge.

At district level a District General Manager (DGM) became accountable for the quality and range of services provided for his district by the health professionals. (Inevitably many of the administrators from the 'consensus management' teams became DGMs, although a few managers from industry came into the NHS at this point to fill these roles.) DGMs then decided upon a management structure for their district, and, when these were approved by the then Department of Health and Social Security, appointed their unit general managers. In a particular district, for example, there might be an acute unit, a chronic unit and a community services unit, each with a unit general manager responsible to the DGM, just as the latter was responsible to the relevant region.

At the time of writing the NHS has continued to be the subject of much concern, and the 1989 Government White Paper discusses the development of some form of internal market within the NHS, with budgets for general practitioners, some hospitals becoming autonomous, and an increase of clinical accountability, with tighter contracts for clinicians, merit awards related to continuing performance and systematic medical audit becoming a professional requirement. There is no lack of signs of continuing agonizing over how to manage so complex an organization and how to measure the performance of a service provision with no unequivocal performance indicators. Giving the Radcliffe–Maud Memorial lecture in 1987, Sir Roy Griffiths said that:

> The triumph of the 1940s was the vision to create the NHS. The triumph of the 1980s and 1990s will be to create the wealth to secure it and the will to manage it,
> (Griffiths, 1987)

thus neatly encapsulating current concerns.

To anyone thinking about the special nature of the NHS as a management problem it is obvious that it has a number of characteristics which do not apply to industrial companies, and which make its management problematical. These are problems beyond those associated with the immense size of the organization, itself a source of many difficulties.

Firstly, there is the problem of thinking of the NHS as an entity, in the same way that the Ford Motor Company, the National and Provincial Building Society or the Union Jack Rubber Company can be, and are, thought of as single entities. During an industrial dispute in the Health Service some years ago a Mr David Grime wrote a letter to the *Guardian* newspaper in which he pointed out that health workers do not think of themselves as working *for* the NHS. A million people work *in* it, but no one works *for* it, not in the same sense that employees work *for* the Union Jack Rubber Company. This is a dramatic way of saying that there is no demonstrable unitary power structure in a managed NHS, for all that the general management initiative following the Griffiths Report tried to push the Service in that direction. Health care provision in the UK is to a major extent in the hands of the NHS, obviously. But the parts of the Service are themselves locked into a complicated network of autonomous and semi-autonomous groups concerned with health matters. The provision of health care emerges out of, rather than is routinely delivered by, the professional activity of these autonomous and semi-autonomous groups. The network includes not only the parts of the NHS but also such groups as local authority social services departments (with local authority boundaries not coinciding with those of health districts) and voluntary and charitable organizations such as Age Concern or MIND (National Association for Mental Health). It is obvious that in such circumstances managing, say, services related to care of the elderly in a given geographical area, is not a straightforward task. Networks such as exist in health care provision cannot be managed on the same lines as the Ford Motor Company.

A second characteristic which makes their professional life complicated for Health Service managers is that actual delivery of health care is in the hands of clinical professionals rightly concerned to protect their autonomy as professionals. If, as is often the case, there is more than one way of carrying out a surgical operation, a DGM cannot instruct a surgeon to adopt the cheapest way, that must be left to the surgeon's professional judgement. In fact if the DGM role is examined through the frame provided by the classical management texts of yesteryear, with their talk of defining budgets, responsibilities and spans of control, etc., then the district (and the unit) manager role in the NHS appears to be technically impossible! The difficulties of NHS management can be understood starkly by imagining what it would be like to manage an engineering works in which the shop-floor workers could never be *instructed*, but had to be left to exercise their own professional judgement. It would not be impossible to run the engineering works, but it would not be easy.

From the two characteristics of managing in the NHS just described, a third arises. This was first pointed out to the authors by the perspicacious Chairman of a District Health Authority who had come into that role after a career in industry (Wood, 1985). He argued that when planning a project in the NHS it is never enough to think: What activities do we need to carry out to achieve the objectives of this project? It is also essential to devote at least as much thought (and often more) to the question: What *enabling* activities will also have to be carried out in order to be able to do the main activities of the project? This kind of thinking is of course very relevant in industry too, but it assumes very much greater significance when dealing with the network of autonomous and semi-autonomous groups of which the NHS is a part. This has been confirmed for Checkland in a dozen systems studies carried out in the NHS, and it has been found useful to ask of any health care project such questions as: What relationships in the network will be affected by doing this project? What is the current state of those relationships? How will they be affected? What second-level (enabling) activities are therefore required? These questions, useful in industrial projects, are often crucial in projects in the NHS.

The arguments developed above all suggest both that SSM, with its emphasis on multiple perspectives, could be an appropriate approach to tackling problems in the NHS and that that arena should provide a rich testing ground for the approach. A number of systems studies have been carried out in the NHS in recent years. Described below is one such study concerned with the tricky problem of performance evaluation in community medicine.

Context—General: Community Medicine

Community medicine is the medical speciality concerned not with the health of individual patients but with public health, 'the health of the nation'. In 1988 the Acheson Committee of Inquiry into 'the future development of the Public Health Function' expressed the hope that

> our recommendations will improve the surveillance of *the health of the nation*, clarify roles and responsibilities, show how each particular skill may be brought to bear at the appropriate point in the NHS within the framework of general management and, taken together, will provide a structure conducive to *better health for all*. (Acheson, 1988, p. i; authors' italics)

The idea of taking action at the level of society as a whole to try to ensure the health of a population (rather than of individuals) is, in England, more than 100 years old. In spite of that the specialty of community medicine emerged only in the 1970s. The Todd Report on Medical Education, 1968, recommended the establishment of the new medical specialty, and a Faculty of the Royal College of Physicians was established to oversee training and standards. In 1972 a further report (Hunter, 1972) suggested bringing into the new specialty the former Medical Officers of Health (a local authority post), the administrative medical officers of the former hospital boards and academics concerned with public health and social medicine. As a result of the acceptance of these recommendations, community physicians became members of the consensus teams responsible for health service management at region and district level (Acheson, 1988). In principle, community medicine was the very basis for planning to meet the health needs of a region or district, but Sir Donald Acheson is gloomy about the short history of the budding specialty. His report speaks of 'the outdated approach of some community physicians' (p. 6), 'the failure of the specialty to establish its professional standing' (p. 6), 'the decline in credibility of community medicine in some places' (p. 7) and concludes that 'in some cases . . . health authorities, undervaluing the contribution of their public health doctors, failed to give sufficient emphasis to public health issues' (p. 7). The Acheson Committee found in taking evidence that where community medicine is of high quality DGMs 'cannot envisage an organizational structure in which it does not have a central position' but that where the specialty has 'failed to win credibility ... its worth is questioned' (p. 8).

It is not surprising that our study, carried out three years before Acheson reported, found the measures of performance of community medicine to be an interestingly complex issue!

Clearly part of the intellectual base of community medicine will be epidemiology, the study of 'all factors which affect diseases' (Davies, 1984) together with their frequency and spread. In fact it would be perfectly plausible to interpret community medicine conservatively as being concerned at core with *providing information* about the patterns and dynamics of disease in a given population. But it is obvious that the new specialty could also take a more radical view of its role, perceiving it to be concerned with *managing the provision of health care services* which would bring a population to levels of health defined as feasible and desirable.

In the study to be described the Department of Community Medicine in which it was located inclined more to the radical than the conservative view.

Context—Specific: the East Berkshire Study

The study was set up at a meeting organized by Tony Turrill, then head of the NHS Training and Studies Centre known as The White Hart, Harrogate. He brought together Checkland and Dr Jeremy Cobb, District Medical Officer and head of the Department of Community Medicine in the East Berkshire District Health Authority (EBHA). The District serves a population of 350 000 people in the area Slough/Windsor/Maidenhead/Bracknell. It was agreed that a Lancaster team (subsequently to consist of Checkland and two graduate students, Sophia Martin and Chris Caiger) would investigate and recommend ways of measuring the performance of community medicine in the East Berkshire District. The study was to take place over a period of five months.

At the time this study was defined, Dr Cobb had already done some work on the topic. He had organized a weekend meeting at the White Hart at which a number of community physicians from the Oxford Region had discussed the problem of evaluating performance in the community medicine specialty. The papers from the meeting were a starting point for the study described. They included, for example, an analysis of seven relationships which a community medicine department would have to manage. Four were with entities, namely: the DGM, Local Authority Social Services, the Regional Health Authority and pressure groups, while three were with concepts: privatization, information and reduction of environmental hazards. The papers also included rudimentary systems models relating to these domains. The work of the Lancaster team based in Dr Cobb's Department in Windsor built upon this useful initial effort.

Finding Out and Methodology of the Study

The aim of the study was to help the East Berkshire Community Medicine Department (CMD) with the problem of measuring its performance. As mentioned above Dr Cobb saw the practice of community medicine as consisting of more than the provision of epidemiological data; CMD was actively involved in helping to manage the delivery of health care programmes in the District. Hence we needed to understand not only the ideas (and controversy) underlying the different conceptions of community medicine but also the basic mechanisms operating at district level in the NHS in the provision of health care.

The client for the study was interested in the use of systems thinking in the NHS and wanted to see the practical way in which we would use it in the study. We were happy in the circumstances to accept the role of doing the study in continual dialogue with Dr Cobb and his staff, rather than simply facilitating it by providing methodological help. The Lancaster team felt the need to get to know the NHS and its culture better, and this provided a chance to do that. Given Dr Cobb's interest in the methodology it was agreed that a series of 'project notes' indicating progress would be issued, and that these would cover both the substantive work and the use of SSM. Four such notes were issued in the course of the work; they were found to be helpful in disciplining the team to make explicit both their understanding of NHS practice and their use of SSM.

The initial finding-out phase comprised, apart from the obvious reading and interviewing (sixteen interviews early on), attending some EBHA meetings and compiling Analyses One, Two and Three as described in Chapter 2. This study was done during the period when Analyses Two and Three were being developed, and the attempt to apply them, enriching their content over the whole course of the project period, was an important contribution to their development. As an indication of the early use of these ideas, Appendix 1 to this chapter contains an account written by Sophia Martin and Chris Caiger of their early impressions in EBHA structured by the ideas embodied in Analysis One (systems analysis of the intervention), Analysis Two ('social system' analysis) and Analyis Three ('political system' analysis).

While the reading and interviewing were proceeding it was also found useful not only to produce a basic structural picture of the problem situation (Figure 4.1) but also a rather detailed model of a general (primary task) system to plan and organize (but not deliver) health care to a defined

Figure 4.1 A picture of the situation in which the East Berkshire study was done

population. The root definition and concept are shown in Figure 4.2, the model itself in Figure 4.3.

This model in Figure 4.3 has the same status as the EROS model in the previous chapter, though this time the model has the full form of a conceptual model of a purposeful activity system. But its status is ambiguous in that at the same time it represented the start of Stage 3 model building, in a first methodological cycle, while being used also as a source of questions in the initial sixteen interviews. It was found very useful in this latter role, and it also provided a small piece of learning concerning the 'reading' of such models. At one stage in the project, in discussion with people now sufficiently familiar with SSM that models could be tabled in discussion (not something to do lightly, given the culture shock such objects can provoke) the EBHA people present had difficulty with the model because of the way its activities are numbered. They imagined that since they had the lowest number, Activities 1, 2, 3, 4 would be the 'first' activities carried out if the model were operational. 'But', they said 'you couldn't do 1, 2, 3, 4 until 10, 11, and 12 had been done', and that seemed to a problem to them. Now in fact what they perceived is precisely what the model indicates: 10, 11, 12 are followed by 1, 2, 3, 4. In the conventions of model building the arrows and boundaries here indicate that Activities 1–9 are indeed contingent upon Activities 10, 11, 12. We had numbered the activities purely arbitrarily , simply so that we could speak about 'Activity 8' and know what we meant. It would have been better, for public use of the model, if we had numbered Activities 10, 11, 12 as 1, 2, 3, and it is in general a good idea to give the lowest numbers to activities upon which others are contingent. But it is important to understand that this problem cannot ever be solved because models are cycles of mutually dependent activities; they are *not* sequences which could be numbered accordingly. There is no correct sequence of numbering because models are not linear. Through the cycles of monitoring and controlling every activity in a model is in principle contingent upon every other.

Root Definition

> A DHA-owned system, staffed by professional officers accountable to the DHA which, in the light of existing provision of health care (NHS and non-NHS), plans and organizes the delivery of health care to defined populations using current health technology. The system manages the delivery via both ongoing services and specific projects, operates according to principles laid down by DHSS and Region and within the budget allocated. The System responds also to ad hoc issues arising outside the framework described. Its reporting meets the requirements of the 'NHS planning system'.

C   defined populations

A   professional officers

T   population in given health state ⟶ population in improved health as a result of this system's contribution

W   organized provision of health care is feasible and desirable; it can be planned and organized.

O   DHA

E   Structure DHSS/Region/District; 'NHS planning system' as a reporting mechanism; budget

Concept :

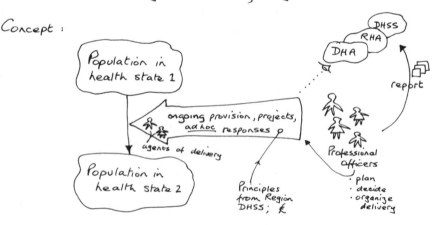

Figure 4.2   Root definition and concept of a system to plan and organize the delivery of health care

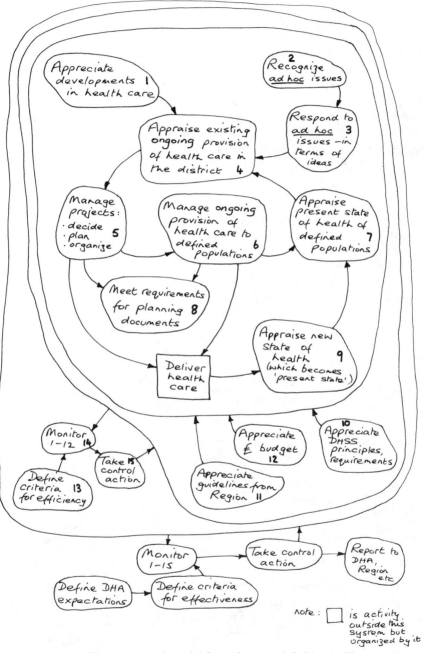

Figure 4.3 The conceptual model from the root definition in Figure 4.2

As the finding out phase unfolded, a specific methodology for the study was created which took the form shown in Figure 4.4. It was clear that the 'more-than-epidemiology' view of community medicine espoused by Dr Cobb and his team meant that his Department would be concerned with all three of the overlapping 'domains' of the NHS with their different underlying values (Smith, 1984): the *management domain* concerned with attempts at rational allocation of resources to meet intended aims; the *policy or political domain* concerned with policies set—ultimately—by politicians and elected represen-tatives; and the *service or professional domain* in which health professionals, capable of and eager for self-governance, provide health care according to professional standards, normally with a focus on individual clients. This

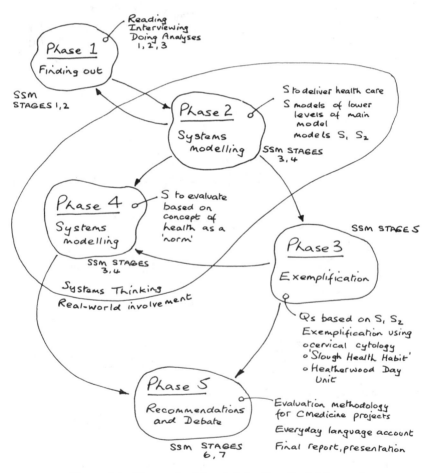

Figure 4.4 The methodology of the East Berkshire study

argued for a very broad finding-out phase, and suggested the relevance of a system to define and organize delivery of health care programmes to a defined population (as in Figure 4.3). It also suggested that some expansion of such a model would feed the concept of 'a system to evaluate' a health care programme. Exemplification would then be attempted by using several current projects in East Berkshire (three were addressed) and finally a version of 'a system to evaluate' would be expressed in an everyday-language version which was of practical value to those with no particular interest in systems thinking or its apparently esoteric language.

The finding out was done at a time of structural transition in EBHA, for the newly appointed DGM was at the time formulating his structure of managed units. We treated Phase 1 of Figure 4.4 very much as a learning phase, and Sophia Martin found the metaphor to describe it, one which in the authors' experience applies to all such phases, when she said that we were 'allowing knowledge to be built up in the manner of a palimpsest', knowing that as we erase and write the new on top of the old knowledge, something of the old knowledge is retained.

In practical terms Analysis Three allowed us to appreciate what was at stake for those undergoing reorganization, and how deeply felt the changes were. In particular, in East Berkshire as in other districts, the existence of community medicine as a discipline was being questioned—although in East Berkshire its eventual restyling as the Department of Planning and Information recognized its active role in the Health Authority. Analysis Three also allowed us to place ourselves in the organization; it made sure that we did not forget that in interviewing and discussing issues we would be perceived as agents of a somewhat radical Department.

Modelling and Exemplification

Since the study sponsors were interested in seeing SSM applied within their situation, models were always displayed to the members of CMD most concerned with the work, especially Dr Cobb and Dr Queenborough, EBHA Specialist in Community Medicine.

The first significant model, that of Figure 4.3, proved to be of interest to them. They approved its focus on planning and organizing (but not itself actually delivering) health care; and they liked its emphasis on continuous process. In the system in Figure 4.3 planning is not an event (producing 'the plan') but is a continuous process in which health is a norm rather than an attainable goal. What is seen as an acceptable norm will change as medical technology, resource levels, lifestyles and public attitudes change; but, like ol' man river, the system of Figure 4.3 just keeps rolling along, seeking an ever new state of health for a defined population in line with current views and values. The model also indicates that helping to define and organize health programmes will entail managing a complex web of relationships.

All this was appreciated, but this initial model was at too high a level of abstraction for useful comparisons to be made. The first thought was to expand each of the activities of the first model into an activity model at the next resolution level, but this led to too much detail and far too many activities for the exemplification phase in which we would use the models to explore current health projects in East Berkshire. Instead it was decided to focus on those activities most relevant to an active (managing) role for community physicians. The key activities were 1, 2, 5 and 7 and this provided a framework for a second set of models (S_1, S_2) which concentrate on the main issues CMD would be involved in: what health care programmes to provide, which agents should provide the programmes, how they should be managed, and programme implementation. The concept, root definitions and models were developed in dialogue with the clients, who felt that they better reflected the 'project' orientation of CMD than did the initial model, useful though that was as a scene setter. But once a project to open a new ward, for example, was completed, the ward became a part of an ongoing service and was no longer a short-term concern of CMD. Figure 4.5 shows the concept and root definitions for S_1, S_2, Figure 4.6 the conceptual models.

In discussion with Drs Cobb and Queenborough, Activities 1 and 5 of system S_1 were further expressed, as shown in Figures 4.7 and 4.8 respectively. Note that Figure 4.7 is not an activity model; rather it spells out our concept of what is meant by the 'appreciation' which motivates the initiation of a health programme. These figures embody both Peter Wood's idea of the need for 'enabling' activities to be thought out as well as direct 'doing' activities (Wood, 1985), and Vickers's idea of managing relationships, a concept which seems particularly apposite in the NHS's network of autonomous and semi-autonomous groups (Vickers, 1965; Checkland and Casar, 1986).

Armed with the concepts expressed in Figures 4.5–4.8 and with the feeling on the part of the East Berkshire community physicians that they were insightful, it was now necessary to test them against the real world by examining some current projects. Discussion with community physicians who had taken part in Dr Cobb's Harrogate meeting identified three projects suitable for exemplification. They covered the three NHS domains and were considered of legitimate interest to CMD. They were: a project to establish a cervical screening service for the District; the Slough Health Habit, a major health promotion campaign to persuade the citizens of Slough to live a healthier lifestyle (giving up smoking, taking exercise, etc.); and the provision of a mental illness day care facility managed jointly by the NHS and the local authority.

Systems to decide upon (S₁) and implement (S₂) health care projects

Concept:

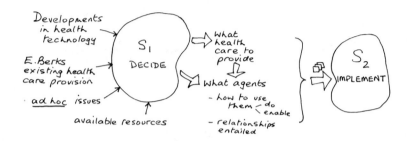

S₁ Root Definition

> A CMD-owned system staffed by CMD and selected agents which, in the light of information on health care provision, available resources, developments in health care technology and ad hoc issues, decides what health care to provide to target populations and what agents should be used, and how to use them. The system attends to what relationships must be 'managed' in providing the health care, and considers both direct and enabling activities.

S₂ Root Definition

> A system owned and staffed by CMD and selected agents which provides the health care decided in system S₁.

Figure 4.5 Concept and root definition of systems to decide upon and deliver health care

Using the models S_1 and S_2 to generate questions, the three projects were examined through interviews with both CMD members and the 'agents' who were connected with implementing the projects. The idea was not to amass a lot of information about each project but to test whether or not S_1 and S_2 led to questions which those directly concerned with the three projects understood to be relevant and judged to be cogent. (The questions took the form of asking for each activity of S_1 and S_2 whether the activity was observable in the project in question and whether or not it was regarded as being well done.

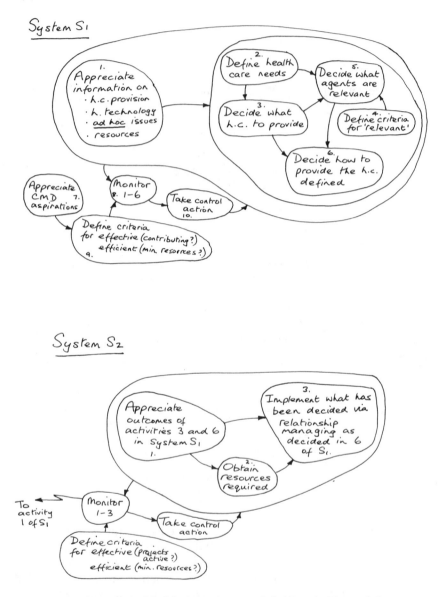

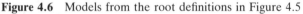

Figure 4.6 Models from the root definitions in Figure 4.5

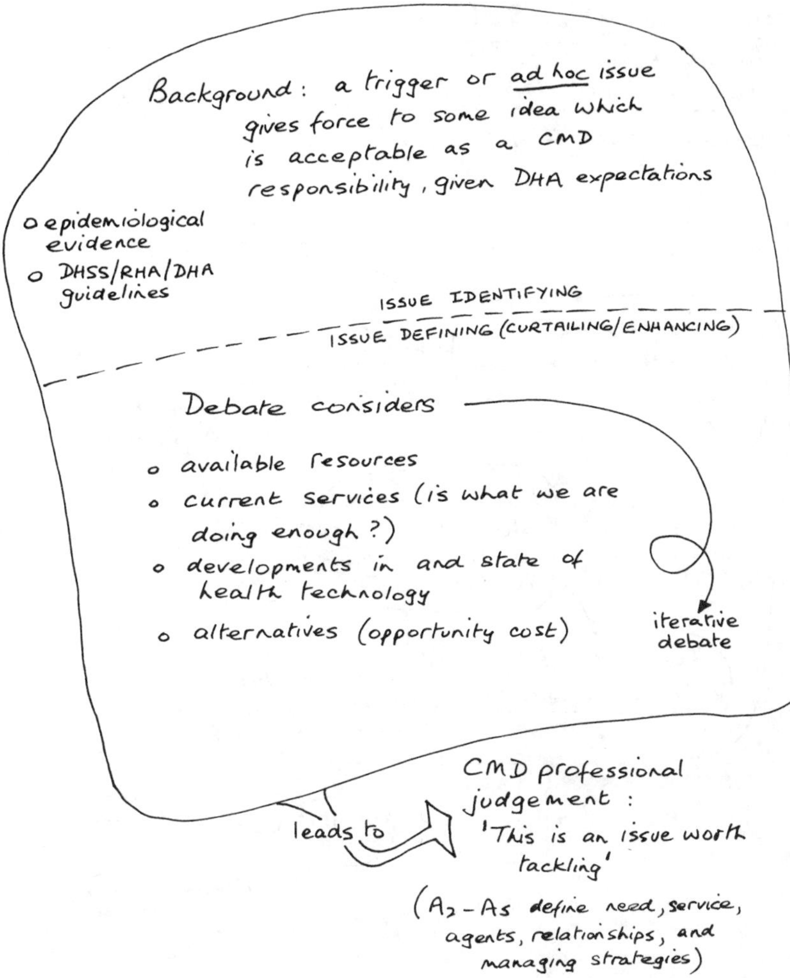

Background: a trigger or ad hoc issue gives force to some idea which is acceptable as a CMD responsibility, given DHA expectations

o epidemiological evidence
o DHSS/RHA/DHA guidelines

ISSUE IDENTIFYING
ISSUE DEFINING (CURTAILING/ENHANCING)

Debate considers

o available resources
o current services (is what we are doing enough?)
o developments in and state of health technology
o alternatives (opportunity cost)

iterative debate

CMD professional judgement: 'This is an issue worth tackling'

leads to

(A₂-A₅ define need, service, agents, relationships, and managing strategies)

Figure 4.7 The concept of Activity 1 of system S_1 expanded

The 'S_1 S_2 questions' are included in Appendix 2 to this chapter.) In addition the questions were tried out on the project of a doctor in CMD who was setting up a breast cancer screening programme as part of his Master's degree. Sophia Martin and Chris Caiger were gratified when he incorporated a number of their ideas into his methodology!

Those concerned with the three main projects found the questions pertinent, and they passed the further test that they actually led to some actions being taken in the projects themselves, though that had not been the main aim of this comparison stage. For example the interviews based on the

Activity 5 of System S1 (Figure 4.6)

(Decide how to provide
the health care defined)

becomes, in more detail :

Figure 4.8 Activity 5 of system S_1 expanded

questions derived from S_1 and S_2 led to a realization on the part of the 'agents' for the Slough Health Habit, the Health Education Department, that they needed help from a professional PR company if they were to have a bigger impact on the people of Slough.

It is important to appreciate the special nature of this comparison stage (SSM Stage 5). Normally the purpose of comparing models with the real world is to generate a coherent debate out of which will come ideas for improving the 'what' and/or 'how' of current practice. In this example any such ideas were a bonus gratefully garnered, but the main aim of the comparison was to test the validity of S_1 and S_2 and the questions derived from those models. The question being asked was: do S_1 and S_2 provide a valid epistemology for critical conversation with professionals about the real project? If so, then our study could move on with some confidence to creating a tool for evaluating CMD work. This would be achieved via the next relevant system, namely a system to evaluate any project to provide a programme of health care. The relevance of this system was now underwritten by the experienced adequacy of the exemplification using real projects, and the nature of the model would build upon the now tested concepts in models S_1 and S_2 (Figure 4.6).

Attention now moved to Phase 4 of our study (Figure 4.4). Based on the experience so far, a system to evaluate health care projects or programmes was built. The concept of it is shown in Figure 4.9, the root definition and model in Figure 4.10. The model is deceptively simple: much simpler than acquiring the experience which gave us confidence in its relevance and usability!

Outcomes

We now had models of both a system which decides what health care to provide through which agents, and a system to provide that care. The models had been tested and shown to be appropriate using three current projects in East Berkshire as examples, and were now incorporated into a model of a system to evaluate any health care project (Figures 4.9 and 4.10). This latter model is based upon the idea of health as a changing *norm* rather than an attainable goal. The norm will be threatened by influences in the environment, and management effort will be required to maintain the health of the population within the tolerable range of the norm, which will itself be determined by societal attitudes, the current state of health technology and the resources available. CMD will contribute to 'health managing' by helping to decide on, plan and organize the delivery of programmes to meet the needs of client groups in the community. It should be in a position to organize the debate between health professionals about what services the district should offer. As a medical specialty it could bridge the gap between the clinical and management domains; and its learning should be relevant not only to its own

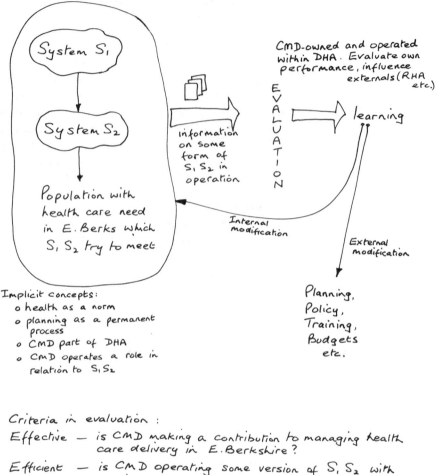

Figure 4.9 A concept of evaluation relevant to CMD

professionals within CMD but also to wider bodies such as the district, the region, the Department of Health and local authorities.

In order to make these now-tested outcomes generally available beyond the group actively concerned with our study, a version of the system of Figures 4.9 and 4.10 was prepared in everyday language. This took the form of a set of questions which should be asked by members of a CMD wishing to evaluate their work. There are four main questions, and these frame the 20-odd derived from models S_1 and S_2 (Figure 4.6). The full set of questions is given in Appendix 2 to this chapter.

Root Definition

A CMD-owned and operated system for the evaluation of projects for the delivery of health care (as manifest in versions of S_1 and S_2) which are themselves part of the achievement of health as a norm in EBHA. The system acquires learning internal to CMD and external in relating to other bodies such as Oxford Region, EBDHA, LA, CHC, etc.

C   CMD

A   CMD

T   Projects ⟶ Evaluated projects (experience)   (experience)   Via questions from S_1 S_2 models

W   The definition of an acceptable norm for 'state of health' can be helped by formal evaluation

O   CMD

E   Organization structure (RHA/DHA, etc.)

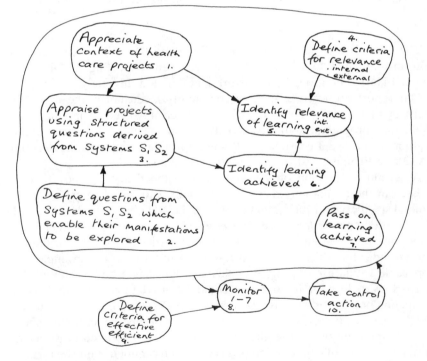

Figure 4.10 A system to evaluate, based on Figure 4.9

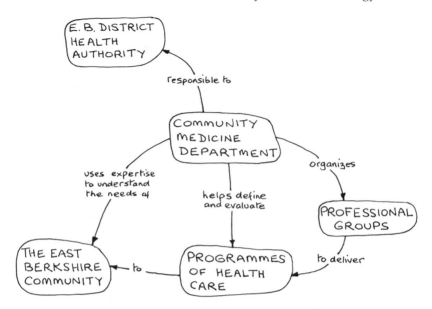

Figure 4.11 A concept of the role of Community Medicine Department

These ideas were acceptable to Drs Cobb and Queenborough, and were now disseminated to CMD and to a reconvened meeting of Dr Cobb's original Harrogate group, whose work had initiated this study.

Our report proposed evaluating community medicine by the use of the ideas in Figure 4.11. Here a CMD is actively involved in mounting and evaluating programmes of health care as an agent of a District Health Authority, a concept consonant with the reformulation of the East Berkshire CMD as a Planning and Information Department.

A second implication of the concepts developed in this study (and expressed in the evaluation scheme in Appendix 2 to this chapter) is of more general significance. Much attention within the NHS has been directed to the idea of defining *performance indicators* (PIs) against which health care provision might be judged. Many people feel particularly satisfied if they can find an indicator which can be expressed quantitatively. For example, a PI applicable to the use of a hospital cleaning fluid might be the area of floor cleaned by a litre of it, and this could be used to distinguish between competing brands. In systems terms, however, PIs—or, in the phrase used in SSM, measures of performance—should never be one-dimensional, should never be plucked out of the air, and should never be defined in a vacuum; they are one part of a description of a *system*, and cannot usefully be regarded in isolation. (The *best* cleaning fluid according to the PI might be the one

Summary of the East Berkshire HA Project

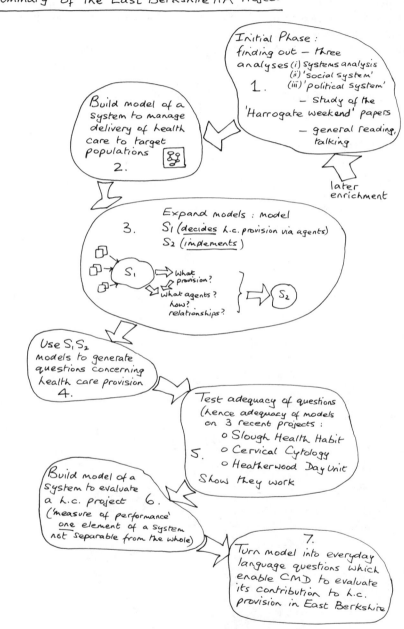

Figure 4.12 The EBHA project as a whole, in summary

which causes headaches in users or it, might be the one for which there is no continuity of supply.) The work on PIs would be more cogent if the indicators themselves were placed in a framework of ideas to do with conceiving and executing purposeful activity, such as SSM provides.

Both in CMD in EBHA and at the Windsor meeting to discuss the work at the end of the project, these ideas seemed to gain a good degree of acceptance among the community physicians present. The study as a whole, summarized in Figure 4.12, had contributed to the thinking of the Department of Community Medicine in East Berkshire Health Authority at a time of management reorientation at district level. Its specific outcome was the everyday language evaluation procedure for community medicine given in Appendix 2 to this chapter, something which is transferable to other departments prepared to take a proactive view of the specialty.

Appendix 1

Use of Analyses One, Two, and Three in the study in East Berkshire Health Authority

This appendix contains the results of carrying out the three analyses used within Stages 1 and 2 of SSM. This was done over a period of several weeks at the start of the study by Chris Caiger and Sophia Martin.

Analysis 1 (based on Checkland, 1981, Appendix 2)

1.1 Client: Dr J. Cobb, District Medical Officer at EBHA.
1.2 Client's aspirations: To develop a way of measuring the performance of the CMD
       : To learn more about SSM.
2.1 Problem solvers: C. Caiger, S. Martin, P. Checkland plus various CMD staff.
2.2 Resources available: SSM. CMD staff, Harrogate study, Nuffield and EBHA libraries, P. Checkland, Huddersfield DHA study, 3 months project time.
2.3 Constraints: time, possibly the management reorganization, political environment of NHS, GM, failure to get support.
3.1 Problem owners: Dr J. Cobb, CMD of East Berkshire, Oxford Region Health Authority Units, the community of East Berkshire (active/ passive), private health companies, Harrogate PI group, information providers, local authority social services, medics, primary care agents, other/all CMDs.
3.2 Implications of problem owner chosen.
    Dr Cobb, while taking a fairly radical view of CM, is looking for practical management assistance so the outcome must be in a form usable by him. Therefore we need access to all EBHA and other relevant staff. We must assess what is culturally feasible as a role for the

CMD and what tools would be useful. If including the ORHA CMDs we must make them feel included in the project. Projects for operationalization must be relevant to their world. Criteria for choosing projects should reflect Harrogate domains and the projects should be within the accepted CMD role and if possible be tackling issues they have experience of. Very difficult to take the community to be the problem owner as we could never identify one voice to validate a model. Also people in general do not voice their needs, but recognize services as useful. It is the responsibility of professionals to diagnose patterns of needs in populations or diseases in individuals.

3.3 Reason for regarding the problem as a problem.

The reorganization has specified annual audits/reviews of activities and requires ways of measuring performance. Current PIs are largely 'hard' and not suitable for measuring the success of the CMD. There is a general awareness of the lack of ability to measure performance in the NHS. Anyone who makes a significant contribution to this area will benefit. Our client feels the orientation of PIs is fundamentally wrong and needs to be rethought using a 'systems approach'.

3.4 Value to the problem owner.

Clearer MoPs would increase bargaining and communication power with the region. It may increase power over subordinates and within the district. Improvements in the effectiveness of the DHA in providing services to the community are the ultimate goal.

4 Problem content.

4.1 To provide MoPs for CM. The problem content could be described by a model(s) of CMD activities from which useful MoPs could be derived by asking the questions 'is the activity done?' and 'is it done well?'

4.2 The CMD activities we are interested in are to do with the deciding on planning and delivery of health care. Nouns and verbs 'relevant to those activities' have been examined by a 'Structure and Process' analysis.

4.3 Constraints on the PC system are the accountability of the study to GM, NHS, etc. (political climate) and acceptability/desirability of our recommendations as practical management tools.

Analysis 2

Roles, norms, values as described by Vickers.

Role: in CMD
Norm: expected behaviour
Value: what constitutes 'good' or 'bad' behaviour in role.

Roles in CMD

CMD roles include maintaining relationships with other role holders, e.g. DGM, nurses, LAs, Treasury, DHA . . . The roles, norms and values of these individual bodies will not be explicitly examined.
The basic role of a CMD member is to contribute to managing the delivery of health care.

DMO possible roles
- analysing and responding to epidemiological statistics and advising on preventative/education implications
- predicting trends from community medical statistics
- receiver and information gatherer, reporting to DGM and DHA
- planner of CM and need definer
- liaiser/coordinator of other resource holders including LA, JCC, voluntary organizations, NHS, FPC . . .
- managing disposal of resources—although cannot fix what the budget is.

Also:
- member of DHA
- Officer in charge of department
- at various times, chairman, member, or other role in committees.

Norms

Generally these are not 'NHS norms', but J. Cobb interpretations. NHS provides little in the way of explicit job description specifications and expectations. At this time the roles and therefore the norms are very uncertain. (The Griffiths Report recommends the discussion of role definition.)

Given J. Cobb's proactive approach a norm could be the balancing of the drive for efficiency against the need for effectiveness.

In the multi-disciplinary and traditional style of NHS he might be expected to be diplomat, smoothing the way, reducing conflict. In practice he sees himself as an instigator of change—sometimes 'spilling blood', 'causing an explosion'. Older members of other CMDs feel he could be readier to accept the authority of his peer group of clinicians, but he believes he is supported by the opinions of younger groups.

Though 'radical' in outlook he uses legitimate committee and bureaucratic channels.

Values

- A commitment to the development of standards of community health.
- The importance of being able to carry out the programmes their research has suggested is relevant.
- A clear sense of purpose.
- Efficiency of resource use but the maintenance of flexibility, the room to cope with new projects.
- Values his autonomy, the ability to take risks, and the credit for being proved right.
- Values up-to-date thinking in health care, also in management schools, and in management literature.
- Community physicians generally do not have high status in the eyes of traditional physicians, e.g. brain surgeons.
- Rewards for most clinicians are more direct; they see the results more easily than do community physicians.

Analysis 3

Disposition of power

- Changing and uncertain at the present while DGM makes arrangements and people settle into new roles.
- Used to be held by clinicians. Medical autonomy still holds but management power concentrated in GMs.
- District management used to be by team consensus, now by accountable managers.
- Nothing to guarantee GMs power. Depends how they grow into new jobs. EBHA favours a team approach: DGM does not want to be too directional but will accept responsibility.
- CM department has been very active and the planning team relatively passive.

Nature of power

- Ability to be able to influence the distribution of resources implies access to information and ability to persuade.
- Ability to enlist support of, influence or direct others. (Rely on cooperation of others to carry out actions, do not even have a specific budget defined for all their work.)
- Ability to define the needs of a community *convincingly*. Use of propaganda.

Process by which power is:

(i) obtained—'grabbed'
      —obtained through
      professional/medical credibility   ⎱ enhanced by previous success

(ii) exercised—taking control of committees by force of personality, giving professional advice to committees. Withholding information, filtering proposals before public discussion; officers reporting what they think fit to meetings.

     exercised—by default because no one else is doing anything. J. Cobb could also claim to have successfully used management concepts giving him an advantage over non-management-oriented people.

(iii) preserved—difficult to maintain, but expectations of CMD in East Berkshire fairly high and proactive. J. Cobb feels supported in being made Director of Planning and Information on DMB.

(iv) passed on—too much change to say. Depends on performance in new positions. Possible doubt of future of CM as a specialty given professionalism of other interpreters of medical community statistics.

Appendix 2

Evaluating a Health Care Project in Community Medicine—A Usable Tool

This appendix contains an everyday language account of an evaluation scheme for the work of a proactive Department of Community Medicine. Its source is the systems models in Figures 4.5–4.10.

(1) Examine the context of the project in terms of:
    (i)   status: time
                       cost
                       does it involve a redirection of existing service, an establishment of a new service, etc.?
    (ii) initiative: own
                       compulsory
                       other?
    (iii) type of care: Is the project to improve or maintain the health, or slow the deterioration of the health of the target group? Is it preventative, or promotional?
    (iv) degree of organizational involvement:
         internal to NHS?
         (which units/services/department)
         external?
           (e.g. LA, FPC, GPs, etc.)
         direct to public?
           (which groups)
    (v) which sources would probably best provide information for an evaluation of the project?
(2) Detailed appraisal is concerned with the extent to which an individual project meets or met certain fundamental conditions:
    (i) was the 'right' health care need identified in the community?
    (ii) was a service to meet that need and its delivery (particularly the quality of management of relationships with the personnel carrying out the service) well-planned?

(iii) were plans actually carried out, and were intended outcomes achieved?

In answering Question 2 use the list of S_1 and S_2 questions which follow Question 4.

(3) Identify learning from the project experience. What strengths and weaknesses of the project are indicated by the answers to the questions?

(4) Who would benefit from this information?
   (a) how could it be used internally in CMD for its future activities?
   (b) could it be presented to external bodies (e.g. health authorities for reviews, requests for funds, recruiting to CM, reports to research or academic bodies, to the public, press or CHC)?

$S_1 S_2$ Questions (derived from S_1 and S_2—Figure 4.6)

(1) What was the trigger for the project—what issues gave rise to it?

(2) In discussing the issues in the final instance, did you consider:
   - quality of existing resources (available finance, personnel facilities, equipment, etc.)?
   - the services already offered? Is what we are doing now sufficient?
   - the status of the technology required (health tech, equipment tech, information tech . . .)?
   - opportunity costs. Was there a debate about what the next best project would be?

(3) How was the background evidence presented and to whom?

(4) Did you have a definable target group for this project with a known range of characteristics?

(5) Did you decide what would indicate that their needs had been met and did it indicate a timescale, or success and completion criteria for the project?

(6) Given the background to the project, why did you select that group?

(7) What were the constraints on the definition of the service?
   - resources
   - available skills
   - costs
   - time

(8) Did you have to go outside the NHS to find people with the skills or resources to provide this service? How clearly did you define the role of agents?

(9) How did you find out about the skills/resources they had, whether inside or outside the NHS?

(10) Why were these agents considered suitable (a) to meet the needs of the target group; and (b) to provide the service you had decided on?

(11) Did you involve other institutions or other parts of the NHS in the planning of the project?

(12) Before this service was brought into operation, were you used to working with these other groups/agents?

(13) Given that you wanted to involve these agents in the project what did you do to enable their involvement?
      Was it necessary:
      - to set up new decision-making structures or committees?
      - to get on to new committees?
      - to establish new sources of information?
      - to learn about other skills?

(14) Was there evidence that the agents understood the plans?

(15) Was the planned service provision actually carried out?

(16) Were the planned resources made available?

(17) Were satisfactory working relationships developed with agents; was there any conflict of values?

(18) Was the stated need met (state 1 → state 2) and was it in the target group identified?

(19) Were project review meetings held regularly and were agents properly represented? Did you consider:
      - use of resources?
      - if plans were carried out?
      - outcomes?
      Was it easy to get information for project reviews on these topics?

(20) Was it possible to achieve changes in the project whenever necessary; what were the difficulties?

(21) Were the issues that gave rise to the project resolved? If not can you identify the problem area as:
      - need?
      - service?
      - agents chosen?
      - relationship management?

(22) How far can you say that it was an issue worth dealing with (evidence)? Could East Berkshire have been better served by some other project?

(23) How efficient was the planning and information gathering process?

(24) Is the project within DHSS and RHA guidelines?

(25) How well did the project serve the interests of the DHA? What are other districts doing about this?

(26) Were sufficient skills available—both within CMD and between CMD and the agents—for the planning and carrying out of the project?

Chapter 5

An Application of Soft Systems Methodology in the Civil Service

Previous chapters have described a 'mainstream' application of SSM in industry and an early application in the National Health Service. This chapter describes a study carried out in the Civil Service. It serves two purposes. Firstly, it illustrates the ideas at work in a different culture—that of the Government service—and, secondly, it provides a link to Part II, in which is described a sequence of studies which were carried out within one (industrial) organization and led to a new view of SSM. That sequence of studies was carried out by the authors, together with a number of their colleagues; the study described in this chapter was the first encounter with SSM for one of the authors (J.S.), an encounter located in a problem situation rather than in a classroom (Scholes, 1987). The nature of this initiation into the use of the SSM was important in leading to the learning set out in Part II.

The Central Computer and Telecommunications Agency

This study was carried out in the Central Computer and Telecommunications Agency (CCTA), a Government organization concerned with the development of computing and telecommunications in the Government service.

In 1972 a body known as the Central Computer Agency was established. It brought together policy, planning and operational support functions which had previously been carried out by various units in the Treasury, the Department of Industry and Her Majesty's Stationery Office, and it operated as part of the Civil Service Department (the department then responsible for managing the Civil Service as a whole).

At this time there were relatively few computers in central government. Computers themselves were large and expensive; departments were inexperienced in their use. Hence there was wide acceptance of the need to pool expertise and exercise control over the development of computing from the centre. During the 1970s many computers were installed in Government departments, the cost of hardware reduced considerably, and computer systems in general became more reliable and easier to use. No department ignored computers completely and some gained considerable experience in the planning and implementation of computer systems.

While these developments were taking place there was a convergence of computer, telecommunication and office machine technologies, and the Agency adapted in various ways in order to provide what it considered appropriate advice and control in a changing environment. The Agency became the CCTA in 1978 when it took on telecommunications work, and it took responsibility for office machines in 1980. Simplified procurement procedures were introduced for small systems; running contracts were introduced for microcomputer procurements; and some departments were given authority to approve their own computer acquisitions within stated limits. By the 1980s the role of the CCTA was increasingly directed towards departmental Information Technology (IT) strategy and the managing of computer projects.

The Agency's aim was set out succinctly, with a certain stately grandeur, in a handbook distributed to departments:

> The CCTA is responsible for the promotion of computing in Government Administration with the aim of improving efficiency and widening policy choices. It provides advice on the development and implementation of computers and telecommunication systems including all technical matters, is responsible for all aspects of procurement and provides advice on possible computer and telecommunication applications.

At the time of the study the Agency comprised a staff of nearly 700 people in three London-based divisions (Projects; Technical Services; Secretariat and Support Services) with the Procurement Division based at Norwich. Projects Division, our chief concern, was mainly occupied with assessing computing projects and approving (or not) the financial expenditure on them. This crucial function was effected through Departmental Liaison Officers (DLOs) who were assigned to each central Government department using or intending to use computers. The DLO, normally the first point of contact between the Agency and a department, was expected to provide direct support and advice, to scrutinize proposed projects and to bring in the expertise of Technical Services and Procurement as necessary. Obviously this would always be a difficult and complex role; and equally obviously it had become more problematical as the technology changed and as departments gained experience and received a degree of delegated authority. Our study concerned the role of the DLO.

Background to the Study

A review of CCTA procedures in 1981 had concluded that the DLO role had become very difficult to fill and recommended that it be examined. The Director of the Agency, aware of SSM as an approach to ill-structured problems, initiated a systems study as a response to the review's recommendation. Initial discussion, more formal than would normally be the case in an industrial company, led to an outline of the study as follows:

> DLOs in Projects Division of CCTA have responsibilities which include advising on and appraising computer, telecommunications and office machine projects. Some departments now have delegated to them the authority to approve projects within various financial limits. In the light of this and other changes the team should carry out a review of the role of DLOs including
>
> 1. the scope of their responsibilities
> 2. relationships with parts of CCTA and CSD
> 3. relationships with departments
> 4. relationships with the computer industry
> 5. induction and training needs
> 6. experience and qualities needed to fulfil the tasks
>
> The team should aim to complete the review within four months and to present a report to the Steering Committee setting out their recommendations.

'The team' referred to had been agreed to consist of: Peter Checkland and Ron Anderton from the Department of Systems at Lancaster; David Culy, a postgraduate student at Lancaster, an engineer in his early thirties; Margaret Exley, a Principal Research Officer (Social Sciences) from the Management Development Division of the Civil Service Department, concerned with the human and organizational aspects of computer and office technology; and Jim Scholes, a Branch Head in CCTA's Projects Division who had himself been a DLO for three years. Of this group, only David Culy worked on the study full time. The two civil servants were seconded to this work half-time, while Checkland and Anderton worked on it at a rate of about one day a week. 'The Steering Committee' mentioned in the study brief consisted of senior representatives from each of the Agency's four divisions and was chaired by the head of Projects Division. Its existence was quite normal in Whitehall: any project of this kind in a government department would be likely to have such a group, their perception of the task very much defined by the name of the committee. The team's hope was that they would see themselves as part-time participants in the study rather than as management policemen.

The Steering Committee asked from the start that the team should produce a final report with recommendations. It was not clear to what extent, if any, the team would be involved in the implementation of change, but given the position of Scholes it was likely that he would be involved whether or not the team continued its work beyond the report stage.

Based on the model of SSM in Chapter 2 (Figure 2.5) a methodological plan for the study was proposed (Figure 5.1), and accepted by the Steering Committee. In the early discussions with the Committee other important points emerged. The study was to take the Agency itself as a given, likewise the Projects Division within it; but CCTA–department interaction was not defined in any constraining way. Finally, it was agreed that many people inside and outside CCTA would be interviewed. The interviews would cover the interests of each of the Agency's divisions, 'customer' departments and other 'supply' departments at the centre of the Government service, as well as computer suppliers. The Committee suggested a list of people and organizations who would have to be contacted. No less than 106 people were seen by the team (some more than once).

> As is obvious from Figure 5.1 the study was conceived as a classic application of SSM. The methodological cycle forms the basis of the project plan; and in fact it provided coherence to the team's activity, which was important with a geographically dispersed

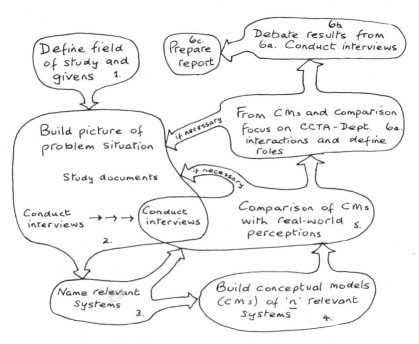

Figure 5.1 The methodological plan for the study

team with only one full-time member. SSM was here used precisely in the way it was normally described at that time, i.e. as in Figure 2.5.

At this stage the impression was of a study taking place in a situation which called for much more formality than is expected in industry, noticeably in the form of a very precise remit and required outcome (a 'final report'), and a process involving formal meetings of the Steering Committee with agendas and circulated minutes. Some of this might be thought to be foreign to the spirit of SSM, given the questing approach of the methodology, but it is natural in an organization always aware of the need to be ready publicly to account for the husbandry of public resources.

The Steering Committee members therefore provided early indication that they saw themselves, as a group, receiving periodic reports and commenting upon them (rather than on the study itself). They would be monitors rather than participants, and this persisted throughout the course of the study.

Finding Out: Stages 1 and 2

Given the study brief, building a picture of the problem situation could have focused on the tasks of the DLO with questions about those tasks addressed to people in many different roles to whom the DLO activity was relevant. However, answers to questions about the DLO tasks would have assumed some implicit model of CCTA, Government departments and the interaction between the two. There was a need to appreciate these tacit assumptions if the role of the DLO was to be rethought, and this led to an attempted definition of factors which could be used to classify both the interviewing and reading which constituted the finding-out phase. This initial structuring is shown in Figure 5.2. With the aid of it the team would know, for example, that a particular interview with person X represented access to a 'large' department regarded as 'competent' in its computer expertise and having 'many' 'innovative' 'computer' projects with 'much delegation'. In this way gaps in the coverage could be identified and filled.

It also became evident that attention had to be paid to the dispersion of the team (interviewees also being widely dispersed) and to the fact that the involvement of four out of five team members was part time. It was necessary to impose some uniformity of approach and to set up some simple mechan-

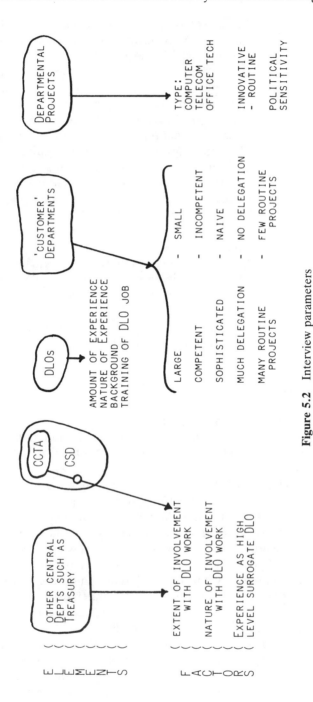

Figure 5.2 Interview parameters

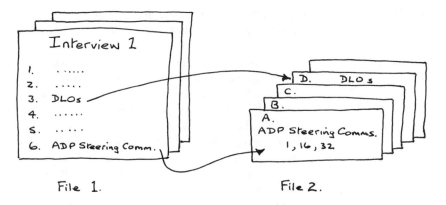

Figure 5.3 Storage and retrieval of interview messages

isms for data capture and retrieval: idiosyncratic interview notes, however meaningful to the note taker, would not mean much to the rest of the team.

A simple structure was defined in which an interview was recorded on two sides of A4 paper with four broad classes: *practices* (interviewee's background, role and tasks, relationship with DLO activity), *perceptions* (views on CCTA, DLO role, projects, etc.) *problems*, and *ideas and comments*. Interviews were numbered in chronological order and the handful of main 'messages' were extracted and recorded on file cards. A second set of file cards in alphabetical order allowed different interview messages on a given topic to be gathered together (see Figure 5.3).

This compilation of the results of the finding-out process continued throughout the project. Nevertheless, as soon as the initial main round of reading and interviewing was completed, several problem themes could be discerned. Interaction between CCTA and departments clearly covered many tasks and required a range of skills—technical, administrative and political. Several strands making up the interaction between Agency and departments converged on the role DLO. Although work on IT policy at a strategic level was going on in the Agency, interviews in departments indicated that the Agency was on the whole perceived as reactive, responding to current pressures rather than promoting new technology, new techniques and new standards, which was the Agency's intent. Senior management in CCTA acknowledged the need to improve the Agency's ability to obtain continuously updated information on computer projects and to learn lessons corporately in the longer term. Finally, although some senior members of CCTA felt that the Agency's prime role was the provision of help with project management, it was accepted that such advice was generally *ad hoc*, dependent upon the experience and knowledge of particular individuals rather than provided corporately in the form of policy, guidelines, standards, etc. This is a summary of the general views expressed in the initial round of interviews.

This example of Stages 1 and 2 of SSM had to be more carefully organized than is usually the case but the intellectual content of it was not dissimilar to that phase of other studies carried out in the early 1980s. The interview classification scheme of practices, perceptions, problems, and ideas and comment is a version of finding out about structures, processes and the relation between the two. The Analyses Two and Three described in Chapter 2 (covering 'social system' and 'political system' respectively) had not then been developed but the interview notes recorded much about the situation as a social system; and the whole study was being carried out in a culture in which there was great sensitivity to political issues. More than once, of a particular high profile project, we were told 'Oh yes, the DLO on that was the Director of the Agency'! This experience, in fact, was one of those which fed the thinking which became formalized in Analyses Two and Three.

As to the time and effort spent on managing the extensive interview programme, it is clear that this was worthwhile both in keeping the whole team in touch with progress (in fact in providing them with a common epistemology) and in providing a means by which the main messages from the interviews could be extracted. It is also clear, however, that the time invested in Stages 1 and 2 was such that it would not have been possible to loop round the methodological cycle more quickly. The Steering Committee's decision that the team should carry out an unusually large interview programme virtually dictated that the project would consist of one careful passage of the SSM cycle over a period of about six months. This was an inevitable response to the external circumstances in the study. What is interesting in retrospect, and especially in the light of the learning gained from the work described in Part II, is that this was in no sense perceived as a problem by the team. Here was an interesting complex study; SSM was simply the methodology for tackling it.

Conceptualizing: Stages 3 and 4

The outcome of the 'finding out' phase was discussed at two team meetings and, from these discussions, many ideas for relevant human activity systems emerged. Eventually a set of a dozen root definitions and CATWOE analyses was derived. The definitions were based on the following ideas:

(1) A system to promote computing in Government administration. This took the official statement of aims as a root definition.
(2) A project management system.
(3) A system to monitor departmental projects (a subsystem of 2).
(4) A system to 'work the CCTA system' (i.e. help departments get their projects 'signed off' by CCTA).
(5) An Agency–department interaction system which helps departments to become less dependent on the Agency.
(6) An Agency–department interaction system which monitors high level computer policy formulation.
(7) An Agency–department interaction system which controls computer activities in Government.
(8) A system which balances the demands made on DLOs by the agency and by departments. (DLOs were conscious of their need continuously to do this.)
(9) A system to provide resources to enable CCTA's work to be carried out (recruiting and training).
(10) A system to facilitate communication between departments and DLOs.
(11) A system for generating procedural guidelines for efficient and effective application of computer and telecommunication technology in Government.
(12) A system to decide the nature of Agency–department involvement in specific projects.

This range from a system which would be operated internally by a DLO (No. 8) to one which does not necessarily assume the existence of CCTA (No. 11) was felt to be broad enough to provide material relevant to the problem themes which the finding out had revealed.

Eventually, although not in one great orgy of model building, twelve root definitions based on the ideas listed above were modelled. Three of the root definitions and models are shown in Figures 5.4, 5.5 and 5.6. Figure 5.4 takes the official aims of the Agency as a root definition and spells out what is logically entailed in that statement if it is taken to be a prescription for purposeful activity. It illustrates that the keyword 'promote' is open to a variety of interpretations and that the aims statement is a mixture of 'whats' and 'hows'. Figure 5.5 is an issue-based definition and model, the issue being the decision on the nature and extent of Agency involvement in departmental

C   Government administration

A   CCTA professionals

T   Govt. admin ⟶ Better Govt. admin. through promotion
                    of computing

W   Computer technology, etc., can improve efficiency

O   High level of Government

E   Civil Service norms ; technology

Figure 5.4   Model treating statement of Agency aim (p. 126) as a root definition

Root Definition

> A CCTA-owned and operated system to decide
> the nature of the Agency-Dept. relationship
> (extent of Agency involvement in Departmental
> computer and telecommunication projects) aimed
> at improving the efficiency of Govt. administration

C   Dept and Agency, hence Govt. admin.

A   Agency professionals

T   need for decision on   need met
    Agency-Dept relation ———→
    in proposed project

W   Different kinds of computer/telecommunication projects
    can be recognized and classified; this is useful
    and feasible

O   CCTA

E   Civil Service norms; technology

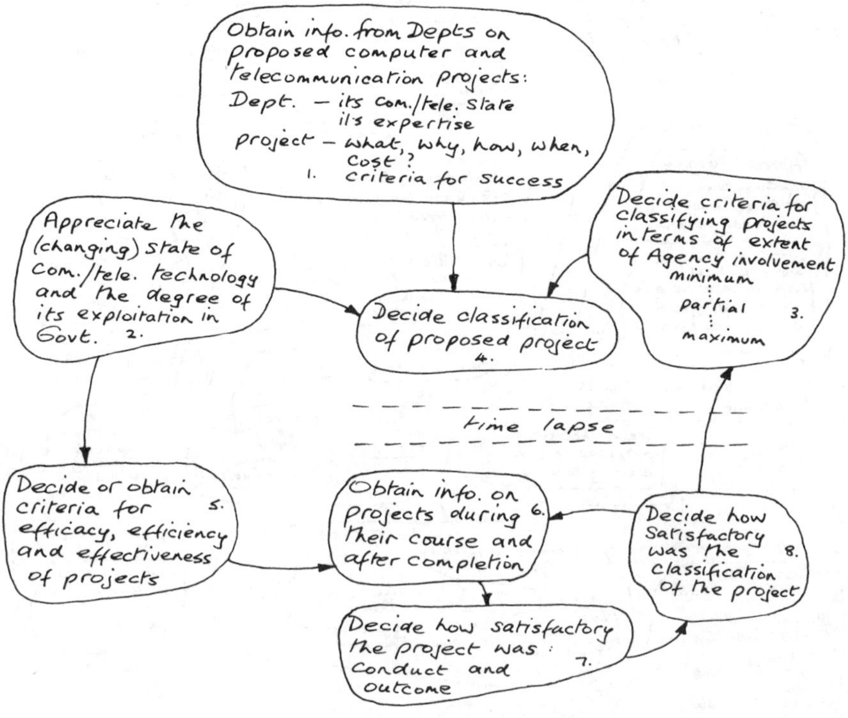

Obtain info. from Depts on
proposed computer and
telecommunication projects:
Dept. — its com./tele. state
         its expertise
project — what, why, how, when,
          cost?
    1.   criteria for success

Appreciate the
(changing) state of
com./tele. technology
and the degree of
its exploitation in
Govt. 2.

Decide criteria for
classifying projects
in terms of extent
of Agency involvement
    minimum
    partial   3.
    maximum

Decide classification
of proposed project
    4.

– – – – – – – – – –
time lapse
– – – – – – – – – –

Decide or obtain
criteria for   5.
efficacy, efficiency
and effectiveness
of projects

Obtain info. on
projects during   6.
their course and
after completion

Decide how
satisfactory
was the   8.
classification
of the project

Decide how satisfactory
the project was:
Conduct and   7.
outcome

Figure 5.5 Root definition and model of a system which decides Agency–
Department relationship on a project

Figure 5.6 Model of a system to help departments towards autonomy in computer and telecommunication projects

computer or telecommunications projects. This notional system would classify projects according to evolved criteria which would determine the Agency–department relationship in specific projects; the system would learn its way to doing this better. (In real life the conception of giving some departments delegated powers to manage their own computer development is a less sophisticated version of this same concept.) Figure 5.6 shows a model of a system which fosters the development of computing in Government by helping departments towards autonomy in the sense of gradually becoming less dependent upon the Agency. The Steering Committee regarded this as a highly relevant 'relevant system'.

> The root definitions and models were developed over a period when interviews were continuing, and they helped to feed and enrich the questioning process. Some were also developed during a week which the team spent together in order to carry out a formal comparison between models and real world (see next section). These models, made and used in 1981, were sufficiently coherent to be useful, even though they do not quite match all the rigour in Chapter 2!
>
> In the model building process the experienced SSM-users naturally took the lead. It was significant, however, that those new to SSM (the two civil servants in the team) being in the middle of a serious use of SSM, had no difficulty at all in appreciating that such models are *not* accounts of part of the real world but are only intellectual constructs (holons in the form of models of plausible, relevant human activity) which may be used coherently to explore real-world purposeful activity in order to improve it. Students in classrooms often do find this crucial fact difficult to grasp, so ingrained is the notion that 'model' necessarily implies an attempt to mimic part of the real world, as in the models of classical management science. In SSM the word covers constructs which the practitioners hope are relevant to a conscious, organized debate about aspects of perceived reality which usually go unquestioned in unreflecting everyday life (Checkland, 1988c).

Comparing and Recommending Changes: Stages 5 and 6

The conceptual models produced at Stage 4 were used in the usual way to define questions which can be asked of the real situation:

- Does this activity, or this relationship, exist in some form?
- How is it done and by whom?
- Is it a source of concern or is it regarded as well done?

Of course many of the interviews already held had to an extent constituted comparisons between models and real world. However, the formal structure of the project and the expectations of the Steering Committee meant that ideas for change would have to be developed and tested in a fairly extensive round of further interviews. The team prepared for this by carrying out as detailed a comparison as possible on the basis of the knowledge available in the team's experience (especially that of the civil servants) and in the material collected in the interview files. Tables were assembled which listed three things: differences between models and real world; potential changes suggested by the differences; and people with whom these issues should be discussed in further interviews. Part of one of these tables is shown in Figure 5.7. Analysis of tables of this kind produced 26 main topics to be raised in further interviews, including, for example: mechanisms for learning from experience at the level of project strategy; planning and policy making in the Agency; Agency links with Management Services; telecommunication and office technology projects in relation to data processing projects; documentation of office technology projects. A matrix of identified interviewees against topics/issues was drawn up to facilitate the interview process and act as an index for information retrieval.

In all 33 'comparison stage' interviews were then conducted, and again the volume of material called for attention to the recording/retrieving process. Interview notes were recorded in a pre-defined format, an example of which is given in Figure 5.8.

While the comparison was being made and the further interviews were being conducted, the team was continually striving to grasp as a whole the picture emerging in the study. In pursuit of this aim one strategy adopted was to extract from the models any activity which might be expected to be done by anyone concerned with Agency–department liaison. The idea was to collect all these activities and see whether they could be structured into a coherent conceptual model. If so, then its root definition could be inferred, and the combination of definition and model would give a rich account of the liaison role. [This was an early emergence of the process which Wilson later termed 'model condensation' (Wilson 1984, pp. 87–89) in which activities in models of which there is some real-world manifestation are extracted and assembled in the search for a would-be neutral primary task definition.]

<RD 5> Differences: Model—Real World	Potential Change	Persons or Institution to Interview
1. Model entails learning at two levels: —project —C & T strategy	(1) Institutionalize learning and development	AW SB 'MGT COM'
2. Relative lack of strategic learning in real world	(2) Redesign policy/planning in the Agency	WA BS 'Mgt Com'
3. RW neglect of cultural aspects (which will be very important in office technology projects)	(3) Link Agency into Mgt Ser network	NG BN JK
	(4) Relocate OT Group	
	(5) Better documentation for telec and OT	CS BR
	(6) Educate the Agency in implication of t/OT work	CG MOF

NOTE: Numbers, thus: 3, signify reference numbers for model—real-world difference and thus: (5) signify reference numbers for topics of potential change. t = telecommunications.

Figure 5.7 Part of a table recording the outcome of a comparison between model and real world

CCTA Project: Role of DLO .

Interviewee: J.J.

Date: 02/10/81

Interviewers: J.S. and D.C.

Topic Nos. (7), (8,9,10), (15), (19)

Questions:

a) Agency as resource pool
b) ADP steering committees (membership T.O.R.'s info. flows)
c) Reduce delays in resource acquisition
d) Classifying projects to define CCTA involvement
e) Selection and training of DLOs

Interview Sheet 'Comparison Stage'

Record of responses

a) Require TS support, detailed knowledge of new developments—need to be outward looking. Experts on project management—not possible with present setup—experience with PROMPT would be useful. Need 'Project Determination Group', assessing project worthiness, challenge fundamentals—establish cadre of 'Technocrats': fewer but better people, more experienced; higher grade (Young A/S).

.   .

etc

e) DLO 'work the system' is vital—one point which Departments can contact. Someone to test one's ideas against. Present DLO tasks impossible—lack grade/skill/experience. Technocrat idea would need career development path—high powered group in CCTA—might bring in outsiders. Need for CCTA and Depts to identify people to develop for this.

Other points

CCTA should promote aids such as PROMPT. Investment Appraisal—etc.

Investment appraisal vital. Not handled well by CCTA—effort should be related to size/complexity of investment being appraised. How to deal with sunken cost? Ongoing review of project investment. CCTA need to match forward projection of technological developments with good financial practice. Buy V Rent etc.

Need to be looking ahead at direction of developments—rather than being re-active. Need planning function.

Figure 5.8 Part of an interview record from the comparison phase

In the case of the DLO role it was impossible to assemble the extracted liaison activities into a coherent conceptual model, although the assembled activities were agreed to be a reasonable account of what actually existed. This illustrated the multiple disparate roles the DLOs were required to fill.

A concept of liaison between two parties was formulated in which liaison was defined as 'identifying and taking actions to eliminate perceived shortcomings in meeting each other's needs'. This catered for the mismatch between perceptions and expectations of the parties involved; and if interactions were perfectly engineered there would be no need for 'liaison'. This concept enabled the team to gain a clearer grasp of the current complexity of liaison activities. Expressed in these terms, liaison could be seen as *the concern of the whole Agency*, and comprised at least five roles: inspector, consultant, communicator, representative, promoter. The difficulty in the DLO's situation could be summarized as arising from the fact that at different times, in different circumstances, to different extents, all these roles might be expected of the DLO; yet some would be fulfilled by other Agency colleagues simultaneously.

After completion of the comparison interviews the team met for two days to write a draft message of the study. Its essence was as follows:

- Analysis of the DLO role reveals at least five components: inspector, consultant, communicator, representative, promotor.
- It is unrealistic to expect this role to be filled as it is at present by relatively inexperienced, middle-grade, temporarily seconded people working in a wide-ranging, rapidly developing, highly technical area. There is a mismatch between the task as currently defined and the capacity of those expected to do it.
- Something can be done to improve the capacity by training, by selection, by better support. But there are limited possibilities here.
- Therefore the job must be reduced; not, the team suggests, by reducing the technical coverage, but by reducing the range of responsibility.
- It follows that some responsibilities currently those of the DLO must be placed elsewhere.
- This could be done in a number of ways: by more delegation; by having several types of specialist DLOs; by change of CCTA internal organization. (Possibilities were then spelled out.)

The members of the Steering Committee considered the team's message in the form of a draft report and subsequently discussed it at a meeting between the Committee and the team. Although the team hoped for the involvement of the Committee in the debate about change, the Committee chose to focus its attention on the *product* of the study (the report) rather than its process and content. The outcome of the meeting was an agreed set of amendments to the draft report rather than a shared commitment to change. Errors of fact were corrected, the legitimacy of statements regarded as provocative was tested, and comments outside the study's terms of reference were deleted. This cleared the way for issue of the final report which had been defined at the start as a required outcome of the study. It has already been remarked that work to implement change is the real aim of SSM, but a final report is a legitimate outcome of a study constrained to end at Stage 6, as was this one.

The final report itself, addressed to the head of Projects Division (who chaired the Steering Committee), gave a fairly full account of the study. It argued for seeing 'liaison' in the sense defined above as the role of CCTA as a whole, and making changes both to the defined role of the DLO and to internal Agency arrangements in order that that liaison could be carried out appropriately at different levels and in a way which would lead to corporate learning.

By this stage the study could be seen as a fairly straightforward passage through the stages of SSM in sequence, with some internal reiteration as, for example, further comparison-stage interviews fed further modelling. As required by the initial terms of reference, the nearest thing to Stage 7 action was here the somewhat unsatisfactory one (from the team's viewpoint) of producing a final report for consideration by the Agency seniors.

Compared with most of the studies carried out in industry, this one was more formal in its arrangements and requirements. This was accepted by the team at the time but with hindsight it can be seen that the study was flawed in two ways. Firstly, the terms of reference for the work were narrow. They restricted it to a study of the DLO role. Carrying out the work repeatedly led to issues and interim outcomes which suggested that a systems study of the Agency itself would be more appropriate, but it did not seem possible to bring about that change.

Secondly, the study was carried out within an organizational arrangement (that of the Steering Committee) which focused on the *product* of the study (the final report) rather than on its *process*. The two civil servants who worked in the team, Exley and Scholes, became committed to the idea of SSM as a process of enquiry, but inevitably members of the Steering Committee, in that role, saw their task (correctly) as an editorial one. Their job as a Committee was to make sure that the final report was in an acceptable form. Individually, members of the Committee were very helpful in interview and contributed significant ideas to the study. *As a Committee* their concern was with the final product. From the point of view of SSM as an enquiry process, we had here a situation in which ownership of the *process* of the study was deficient.

The experience indicated the extent to which arrangements for carrying out a study can influence its content; useful learning was thus gained over the need to pay very careful attention to the 'engineering' of the 'problem solving system'. At a more general level this experience indicated the need for more explicit analysis of the social and political aspects of a situation under study. It was one of the experiences which led to the emergence of Analyses Two and Three, described in Chapter 2.

Outcomes

Given that the 'taking action in the situation' in this instance necessarily took only the weak form of producing a final report, this discussion of the outcomes of the study is an account of what happened following submission of the report rather than a conventional account of Stage 7. It is included here both for completeness and because it enhances the learning derived from this experience.

When the team's report was available, the Steering Committee prepared a commentary on it which set out the Committee's position on possible changes. The commentary made a number of points which included the following: the report was helpful in identifying the difficulties of the DLO role and providing a coherent framework for discussion of it, without being complete enough to enable immediate decisions to be made. It was accepted that the present DLO role was unrealistic and that responsibilities for project liaison should be more clearly differentiated from departmental liaison, which should be dealt with at a more senior level as advocated in the report. Actions would be taken to deal with some specific concerns (e.g. training and information handling.)

The report together with the Steering Committee's commentary was passed to the Director of the Agency, and a presentation of the report's findings was made to DLOs. What followed was a process spread over some months in which the study and its messages were digested and assimilated into the thinking of all the relevant senior civil servants within the Treasury. That such a process should happen over a protracted period of time is more typical of the culture and expectations of the Government service than industry, and presumably reflects recognition of the axiomatic need for public accountability.

Some months after submission of the report the Deputy Secretary in HM Treasury responsible for CCTA chaired a meeting attended by the Director of the Agency, some of his senior colleagues and three members of the study team, Anderton, Checkland and Scholes. The team's findings were outlined, and discussion developed in which it was agreed that changes should be made in the Agency—for example to create procedures for departmental liaison more independent of liaison on specific computer or telecommunication projects—and that proper attention should be paid to planning and managing the implementation of the changes. These changes were to be part of an 'autumn initiative' covering wider reorganization within CCTA. In the event freeing resources to manage the change posed a problem and progress was slow.

At the beginning of 1983 a newly appointed Director, Dr Paul Freeman, began a fundamental review of the Agency's role and organization. Scholes, who by then was working in the computer company ICL on secondment from the Civil Service, met the new Director and was asked to contribute ideas. The report on the DLO role was discussed and further meetings took place. During this period SSM was used to help structure discussion on topics the Director judged to be important. A root definition, model and comparison table from this work are shown in Figures 5.9 and 5.10.

A few months later, in discussion with Checkland, the Director outlined his ideas on reorganization of the Agency. These were discussed in relation to the analysis in the report of the DLO Review team, fairly close mapping being revealed.

The Director's report, a *Review of the Central Computer and Telecommunication Agency*, was submitted to Treasury Ministers and published by HMSO in July 1984; its recommendations have now been implemented. The Freeman Review described the main functions of the Agency as being to authorize departmental expenditure on IT, to provide advice, support and infrastructure for Government IT, and to procure IT goods and services for Government departments. As part of the review, departments were asked whether they would prefer there to be no central agency at all. The response was clear:

> Virtually all departments consulted wanted there to be fundamental changes in CCTA's role, in particular to make the Agency less overbearing. . . . Departments also wanted the Agency to help and support them whole heartedly rather than to nanny and 'second guess' the judgements of departmental managers (para 3.4).

Root Definition

> A Director-owned, CCTA-expert-manned system which
> collects relevant information and projects the probable
> long-term direction of IT in Government; and recommends
> appropriate action for CCTA to take in order to match
> likely departmental needs.

C   Director of CCTA (or CCTA management team)

A   CCTA experts

T   Information on current ⟶ recommended courses
    and future IT trends   of action for CCTA

W   CCTA needs to look ahead strategically

O   Director of CCTA

E   Ongoing Agency – department interaction on IT

†
eg.
technology trends
departmental IT plans
external examples,
etc.

Figure 5.9   An issue-based model concerned with CCTA strategy

Model	Real World
Define 'Relevant' information	Does not exist currently—but CCTA does not look like this model. Given W is 'think strategically' then need would be to identify 'strategic' info. so that projections could be made from current base of knowledge (which would also need to be formalized).
Establish info. collection mechanism	CT(P) and CT(TS) networks already exist. Would need to widen to cover *IT* effectively. Implies pulling in Man Services info. + possibly pooling info. with DI and MPO. Also need to look outside, e.g. market research, commercial trends, etc.
Collect relevant info.	Largely on *ad hoc* basis at present. Much info. collected on DLO net but not pooled. External info. *ad hoc*—not pooled (more *product* orientated rather than application). But CT(P) + TS should be able to get most of what is needed.
Identify and evaluate possible futures	Not currently done. Needs skills/training. Akin to market planning.
Project possible futures	Not currently done. Needs skills/training. Akin to market planning.
Assess what CCTA can/ should do	Not currently done. Needs skills/training. Akin to market planning; also take corporate view of strengths and weaknesses of CCTA and depts.
Recommend appropriate actions	Not currently done. Needs skills/training. Akin to market planning; also take corporate view of strengths and weaknesses.
Monitor and control	Need to establish some measures (e.g. value to 'customer').

Figure 5.10 Comparison table for the model in Figure 5.9

In spite of these reservations, however,

> . . . not one department preferred the 'zero option' of no central agency (para 3.4).

The Review went on to argue for a fundamental shift in the nature of the relationship between CCTA and departments, the prime recommendation being that

> Departments should have the responsibility to identify, develop and manage their own IT projects, subject to the general procedures established by the Treasury (para 3.21).

The main aim of 'the new CCTA' was summarized as

> to bring added value to the development and application of IT in Government administration in ways which are best provided from a Central Government body (para 7.4).

This fundamental change in the role of the Agency was accepted by ministers and implemented. A *CCTA Progress Report* in August 1985 considered the progress in 'the programme of work set for CCTA following the publication of the Freeman Review'. Its description of CCTA's new central aim usefully summarizes the new thinking:

> The central aim of CCTA is to add value to the development and application of IT in central government by assisting departments towards the effective use of IT.

Thus the final function of the systems study and report on the role of the DLO, carried out two years previously, was to make its contribution to the review of the Agency by the new Director. If SSM is viewed simply as an action-oriented process of analysis, then this seems an unspectacular outcome. But the history of this study in its context offers a good example of the continuous evolution of what Vickers (1965) calls the 'appreciative settings' of an organization, the corporate readiness to notice as significant certain aspects of reality and to discriminate and judge them by standards which are themselves created and renewed in the ongoing flux of events and ideas which constitutes organizational life.

The systems study made a contribution to the thinking about the Agency and its role. The Freeman reforms introduced significant changes in the Agency and also changed not only the way in which departments perceived and judged the Agency and its activities, but also changed the way the Agency perceived and judged itself, all this taking place over a period of many months. Vickers formulated the concept of an appreciative system in an attempt to make sense of his 40 years' professional experience in the world of affairs (Checkland, 1983b; Blunden, 1985). SSM can lead not only to purposeful action to improve a problem situation; more generally, it also helps to orchestrate the process of 'appreciation' in Vickers's sense (Checkland, 1981, Chapter 8; Checkland and Casar, 1986), sharpening views and making choices of action more explicit. This example of the process of appreciation complements that described in Chapter 3 in an industrial setting.

Conclusion

As has been mentioned earlier, this application of SSM yielded useful learning about the need to explore the problem situation both as a 'social system' and as a 'political system' (using these phrases in their everyday-language sense) and this was later formalized in the Analysis Two and Analysis Three described in Chapter 2. The DLO study also indicated the importance of paying attention to the carrying out of the study itself. Using the language of SSM, we can say that one choice of 'problem owner', named during Analysis One, can (perhaps should?) always be those who will carry out the study. This choice leads to an obvious naming of a relevant system: *the system to do the study*. It is often useful in complex situations to model a version or versions of this system in order to ensure that aspirations are clear, that resources are appropriate and that a plan of action is available from the start, even though carrying out the study will almost certainly lead to its modification during the course of the work.

In the present case the organization of the study itself *was* taken very seriously, and this led to the two civil servants being seconded to work in the team. With hindsight, though, even more effort might have been appropriate, aimed at persuading the Steering Committee members to become part-time participants in the process of the study rather than simply acting in the Steering role. In fact it was subsequently clear that the Committee had in their heads a model of the consultancy process in which outsiders enter a situation, carry out their work using data from the situation, and present their recommendations in the form of a report. The efforts to get civil servants included in the team, though successful, did not modify fundamentally this model of the consultancy process which their previous Civil Service history made inevitable for the members of the Steering Committee.

A consequence of this, we now surmise, was the period of time required to consider the report at senior level in the Civil Service Department prior to taking action. The point is that the experience of actually carrying out the study is so very much richer than any account of it in a report can be—just as all the experiences behind this book were very much richer than the book itself can convey. Experience will always motivate much more strongly than any written analysis and argument, and this reinforces and underlines the view that strenuous efforts should be made to use SSM participatively. It is perfectly feasible to use SSM to conduct an outsider's study of a problem situation, and it has often been used in that way, but its richest role is to structure an exploration which both changes appreciative settings and moti-vates participants to take purposeful action.

The other significant feature of this study, when we view it as part of a sequence in which the learning was continuous, is the fact that the three members of the team to whom SSM was familiar simply took it as given that SSM was the way to conduct the study. That was the inevitable model in *their* heads as a result of *their* history. This assumption came to be questioned in the sequence of studies which the present authors carried out after working together on the CCTA study. It provided the new insights which are the subject of Part II.

Part II

Soft Systems Methodology in Action— Learning Through Use

Chapter 6

Two Studies in a Product Marketing Division

Part I has given a view of SSM as it is in the late 1980s, after nearly twenty years of development, and has described three applications of it in different circumstances: in industry, in the National Health Service and in the Civil Service. In the last of these studies, in the Central Computer and Telecommunications Agency, the authors worked together, Scholes as a civil servant seconded part time to the study team, Checkland as an outside consultant brought in to lead the team. Part II now describes a sequence of studies in one organization on which the authors subsequently worked together. The sequence led to useful learning about SSM, and Part II attempts to re-create the experience of that learning.

Following the CCTA study (Chapter 5), Scholes joined the UK's major computer supplier ICL, initially on secondment, later on a permanent basis. The studies described in Part II were carried out in ICL and formed part of the managerial work of Scholes in various Company roles. Their context therefore differs from that of the studies previously described. In the studies in industry, the NHS and CCTA, described in Part I, there had in each case been a corporate decision that a highlighted study involving a team of people should be made, and that SSM should be used to structure it. In the sequence of studies in ICL, now to be described, SSM is not always used as the methodology for a special study somewhat outside the normal run of day-to-day managerial work; it is used also as a managerial aid by a manager going about his normal work.

NETWORKED PRODUCT LINE SECTOR

Context

ICL (now part of STC) is the UK's largest computer company, employing around 20 000 people, with operations in 80 countries. The Company's history in a new and fast-changing industry has not been placid. In the early 1980s following Government intervention at a time of serious financial difficulties, a new Chairman (Sir Christopher Laidlaw) and a new Managing Director (Robb Wilmot) were appointed. Robb Wilmot, a charismatic figure, made many changes in the Company involving not only its products and organization but also its management processes and style. One of Wilmot's main strategic thrusts was the development of the 'Networked Product Line'. In ICL's Product Marketing Division a Networked Product Line Sector was set up, and it was here that Scholes had his first Company role, as Business Planning and Control Manager for that sector.

In order to understand the complex background to this particular study, it is useful to place it in relation to the general problem of organizational structure which any company like ICL will face. The problem for any international business based on a fast-changing technology is to find an evolving structure which enables the company and its managers to think about the businesses they operate in several dimensions and hence to continue to *learn* both individually and corporately. Three obviously relevant dimensions are products, markets and geography. It would be unwise to set up a structure based on only one dimension, such as an exclusive set of product divisions, which in ICL's case might be from mainframe computers to microcomputers or computer peripherals, since this could lead to neglect of other important considerations—such as the nature of the needs of a particular market, or a particular geographical area. There will in fact be no single way of carving up company activity into organizational blocks which make those blocks efficiently autonomous. A company like ICL will have to operate some kind of complex matrix in which managers with specific expertise in one dimension can communicate and collaborate with others whose expertise lies elsewhere. This is because any particular *piece of business*, which is what matters to ICL, will entail meeting the needs of a particular market in a particular place by means of particular products. Making a successful business out of meeting, say, the *information processing needs* of the *mail order business* in *the UK* requires that three inter-related bodies of expertise in ICL are in communication with each other.

In addition to these considerations there will be other functional activities which a company like ICL has to carry out in order to survive and prosper in the long term, activities such as obtaining and developing human resources, financing the business operationally and strategically, and doing research and development. And finally, in a science-based business whose products are ubiquitous, it will be essential to find an evolving organizational structure which does not inhibit managers but allows and expects them to develop their own ideas and to take initiatives.

At the time of the Networked Product Line study, ICL's response to these complex organizational issues was the structure shown in Figure 6.1. Here the

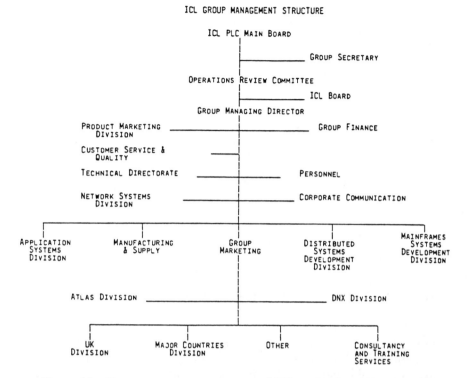

Figure 6.1 Group management structure of ICL at the time of the studies

activities of developing, making and selling products are shown in the divisions shown horizontally ('Group Marketing' being the selling operation, organized geographically), with support from a number of functions shown vertically, such as Personnel and Group Finance. One of these functional divisions was Product Marketing Division (PMD), a group of 280 people headed by a Director. The formal mission statement for this function, expressing its aspirations, was as follows:

> The organization purpose of PMD is to formulate worldwide marketing strategies which support the achievement of corporate objectives. It translates these strategies into practical business plans which are clearly visible and understood throughout the Company. It drives the execution of these plans by the appropriate division, and monitors their progress and authorizes changes in strategy. Product Marketing Division earns ICL the reputation of being an aggressive and competitive multinational noted for its excellence in marketing.

The organization structure of PMD itself, shown in Figure 6.2 reflected a response to the same issues as those facing ICL as a whole. It consisted of a

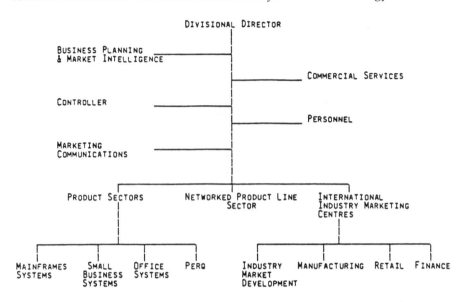

Figure 6.2 Structure of Product Marketing Division at the time of the studies

number of units (horizontally arrayed in the figure) carrying on the activity of formulating marketing strategy, supported by a number of functions such as Business Planning and Market Intelligence, and Personnel.

Networked Product Line Sector was one of the units formulating marketing strategy but its task was more ambiguous than those of the Product Sectors (such as 'Office Systems') or the Industry Marketing Centres which in the main focused on a particular target market such as 'Manufacturing' or 'Retail'. Finally the structure of Networked Product Line Sector itself was the last reflection of the Company matrix (Figure 6.3); it consisted of five units mainly focused on aspects of the networked product line concept and two supporting functions in the shape of a 'Programmes Manager' and a 'Business Planning and Control Manager'. Scholes was in this latter role.

The Networked Product Line Sector has been described as 'ambiguous' above, because its logical status is less clear than that of sectors which address particular industries or particular product groups. 'Networked Product Line' is an abstract concept rather than a product or a market. It expresses the *idea* that ICL products should be capable of being linked to form networks. The Managing Director pushed the idea enthusiastically, but inevitably the concept was not clearly understood throughout the Company. Discussions with several managers within the sector and with others in other PMD sectors, as well as with managers having development and sales responsibilities, revealed a number of perceptions of the 'Networked Product Line'. It was variously

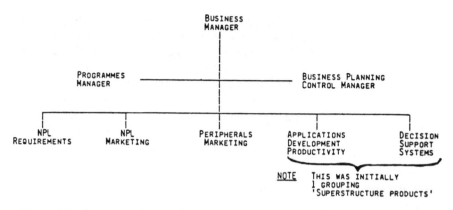

Figure 6.3 Structure of Networked Product Line Sector at the time of the study

seen as a product strategy, as a set of technical standards, as a technical capability, as the name of the total ICL product family, as a selling point for the sales force. Clearly there was no single unambiguous 'deliverable' which could be termed the Networked Product Line; and for the 'Business Planning and Control Manager' there was no clearly defined *business* to be planned for and controlled.

An obvious response to the situation described might be simply to assume that ambiguity is intrinsically a bad thing and hence work to remove it, seeking an unequivocal definition of 'Networked Product Line' (NPL) which could then be used to define necessary action. However, the ambiguity of the NPL concept, though recognized, was not regarded as a particularly serious problem by managers in the sector; and in any case a degree of ambiguity can be helpful in releasing energy and in making room for initiative to be shown. *Too* tight a definition of objectives and tasks within a marketing function can inhibit new thinking and creativity.

In fact the opportunity for a systems study came from the view, widely held within the sector, that the most significant problem was that of obtaining necessary information relating to all the sector's business responsibilities. Having taken part in the CCTA study described in the previous chapter, Scholes decided to use SSM to examine this problem situation, and the Sector Manager agreed that a study should be carried out by Scholes in association with Checkland. The study then started as a part of normal managerial work.

It was initially imagined that the methodological cycle would follow the same pattern displayed in the CCTA study, though the scale of this study was much smaller. The main difference between the two, at the start, was that there was here no fanfare, no setting up of a study team with formal terms of reference, no expectations of work aimed at creating a final report which signals the end of the study. Here was a manager fulfilling his normal professional role as an organizational problem solver, and doing so in this instance by using a particular approach. What were looked for were useful outcomes, conceptual or practical, which would ease a situation regarded as problematical.

Exploring and Expressing the Problem Situation

At the start of any study it is useful to acknowledge what is being taken as given. In this case the givens included the accepted need for the NPL Sector (though its form was open to question) and a focus on it rather than on Product Marketing Division as a whole, or other parts of the Company.

The appreciation of the problem situation came from both documents and discussions. Numerous sector documents were available, including one defining sector roles, as well as statements from the Managing Director concerning, and showing his enthusiasm for, the NPL concept, together with a number of marketing papers. Open-ended discussions with all the managers in the sector were held, and such interviews continued throughout the study.

A number of problem themes emerged early on, and can be seen in comments recorded at the time. These included the need for a clear sense of the sector mission, the need for the sector to have available to it particular information from other marketing sectors and from the development and sales functions. This information was that needed if the sector were to ensure that the NPL concept informed all product definition and development. Using such information implied considerable sector knowledge of the realities of what is involved in planning, implementing and managing computer networks; and it was also clear that the relation between the NPL idea and other concepts such as 'Decision Support System' needed to be thought through.

It is worth commenting on the absence of 'rich pictures' of the problem situation literally in the form of pictures or diagrams. Within SSM the ideas of representing the complexities of a human situation in picture form has been a powerful one. (Figure 4.1 shows such a picture in the NHS study described in Chapter 4.) It is an efficacious way of recording the finding-out phase because relationships and interactions are more briskly captured in pictures than in linear prose. However, the fundamental requirement is to gain a discussable appreciation of a problem situation; pictorial representation is simply one means of doing that which has been found useful. But it is not an axiomatic requirement. The guideline is: do what *you* find to be insightful and comfortable. Stating this guideline raises methodological questions concerning the description of SSM itself in terms of constitutive and strategic rules as originally suggested by Naughton (1977) and endorsed in modified form by Checkland (1981, pp. 252–254). It can be argued that if constitutive rules are expressed in terms of a concrete output—a literal 'rich picture', say, rather than an abstract 'appreciation'—then this may have the effect of reducing methodology to method, and lightening the looked-for emphasis on change accomplished rather than reports written. This whole issue of constitutive and strategic rules is taken up in the final chapter.

Choosing Relevant Systems

The appreciation of the situation gained through interviews and reading led the authors initially to conceptualize the role of a business planning and control manager as a *service system* to a core business in which raw materials (concrete or abstract) are converted into products designed to meet the needs of defined markets. The service is concerned with planning for the total business activity. This entails defining criteria for measuring business performances, making business plans, monitoring the business activity using the criteria, and taking controlling action, either directly or through plan modification. This basic instrumental concept led to the framework set out in Figure 6.4, in which the role of a business planning and control manager is seen as being to *conceptualize* the appropriate business and *enact* the planning and control, with 'appropriate' here defined by the applicability of the NPL concept. Such a picture, whose status is similar to that of the 'EROS' model in Chapter 3 (Figure 3.1), immediately suggests relevant systems which might usefully be modelled for detailed comparison with real-world activity.

In the NPL study about a dozen relevant systems were discussed and, to different extents, worked on at this stage. They included, as well as the systems in the figure: systems to provide the information flows defined by Figure 6.4; a system to produce plans to meet ICL's desire to move into a 'total solution' business; a 'competition awareness' system; a system expert in networking computers; and a system to correlate potential products and potential markets. Not all of these possibilities were modelled in detail via

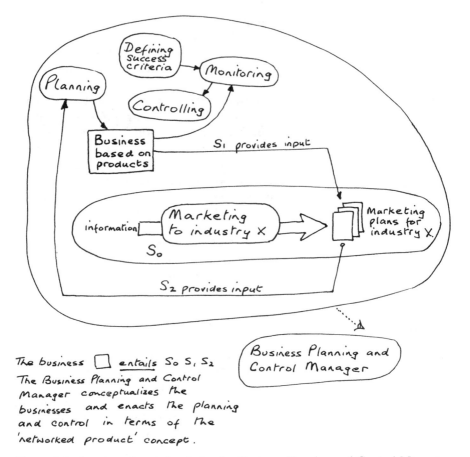

Figure 6.4 represents a hand-drawn diagram with the following labels:

Defining Success criteria

Planning

Monitoring

Controlling

Business based on products

S_1 provides input

information

Marketing to industry X

S_0

Marketing plans for industry X

S_2 provides input

The business □ entails S_0 S_1 S_2

Business Planning and Control Manager

The Business Planning and Control Manager conceptualizes the businesses and enacts the planning and control in terms of the 'networked product' concept.

Figure 6.4 Levels and meta-levels for the Business Planning and Control Manager

root definitions, but attention was given to a number which reflected the expressed concern in the sector for clarifying its mission and for ensuring that the NPL concept influenced marketing plans.

Building Conceptual Models

A number of detailed models were built from root definitions, and these were used in further interviews to shape discussion with managers in the sector. These discussions led to further ideas for relevant systems and new root definitions were formulated. Not all were modelled in detail. They were used to the extent that seemed useful in the light of immediate feedback. Figures 6.5, 6.6 and 6.7 show three of the detailed models which were used at this stage.

Root Definition

> A system owned and manned by NPL Sector which
> takes as given corporate objectives, receives information
> on marketing and development plans and influences
> those plans and marketing and development operations

C Marketing and Development
A NPL Sector
T Plans [operations] ⟶ Influenced plans/operations
W NPL concept is powerful enough to bring about change
O NPL Sector
E Corporate objectives, structure

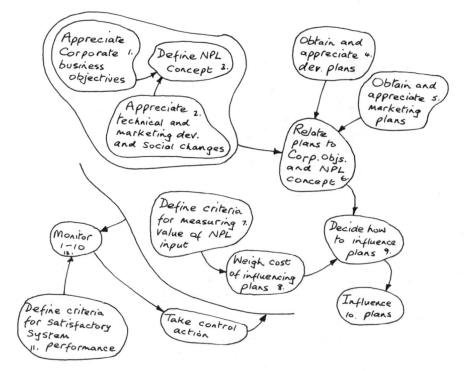

Figure 6.5 A system to influence marketing and development NPL plans and operations

Root Definition

> A Sector-owned and operated system which, in
> the light of corporate business objectives and
> external factors (tech. mkt. social and legislative,
> for example), defines concept X relevant to the
> Sector's achieving its mission.

C Sector

A Sector

T Need for concept relevant ⟶ Need met
    to Sector mission

W A mission-related concept is necessary, feasible

O Sector

E Corporate business objectives, external information

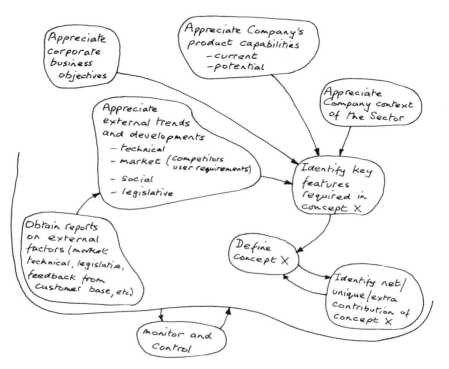

Figure 6.6 A system which thinks out the driving concept of a Marketing Sector

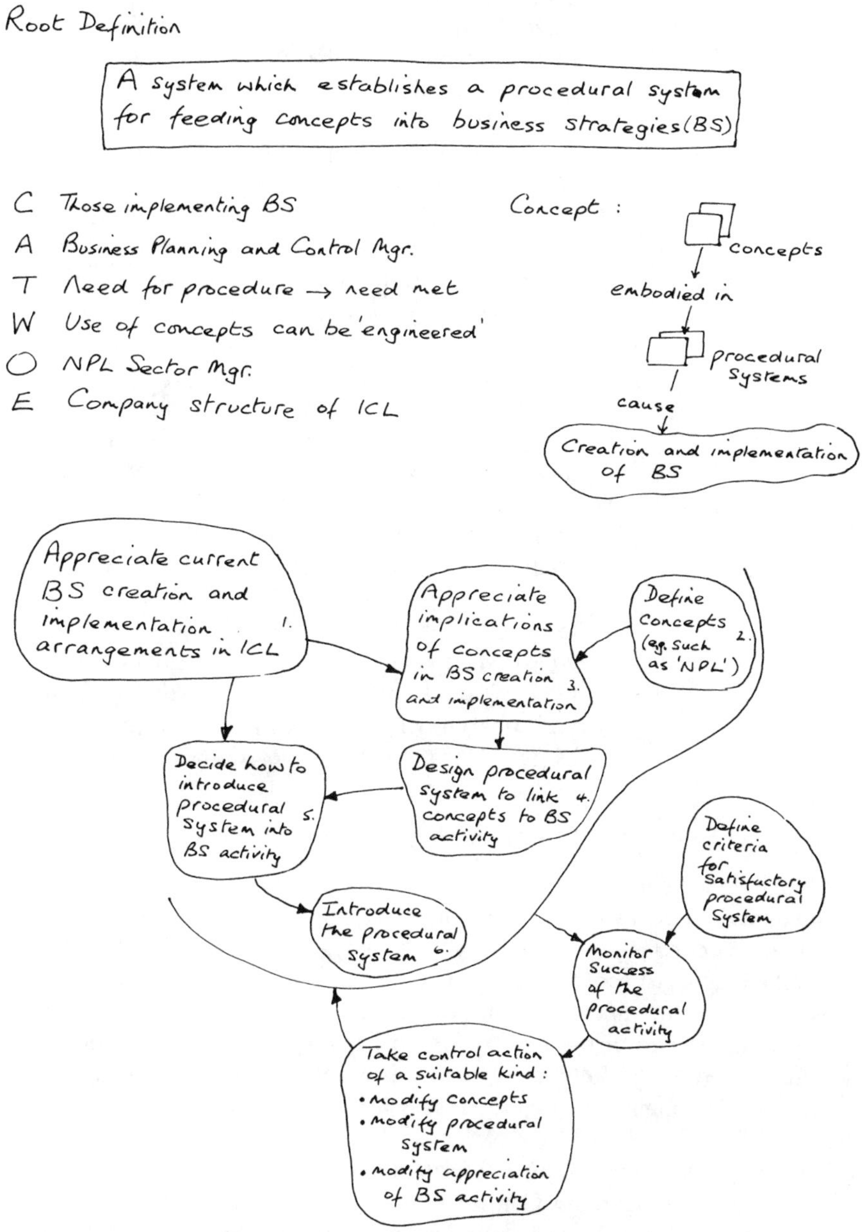

Root Definition
A system which establishes a procedural system for feeding concepts into business strategies (BS)
C Those implementing BS
A Business Planning and Control Mgr.
T Need for procedure → need met
W Use of concepts can be 'engineered'
O NPL Sector Mgr.
E Company structure of ICL
Concept :
concepts
embodied in
procedural systems
cause
Creation and implementation of BS
Appreciate current BS creation and implementation arrangements in ICL 1.
Define concepts (eg. such as 'NPL') 2.
Appreciate implications of concepts in BS creation and implementation 3.
Design procedural system to link concepts to BS activity 4.
Decide how to introduce procedural system into BS activity 5.
Introduce the procedural system 6.
Define criteria for satisfactory procedural system
Monitor success of the procedural activity
Take control action of a suitable kind:
• modify concepts
• modify procedural system
• modify appreciation of BS activity

Figure 6.7 A system by which the Business Planning and Control Manager exercises influences on business strategy

This is a point at which a description of the study stage-by-stage according to the methodology tends to break down. What was underway here can be described in methodological language as a series of very rapid iterations round Stages 2, 3, 4 and 5 of SSM. It is not easy to give a blow-by-blow account of such a process, not least because some iterations occurred *in the course of a single discussion*. This mode of operation is in marked contrast to that used in the CCTA study with its more stately formality. In their own everyday language, what was being said to the NPL sector managers who participated at this stage was essentially this: 'Look, these conjectured purposes seem relevant to our problems. Taking them seriously leads to this set of necessary activities linked in this particular way. Let's compare this with what we're actually doing at the moment.' A number of managers who took part in this process later confirmed that what they found useful was SSM's distinction between the real world of complex action (Stages 1, 2, 5, 6, 7) and the worked-up intellectual constructs (Stages 3, 4) which enable managers mentally to stand back from the everyday hurly-burly and coherently interrogate the usually unquestioned assumptions behind the everyday work.

Comparing Models with Reality and Implementing Changes

Although much comparing of models with perceived reality took place during individual discussions with sector managers, a more formal comparison stage (though one conducted in the everyday language of the sector rather than in analyst's language) can be seen in meetings of the sector at which the work was discussed. An important point was reached about a month after the study had begun. At one of the regular monthly management meetings the study was described to the full management team and the emerging themes were discussed. The main thrust of the discussion, recorded in notes at the time, was as follows.

The study set out to establish information needs, from the point of view of the Business Planning and Control Manager; there are identified problems concerning the availability, accuracy and timeliness of such information; defining the required information in detail requires a clearer concept of what the sector is seeking to achieve, which would enable both necessary activities and information flows to be defined.

Following a long and generally constructive discussion at this meeting, the NPL Sector Manager asked for a summary of the work so far, since it seemed relevant to his own setting of individual objectives for members of his team. Subsequently he took two initiatives based upon the discussion at the monthly meeting and the interim account of the study. Firstly, he organized a full-day meeting with interested parties outside the sector in order to consider the future shape of ICL's business and determine how the 'product' and 'industry' elements within the Division's sectors should be pulled together. Secondly, an NPL Management Team meeting was organized specifically to define the NPL concept in terms which could be used throughout the Company. Meanwhile some individual managers in the NPL Sector had already taken action within their own areas as a result of taking part in comparisons between their present activities and such 'logical machines' as those represented in Figures 6.5, 6.6 and 6.7.

> These happenings within and as a result of the study illustrate the fact that in a study not isolated and highlighted as such, but fitted seamlessly into the work of a section of a company, it is sometimes necessary to map the happenings which take place on to the methodology—which provides a way of making sense of them—rather than think simply in terms of 'following the stages of the methodology'.

A month after the interim account of the study was produced, a further account covering the full three months' work was written. This account was copied to the Director of PMD, an act which helped eventually to lead to further studies using SSM which are described in Chapters 7 and 8. Thus the outcome of this particular study, occurring both during and after it, can be summarized as: changes to the way NPL Sector as a whole thought about its task; changes to the way in which individual managers in the sector did their jobs; changes to Divisional and Company-wide thinking and action concerning a Networked Product Line; and the initiation of a sequence of further studies using SSM.

> From the authors' point of view the study provided a useful contrast to the study in CCTA on which they had previously worked together. The contrast is that between using SSM to *do* a study (as in CCTA) and doing a study *using* SSM (as in NPL Sector). This is not simply a play on words, we believe. We are using the two phrases to try to mark what we now see as a real distinction, although not one which would have

been recognized at the start of the respective studies. If SSM is used to do a study, then it offers a design template for the activity of the study, and the outcomes of the supposed 'constitutive rules' are likely to form milestones for the project journey (Checkland 1981, pp. 151–254). If a study is done which happens to use SSM, then its use is likely to be very flexible, *and at more than one level*. Its value is then as an epistemology by means of which an analyst can make sense of the flux of project activity. It is of course much easier to describe serially studies like those in Chapters 3, 4 and 5 in which a special study is set up and carried out by a study team, than it is to describe a study like that in the NPL Sector which was done as a normal part of managerial work. Nevertheless, it is the more flexible use of SSM in studies of the latter kind which has characterized some of the richest learning experiences in the last five years of the development of SSM. Another brief study of this kind will conclude this chapter.

WHAT DECISION SUPPORT SYSTEMS COULD DO—A CONTRIBUTION FROM SYSTEMS THINKING

Context

This short contribution to the work of the Networked Product Line Sector was in a sense an outcome of the study just described. That work created interest in the systems approach within the sector; and in particular the Marketing Manager responsible for Decision Support Systems (DSS) asked for help in thinking through the DSS concept in order to define possible user requirements which ICL might satisfy with its products. At the time both the concept of DSS itself and the likely impact of the networking concept upon such systems were ambiguous.

The work to be described was not a major study of DSS; it took place over a three-week period in three meetings and a number of brief discussions. Methodologically the idea was to appreciate some dimensions of the 'DSS problem', to build models relevant to it and to use the comparison to identify lessons for ICL's development of DSS. An initial view at the start was that it would be useful to have a usable definition of DSS so that possible ICL products could be identified, together with some idea of the implications of marketing such products. Setting up the study occupied about an hour's discussion.

At the time of this work the rich but ambiguous phrase 'Decision Support System' (Keen and Scott Morton, 1978) was very much in vogue, and it is worth dwelling briefly on this context. It is obvious that as soon as information systems aspire to serve managers at levels higher than that of the clerical (at which level the computer simply does what clerks previously did on paper) there arise many problems concerning both definition, selection and creation of information from the plethora of data available, and the intentions behind its provision. This nest of problems has in general been tackled by adopting a particular view of the management process and conceptualizing the role of information systems in relation to it. The view of the management process which has been taken as given in this field comes from the work of Herbert Simon (1960, 1977; Simon and Newall, 1972), and the attractiveness of its simple clarity has made it the conventional wisdom in spite of its obvious limitations and the fact that it is very one-dimensional (Ciborra, 1984, 1987; Checkland 1985a). Simon's simplification is that managers are *decision makers* who proceed through a three-stage process of: intelligence; design; choice. In the first stage problems are identified and data are collected; 'design' consists of planning for possible alternative solutions; 'choice' entails selecting an alternative which is good enough, and monitoring its implementation. This is a very limited view of what managers do, of course, but it has the apparent virtue that once adopted it makes 'obvious' what information systems are, namely: systems which provide the necessary information for decision making, especially for the crucial 'design' stage in which alternative problem solutions are generated. Thus arises the conventional wisdom that although computers cannot take over the human role, they can provide *support to decision making*. Ahituv and Neumann (1982), taking Simon's model as given, succinctly express this conventional view:

> It is important to note that design-aid systems never replace human decision making! They are capable only of supporting decision-making processes. At this level of management there are always additional factors that cannot be computerized, such as morale and ethics. Therefore, we often call them *decision support systems* (DSS).

This is not the place to pursue this discussion, but enough has been said to indicate the intellectual context of this short study—one in which computer companies were asking themselves how they could offer products under the DSS label. The Marketing Manager for DSS assumed that ICL would offer such products, that the necessary organization to support such a development would exist, and that there would be a welcome for such products from customers.

Exploring the Problem Situation

In ICL at the time, the view was that DSS were concerned with business problems, with organizational issues, with attitudes and behaviour, and with helping managers cope with the many different kinds of issue which compete for their attention. In this study some actual pictures of the problem situation were drawn, and a table was compiled of the several levels at which management effectiveness could potentially be improved by information processing, including those of: interacting with the external environment; meeting the competition; dealing with the internal aspects with respect to tasks; and dealing with the internal aspects with respect to people in the situation. Building on the picture which this represented, it was agreed with the Marketing Manager that it seemed sensible to examine systems which expressed 'what DSS might support' rather than work from either 'universal' or 'ICL-specific' definitions of DSS. This would ground the Marketing Manager's thinking in the everyday world of the manager rather than in the world of academics with a professional interest in clarifying DSS as a concept.

Choosing Relevant Systems and Modelling

Five root definitions and models were developed at one of the meetings which comprised the study; three of the models are shown in Figures 6.8, 6.9 and 6.10. Each of these models was built in a few minutes.

It is worth noting that full root definitions with CATWOE analyses were not produced. It will be seen that the models in Figures 6.8, 6.9 and 6.10 could be more tightly constructed, and that CATWOE statements *could* be produced by assuming that the systems in question were relevant to some notional business organization which interacts with its environment and makes decisions. These models were deemed 'good enough' in this two-man exercise whose measure of performance was only that the Marketing Manager for DSS found it relevant to his concerns, and helpful. (To a reader who may feel that he or she can think of relevant systems which are 'more relevant' than those selected—such as a system to carry out market research on DSS, or a system to define the role of the Marketing Manager, perhaps—we may say that they might indeed seem to be more relevant *to you*. And they might indeed

'Rough' Root Definition

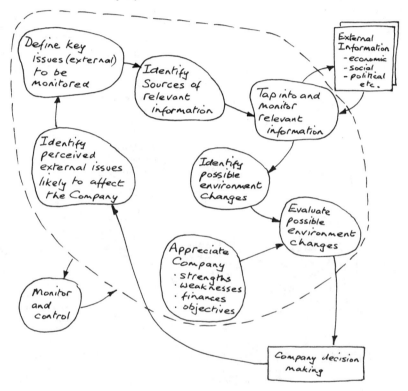

A system which identifies and evaluates possible changes in the 'macro' environment which are likely to impact the business operations of the Company.

(No CATWOE analysis)

Figure 6.8 A rough model used in the DSS study, externally oriented

have been shown to be highly relevant in the study itself, had you been there to argue for them. The important point is that SSM makes the course of such debate more open and explicit. What is being argued about is less confused than usual; shifts of position are discernible; reaching a cul-de-sac can be followed by an orderly retreat and an exploration of new directions; the problem-solving *process* is better articulated.)

'Rough' Root Definition

> A system which provides information to decision makers
> on possibilities to increase return on investment and the
> productivity of the organization, and to reduce costs

(No CATWOE analysis)

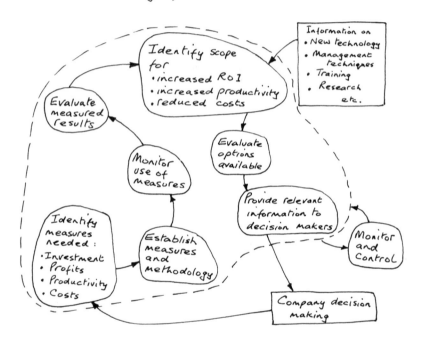

Figure 6.9 A rough model used in the DSS study, internally oriented

Comparing Models with Real Possibilities

At a further meeting the comparison stage of SSM was used to provide a structure for the debate. The five models became the source of five tables which listed activities from the models and showed what it was regarded *would* or *might generally* happen in real situations with respect to those activities, together with what a DSS might contribute. Part of one such table is shown in Figure 6.11 (related to the model in Figure 6.8). The right-hand columns from the tables contribute a flow of ideas to the development of products based on the DSS concept.

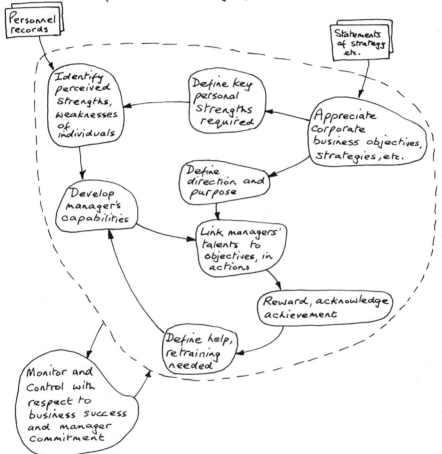

'Rough' Root Definition

A system which enhances the success of a business by harnessing individual managers' talents in a way which develops their capabilities and commitment to the organization.

(No CATWOE analysis)

Personnel records

Statements of strategy etc.

Identify perceived strengths, weaknesses of individuals

Define key personal strengths required

Appreciate corporate business objectives, strategies, etc.

Define direction and purpose

Develop manager's capabilities

Link managers' talents to objectives, in actions

Reward, acknowledge achievement

Define help, retraining needed

Monitor and Control with respect to business success and manager commitment

Figure 6.10 A rough model used in the DSS study, manager-effectiveness-oriented

ACTIVITIES IN MODEL	REAL WORLD	WHAT DSS COULD DO
Identify external issues likely to affect the company	Decision makers do based on past experience of events. Intuitive. Comparison with others' *weakness*. Heavily reliant on knowledge of individuals/ groups (memory). Not formalized. Past experiences may not be valid?	Knowledge base of experience. Actual facts not memory. Modelling of organization in its environment. Express comparisons with others in directly usable form (+ history of comparisons), i.e. *knowledge* of what others have done in comparable situations.
Define key issues (external) to be monitored.	Limited by practicality of what can easily be done. Does not differentiate between key and dependent issues.	Highlight dependencies and bring out key issues.
Identify source of relevant information.	Personal contact, 'Library', commission research, etc. Information suppliers, adverts, etc. Experience, consultancy. Trade— business bodies, etc.	Hold knowledge of external sources of information—a register.
Tap into and monitor relevant information.	Set up subscriptions. Pay for/do research. Employ staff to do. Read tomes of information. Set up education/training seminars.	Directly interface with information suppliers. Could cut out superfluous/noise information. Focus on key *changes* rather than ongoing information on non-changes. Pull in information automatically. Speed up processing—availability of information. Compare information and highlight inconsistencies.

Figure 6.11 Part of the comparison stage based on the model in Figure 6.8

Identifying Outcomes

As the 'client' for the study was also a 'problem owner' and one of two joint 'problem solvers', there was no requirement to produce a written report on the study. However, the fact that learning from this study had occurred was evidenced by subsequent actions taken in the company. And the study was successful in the sense that the 'client' had perceived sufficient value and learning coming from taking part in it that he was cheerful about investing his time. At a later stage the work done in this short study was instrumental in helping set up a systems study of the then newly established Management Support Systems Sector in the Product Marketing Division.

> This short study was initiated by the request for help from the DSS Marketing Manager. SSM was used to provide a coherent framework for discussion, thinking and subsequent action. Compared with a major study such as that in CCTA (Chapter 5), this exercise was on a very modest scale. It was useful, however, as was the NPL Sector study, in indicating that SSM is not restricted to any particular scale (or time length) and can be used not only to do a study which is then formally structured according to its stages, but also to provide quick help for a client not required to learn the language of the methodology.

Chapter 7

Soft Systems Methodology in an Organizational Change Programme

In the previous chapter SSM was in action not as a way of structuring a special 'one-off' study, but as a way of carrying out a manager's job in the normal run of day-to-day activity. As has been shown, this entails a looser, more flexible use of the same ideas than in classic applications of SSM—such as those described in Chapters 3, 4 and 5—but with much more recycling round the different stages, often done rapidly, and with less scrupulous attention to the formal niceties of formulating root definitions and building models. To some extent SSM was veiled in the studies of Networked Product Line Sector and Decision Support Systems. It was simply the way a manager chose to structure his thinking in relation to a couple of professional problems.

Nevertheless, in the history being recounted here it was the Networked Product Line study which led directly to further uses of SSM in ICL, both formally and more flexibly. A presentation of the NPL study to the Managing Director and the Director of Product Marketing Division led to the authors' involvement in a major programme of organizational change then underway in ICL, an initiative known as 'the ICL Way programme'. The origins and nature of this programme are described in the next section. Scholes was made responsible for the ICL Way programme in Product Marketing Division. SSM was used to help define, plan and manage that programme, which was not itself intended to be 'a study' of any kind, but rather an initiative to bring about strategic change in the company. Then within that programme, as one of the several parts of it, two more formal studies were carried out. One was a systems study of a newly established Management Support Systems Sector within the Division. This itself then contributed to a bigger systems study of Product Marketing Division as a whole. These three uses of SSM in the organizational change process are described in this chapter.

Since the two more formal studies included postgraduate students from Lancaster University's Department of Systems and Information Management, as had the CCTA study, this provided a good opportunity to reflect upon and enrich the learning recently gained at that time from the contrast between the work in CCTA and that done in ICL. This is a good example of the way in which an ongoing programme of carefully structured and organized action research can provide good learning opportunities for what Schon (1983) in his account of 'how professionals think in action' calls 'the reflective practitioner'.

USING SSM TO HELP DEFINE, PLAN AND MANAGE THE 'ICL WAY' PROGRAMME

Context

During his Christmas break the Managing Director Robb Wilmot prepared a 'New Year letter' to all ICL managers. This aimed to set the tone for the further development of the Company, and did so in terms of 'basic commitments' expected of all employees and 'management obligations' laid upon every manager. The *commitments* defined a company culture: customer oriented, ready for change, striving for achievement through teamwork, committed to developing its people and to making the company itself a 'productivity showcase'. A manager's *obligations*, as a manager of both people and a business, were to provide direction to achieve high-value outputs through teamwork and development, adopting a 'can do' attitude, innovating, and facing up to and dealing with difficult issues. Finally, the manager had an obligation to reflection and self-measurement in order to improve his or her performance over time.

In the spirit of the culture it was trying to create, the letter did not spell out exactly how the desirable ends were to be achieved. Rather, each ICL Division had the authority and responsibility to define and implement its own programme aimed at achieving the 'ICL Way' of doing things.

Subsequent discussions with the Managing Director and other senior managers over a period of time filled in the background to the launch of the ICL Way programme. Following the financial difficulties of the early 1980s the company had undergone rapid change. New top management had been appointed, staff numbers reduced, old products discontinued, new products acquired and/or developed. There had been a major reorganization to achieve all this, and a shift in strategic direction away from that of supplying mainframe computers to becoming a supplier of 'systems solutions' to problems in selected industry or market segments. These changes had imposed considerable demands upon both company management and workforce, but the Managing Director felt that the change process was far from complete. In a volatile environment he sought some unifying theme which could persist as

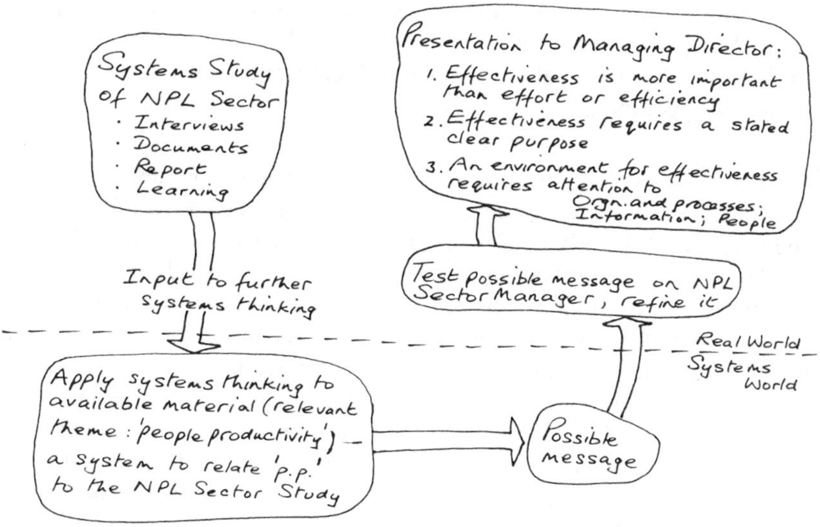

Figure 7.1 The approach to a presentation to the Managing Director based on the NPL Sector study

further changes occurred and as different organization structures evolved. Establishing the idea of doing things the 'ICL Way' was an approach to this problem.

The authors' involvement arose as a result of the earlier systems study of the Networked Product Line Sector.

The Managing Director held quarterly day-long reviews of the work of each Division. These included a series of presentations from the Divisional Director and his managers. At one of these Scholes was asked to make a presentation based on the NPL Sector study, focusing on 'people productivity'. The approach to preparing and making the presentation can itself be seen as a kind of minor systems study. The thinking about the presentation, which maps the shape of SSM (deliberately!) is shown in Figure 7.1.

Following the presentation, discussions with the Director of Product Marketing Division (PMD) took place in which he outlined a number of areas for improvement within PMD. These included:

- *'automation'* of the Division: making it a 'productivity showcase' using ICL products;
- *decision support*: providing divisional management with information support;
- *'house style'*: establishing a PMD norm for organization, meetings, progressing actions, and security;

- *corporate leadership*: providing a lead to the rest of the company in the ICL Way programme; and
- *balance*: achieving a balance between direct support of sales and work on marketing planning.

After this discussion, a proposal was prepared for a programme of work to tackle these issues. Preparation of it was the first phase of thinking to help define and plan PMD's ICL Way programme. This was done using SSM in its 'everyday managerial' mode, as opposed to its 'special study' mode.

This time no particular version of the methodological cycle was used as a basis for a plan of the project, against which an account could be given of the progress made and the lessons learnt. In this application an *internalized* SSM was being used, and reflection on its use came after the methodological cycle had been applied. What follows, then, is more an account of the work done than a stage-by-stage account of SSM in action. Reflections about the work done are included in the usual way, and they do relate the work done to the stages of methodology. But it is important to appreciate that this is the result of subsequent reflection. While doing the work, thinking and action flowed together. That it is possible to cast the later reflections in the language of SSM stems from the consideration that the thinking had in fact been based implicitly upon the methodology.

Appreciating the Problem Situation

Based upon the presentation material used for the Managing Director's review, the 'New Year letter', and the Divisional Director's expressed concerns, initial thinking consisted of the assembly of the picture shown in Figure 7.2. Two of its themes were then further developed: the notion of acting as 'catalyst' of a programme of work which would have to be undertaken actively by PMD managers if it were to bring about change, and the idea of questioning the nature of the activities necessary to bridge the perceived gap between the real-world ICL and the vision of the company it aspired to be.

These themes were developed in the form of the conceptual model shown in Figure 7.3. This led to further thinking about its two levels of activity and to the development of a proposal for an ICL Way programme in PMD which was discussed with the Director of the Division. The proposal covered the areas to be tackled, the duration of the work, Scholes's role and, especially, the overall approach to be adopted. This stopped short of any detailed account of what should be done since a principle was that this would have to be done collaboratively with PMD managers in order to stimulate initiatives and commitment from them.

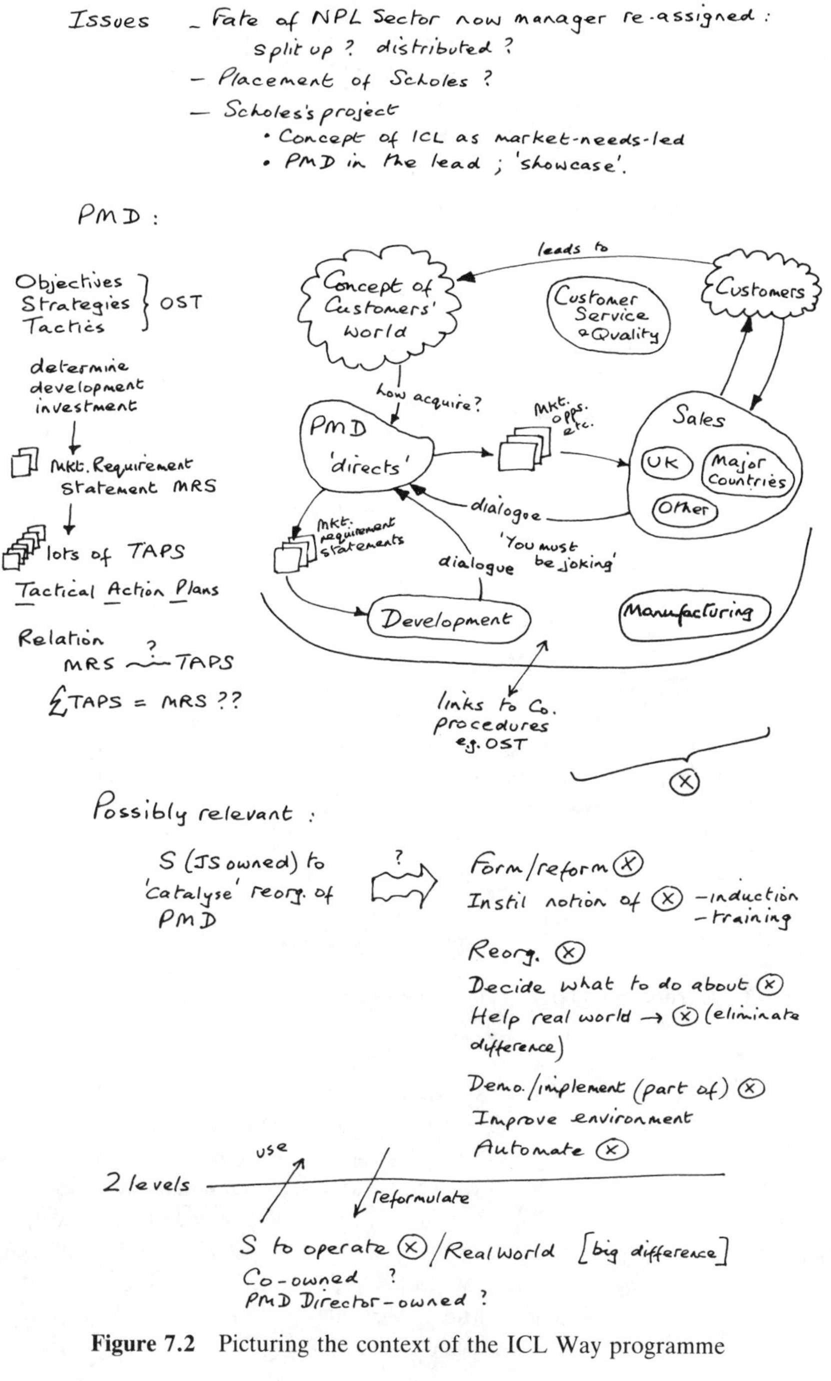

Issues — Fate of NPL Sector now manager re-assigned:
split up? distributed?
— Placement of Scholes?
— Scholes's project
• Concept of ICL as market-needs-led
• PMD in the lead; 'showcase'.

PMD:

Objectives ⎤
Strategies ⎬ OST
Tactics ⎦

determine
development
investment

Mkt. Requirement
Statement MRS

lots of TAPS
Tactical Action Plans

Relation ?
MRS ⁓ TAPS
∑ TAPS = MRS??

Concept of
Customers'
World

leads to

Customer
Service
& Quality

Customers

How acquire?

PMD
'directs'

Mkt.
opps.
etc.

Sales

UK   Major
Countries

Other

mkt.
requirement
statements

dialogue

'You must
be joking'
dialogue

Development

Manufacturing

links to Co.
procedures
e.g. OST

Ⓧ

Possibly relevant:

S (JS owned) to
'catalyse' reorg. of
PMD

?

Form/reform Ⓧ
Instil notion of Ⓧ —induction
—training
Reorg. Ⓧ
Decide what to do about Ⓧ
Help real world → Ⓧ (eliminate
difference)
Demo./implement (part of) Ⓧ
Improve environment
Automate Ⓧ

2 levels _____

use

reformulate

S to operate Ⓧ/Real World [big difference]
Co-owned ?
PMD Director-owned ?

Figure 7.2 Picturing the context of the ICL Way programme

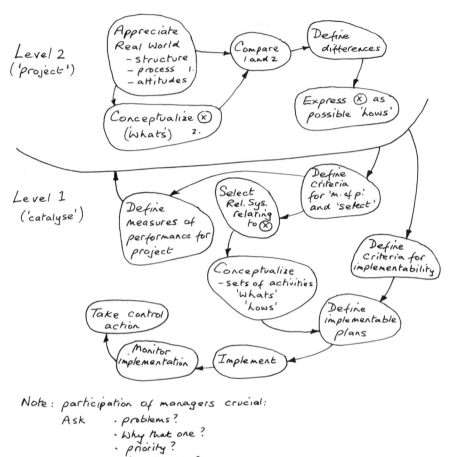

Figure 7.3 A crude model of project activities deriving from the picture in Figure 7.2

Figure 7.2 could be taken to be one of SSM's Stage 1/2 'rich pictures', a pictorial account of many aspects of a problematical situation. It was produced not as a formal outcome of a conscious Stage 2 of SSM but simply in response to a felt need to grasp what was here a much more variegated situation than those in the NPL and Decision Support studies. Those studies were defined more clearly in terms of tasks explicitly acknowledged as needing to be done.

The model in Figure 7.3, alas for the purists, did not have a root definition or CATWOE analysis, though it would not be difficult to provide them and so 'tighten up' the model building. The feeling that progress was being made and was explicitly captured in the picture and model removed any urge to do this at the time.

Agreeing the Approach

The thrust of the proposal to the Director of PMD was that this work should be the Division's contribution to providing a corporate lead in translating the ICL Way into action. Its argument was that positive action would be needed to bring about change, which would require the catalysing of a programme which managers would themselves carry out. But improvements achieved through such a programme would never be absolute. The important thing was to inculcate a *process* which should become PMD's version of the ICL Way. To achieve such an end the Director's personal sponsorship was essential, and to avoid boredom and/or cynicism the programme should have a limited duration before reappraisal—after which 'ownership' should lie with the Division staff.

The proposal made suggestions for work in five areas: productivity; planning; 'people development'; the working environment; corporate learning. There would be a series of discrete modules of activity, some tackled in parallel, some serially. Each had its contribution to make to the programme as a whole but care would be taken to see that the whole did not stand or fall on the success of any individual module.

It was agreed by the Division Director that the programme should be undertaken on the lines proposed.

Implementation

The agreement to go ahead with a six-month programme made Scholes responsible for defining, initiating and coordinating the activities. The Director agreed to sponsor it and be its ultimate controller. The programme would be steered (shades of CCTA!) by a group consisting of the Director himself and four PMD business managers, and this would be a forum for capturing corporate learning. The first phase of the programme would consist of the detailed definition of the rest of it, based on contributions by people in PMD. The second phase would focus on carrying out the modules and trying to establish mechanisms by which learning from the experiences could be gleaned.

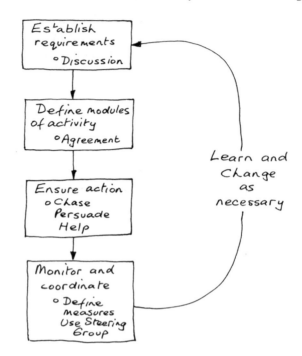

Figure 7.4 The agreed way of carrying out the ICL Way programme

Each of the activities in the programme was to be treated in the manner shown in Figure 7.4, which is clearly a skeletal model of a purposeful activity system.

The main difficulty in getting work on the ICL Way programme in PMD underway was to avoid thinking from the start in terms of the content of various modules of activity. The flexible informal mode of SSM was here used to think coherently about mounting the programme, leaving its content to be filled in participatively with PMD managers, something the spirit of the whole initiative called for.

Thus the picture of the situation (Figure 7.2) and the model (Figure 7.3) both addressed the problem of mounting a participative programme of organizational change. The proposal to the Director focused on the approach, and avoided detailing content. The proposal and the discussions of it which then took place represented the transition from Stage 4 to

Stage 5 of SSM—comparing concepts with perceptions of the real situation. The Director's agreement to the approach, together with its mechanisms and resources, in a sense constitutes Stage 6 of this study. The Director, as a prime 'problem owner', has agreed with the 'problem solver', that a particular action will be taken in a problem situation, the implementation of the programme itself being deemed a 'desirable and feasible' change in relation to the perceived problem of making the ICL Way a reality.

Implementation of the full ICL Way programme in PMD will not be described in detail. The programme itself was not as such a systems study, but those parts of it in which SSM was used are described in the rest of this chapter. However, in order to give the reader a feel for this change programme, it is perhaps useful to indicate by a few examples the kind of work undertaken within the programme's modules.

A programme of office automation was completed, with ICL word processors replacing all typewriters in PMD. This programme was monitored carefully in order to learn about such transitions in ways relevant to ICL's marketing of office machines to external customers. In this way this module aimed both to improve productivity and generate some general lessons.

Common-interest groups were established across the Division, for example a Business Planners' Forum and a Secretarial Development Team. Each of these groups established its own agenda and priorities, contributing—the programme hoped—both to personal development and to improving the way things were done in PMD.

Accommodation was refurbished and exhibitions and displays of ICL products were arranged in common areas (such as entrances) so that staff could better appreciate the range of activity for which PMD was responsible.

Management workshops were organized to encourage the PMD managers to determine priority issues and establish ways of dealing with them.

This implementation of the ICL Way programme, which, with its built-in monitoring and control mechanisms, was itself conceptualized as a learning system, represented Stage 7 of this particular informal use of SSM. The methodology had provided a way of coping with the variety of views about the ambiguous concept of the 'ICL Way', and helped institute a change programme whose content was defined in detail by people in the situation. It was possible to describe to interested parties the learning system which the programme would operationalize (without necessarily using systems' language, if that were felt to be a block) and convey the message that here was a coherent company effort directed at strategic change which urgently required their contributions.

In agreeing to the ICL Way programme, as described above, the Director of PMD also agreed that it should include not only the immediately implementable visible activity—the exhibitions, the management workshops, etc.—but also a more fundamental look at PMD as a whole. A systems study of PMD was initiated. This eventually evolved into the two studies which are described in the rest of the chapter.

A STUDY OF THE MANAGEMENT SUPPORT SYSTEMS SECTOR OF PMD

In agreeing the content of the ICL Way programme and the way in which it would be developed and implemented, as described in the previous section, the Director of the Product Marketing Division welcomed not only the immediately visible actions but also the prospect of a deeper look at PMD and its operations. It was agreed that there should be a systems study of PMD as part of the programme. This provided an opportunity to carry out a more traditional highlighted systems study within the stream of work carried out by the authors.

Starting with the CCTA study (Chapter 5) this had continued in ICL with the less formal use of SSM in day-to-day managerial work (the three studies described in Chapter 6 and the first section of this chapter). Now here was a chance, within the same organization to contrast the classic use of SSM with the less formal mode.

In action research of the kind described in this book it is important to watch out for and seize such opportunities. Action research, not being based upon the hypothesis-testing model from natural science (which is so slippery a concept in the investigation of social phenomena) has to be judged by the even application of two criteria which relate, respectively, to the 'action' and to the 'research': practical achievements in the problem situation and the acquisition of process knowledge concerning problem solving (Checkland, 1981, Chapter 6; Susman and Evered, 1978; Warmington, 1980). It is in general harder to achieve the latter than the former, so it was with some eagerness that the authors embarked upon this phase of the work.

Two mature postgraduates from the Department of Systems at the University of Lancaster, Elaine Cole and Jim Hughes, were brought into the team at this point to help with the formal study. It was decided to work up to the level of PMD itself in two steps, starting with a 'Phase 1' study of the then newly

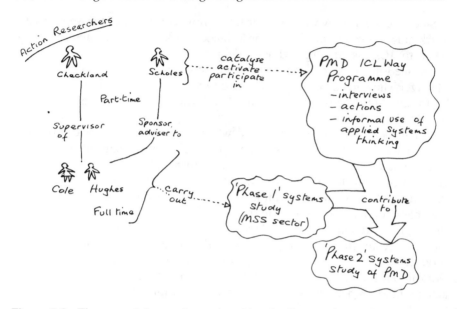

Figure 7.5 The agreed form of team working leading to the study of Product Marketing Division

created Management Support Systems Sector. The second phase would then be the PMD study, with the first study enabling the new members of the team to acquire knowledge of the Company and to learn to read its culture. Then sights could be raised to the greater challenge of a study covering the whole Division.

The network of roles in relation to the various pieces of work was now quite complex, so much so that it was worth making it explicit in order to achieve a shared view within the team of the overall task. This was done by means of the picture in Figure 7.5. Scholes would continue to be responsible for catalysing the ICL Way programme in PMD. This would feed, and be fed by, the two formal studies to which all the team would contribute in their different roles, as indicated in the figure. Cole and Hughes would be working full time in the Company for 20 weeks as part of their Master's course.

Context

The 'Management Support Systems Sector', recently created within PMD but not yet finally named, had largely subsumed what had previously been Networked Product Line Sector. It took over the work of Division Support Systems and Applications Development Productivity, for example.

Since the MSS study was a matter for a group consisting of the appropriate Marketing Manager and the Sector Manager, rather than a formal responsibility to ICL as an organization, detailed terms of reference were not established at the start. The activity was sponsored by the DSS Marketing Manager who had had previous exposure to SSM both in the study of the NPL Sector and in the work done by himself and Scholes on 'What DSS could do' (Chapter 6). He was later to become a manager in MSS Sector, but that eventual name had not been settled when this study began. He provided sponsorship for the study, providing practical support and advice to the postgraduates, and arranging introductions to people they needed to interview. In the first discussions with the team it was agreed that in the limited time available—envisaged as a few weeks—the aim would be to get as far as possible in defining a role for the sector in PMD, provide a possible plan for the implementation of an ICL definition of DSS within the Company, and identify a sector role in defining and exploiting business opportunities for such products. The agreed tentative approach to achieve this and integrate the work into a subsequent study of PMD is shown in Figure 7.6, a rudimentary activity model which the study would make operational.

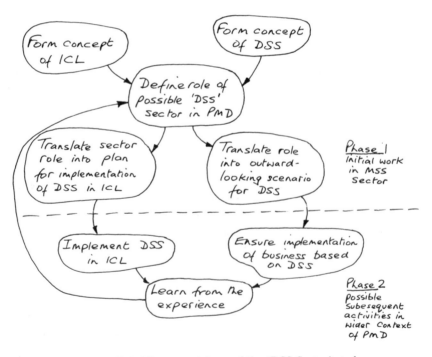

Figure 7.6 The agreed form of the 'DSS Sector' study

This was good enough to get the work underway but it was quickly apparent that too great an emphasis on DSS could unduly limit the value of the work. The Sector Manager saw a wider role for his group than had been realized at the outset: at one stage it was to be called 'Information Engineering Sector' with an emphasis on developing new business opportunities of which DSS was only one. This new understanding led the study team to a revision of the approach to be taken on the lines of Figure 7.7, which is again a rough model of this intervention. (It is expressed, for no special reason, in what engineers would call 'signal flow' form—that is to say, a set of entities with activities on the linking arrows. In the more usual form of models in SSM, the elements are activities with arrows showing contingencies. On the whole this more austere form has been found to be the most useful method of expression for the 'logical machines' which are conceptual models; but the signal-flow form could be used.)

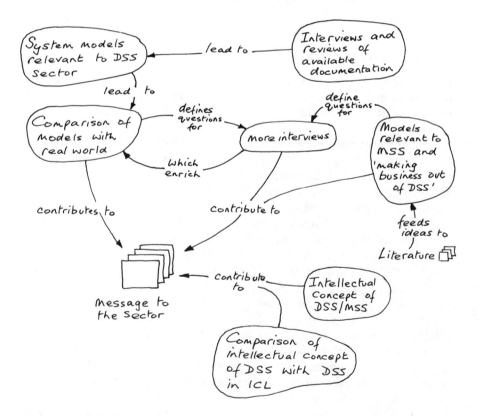

Figure 7.7 The study team's idea of the approach to be taken in the Management Support Systems Sector study

This considerable emphasis on the approach to be followed in the Phase 1 study was due to the desire to fit it into the wider context of the study of PMD as a whole. Methodologically, what has happened here is that the structural analysis of the intervention (the 'Analysis One' of Chapter 2) has initially taken the would-be problem solvers to be in the role 'problem owner', and has modelled—admittedly rather crudely, in Figures 7.6 and 7.7—the system immediately relevant to them, namely *the system to carry out the study*. This is nearly always a good strategy in a formal study as long as it does not become too prolonged or become an end in itself. It provides coherence for team activity and enables a check to be kept on what can be attempted with the resources and time available.

Appreciating the Problem Situation

From a dozen interviews and the reading of many Company documents the postgraduates learned their way into the problem situation. Some important themes emerged from this picture building, and may be summarized as follows.

At Company level a decade had been spent mainly in reactive 'fire fighting' with frequent managerial and organizational changes; many companies had been drawn into the ICL structure, and for many people their cultural roots still lay in their original company; ICL itself was still a 'technical-push' not a 'market-pull' culture; it was more 'personality driven' than 'role driven'; there was a need to link a charismatic MD's leadership to the need for such things as organized strategic planning; corporate functions were relatively weak.

At Division level, communications with other Divisions were not good, so that the PMD 'project management' role was hard to execute; the complexity of the matrix management (Figures 6.1, 6.2, 6.3 in the previous chapter) meant that information flows and necessary feedbacks were not sufficiently coherent; and the Division's image needed to be improved: it needed to be established that a professional marketer was not a 'failed salesman'!

At sector level, planning and development were rather separate cultures, with an emphasis on product-oriented thinking leading to a desire for an unstructured organization in the belief that this would afford flexibility in responding to market needs; DSS were seen as mixes of hardware, software, consultancy and training, as products which would lead ICL away from physical products towards 'end-user solutions'.

Choosing Relevant Systems and Building Models

It was by this stage in the study that it was realized that the emphasis should be on the sector rather than on DSS as one product within it. Bearing this in mind, a number of root definitions were developed covering: organization and structures; new business opportunities; and the concept that the Division should be a 'showcase' for ICL products. Sample root definitions and models from this phase of the work are shown in Figures 7.8 and 7.9. Figure 7.8 shows an issue-based model, one which a Division like PMD would need to operate no matter what its particular structure was. Figure 7.9, also issue-based, shows what would have to be done formally to create a continuously updated 'showcase' for Company products.

Comparing Models with Reality

A comparison between the various models and the team's knowledge of the real situation revealed very little mapping between the two: the models, though agreed by people in the situation to be relevant, were relevant to aspirations rather than to current activities. This emphasized the observation made in the initial finding-out phase that for several years the Company activity had contained much *ad hoc* 'fire fighting'. Gaps between models and current practices were identified and used in additional interviews with people in the sector and elsewhere to analyse the implications of doing or not doing various activities.

From these further interviews, which did not in this instance lead to further modelling, two major themes could be summarized.

Firstly, the existing 'Mission/Objectives/Strategy' (M/O/S) statement for the sector was both at a high level of abstraction and yet very detailed. On the other hand, activities within the sector at the operational level depended very much upon individual initiatives. What was lacking was a structural/procedural framework together with criteria and measures of performance which would ensure a linking between the aggregate of sector activities and the 'M/O/S' statement. Secondly, there was no clear shared concept of what would constitute sector products, such as DSS.

Outcomes

A report was prepared at this stage and discussed by the team and the study's sponsor (the DSS Marketing Manager). It led to a full staff meeting of what was by now officially named the MSS Sector, an opening presentation at the meeting being made by the 'outsiders' Checkland, Cole and Hughes.

The message of the presentation was briefly as follows: the organization, activities and procedures of the sector have been examined, as has the nature

Root Definition

> A system owned by a Division head and operated by Division personnel which seeks to ensure that the need to avoid internal destructive overlaps and conflicts is met, within the existing structures

C   PMD

A   PMD personnel

T   Need to avoid ... etc. ⟶ Need met

W   Internal conflicts can waste energy; it is worth while and possible to avoid/alleviate them

O   PMD head

E   Existing PMD structures

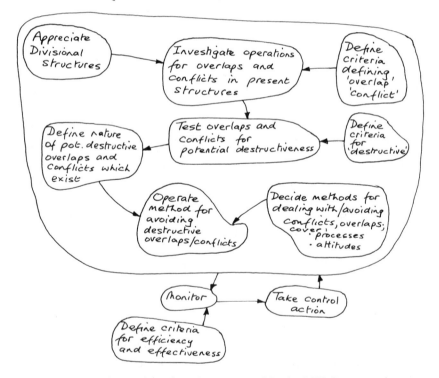

Figure 7.8   An issue-based system used in the MSS Sector study

Root Definition

> A Sector-head-owned system operated within the Company and staffed by Sector personnel, which recognizes a perceived need for the Company, where appropriate and feasible, to do internally those things which it advises its external clients to do, and ensures that the need is met, for the purpose of enhancing the Company's credibility in terms of the advice it gives.

C   The Company

A   Sector personnel

T   Need for the Co. to do what   ⟶ Need met in a way
    it urges its clients to do   which enhances Co.
                                               credibility

W   Credibility depends on doing what clients are urged to do

O   Sector head

E   Existing Company structures and business type

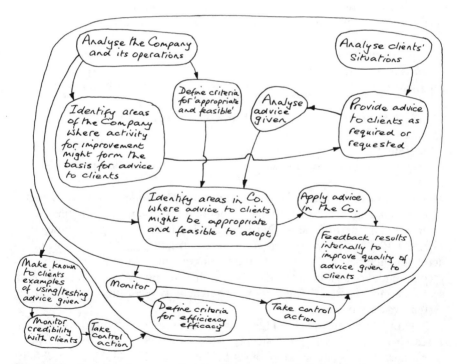

Figure 7.9   A system to ensure that the Company is a showcase for its own products

of such products as DSS. The conclusions reached are: that some positive 'engineering' is required to link operational activities to the Mission statement; and that there is need for a research programme to develop such products as DSS and to ensure that they are targeted to new business.

A good discussion developed at the staff meeting. The basic message of the presentation was accepted, and the approach displayed was thought sufficiently promising to justify further work to take this project further. It was therefore decided with the agreement of the head of MSS Sector that a second phase of this study should be carried out by one of the postgraduates (Hughes) while the other (Cole) moved on to the PMD study.

Hughes's further work entailed trying to fill the observed gap between the sector's Mission/Objectives/Strategies statement and its operational activities. To help in this, root definitions and conceptual models were developed from the M/O/S statement. Separately, interviews with sector members sought a consensus primary-task root definition and model relevant to the sector as a whole. However, two problems emerged as this work was pursued. The models developed from the M/O/S statement revealed some internal inconsistencies between the M, the O and the S: each element in it did not lead to, or derive clearly from, the others. Secondly, a consensus model of the new sector proved very elusive, so different were the conceptions of it by its members.

In the event it proved more useful to bring out a series of themes from the consensus-seeking interviews and feed them into the comparison stage based on the M/O/S models. From this comparison three relevant systems emerged:

- a system to identify new business opportunities and ensure their exploitation;
- a system to manage projects;
- a system to produce a message concerning MSS of corporate significance.

Models from these fuelled further interviews and comparisons with sector activity, leading eventually to recommendations discussed with the Sector Manager. This part of the further work in MSS is shown diagrammatically in Figure 7.10. It made its contribution to the structuring of the sector and its activities then being worked out, and within two months of its completion its recommendations had made their mark; job descriptions were created for all sector members and a simple computerized activity recording and monitoring system was established.

In parallel with this part of Hughes's work, a second stream focused on the research needed for a firm conceptual basis for DSS; this work was not a direct part of the systems study but it helped to build up Hughes's credibility in the sector. It led to a report to the Sector Manager which contributed to the formulation of the sector's research programme.

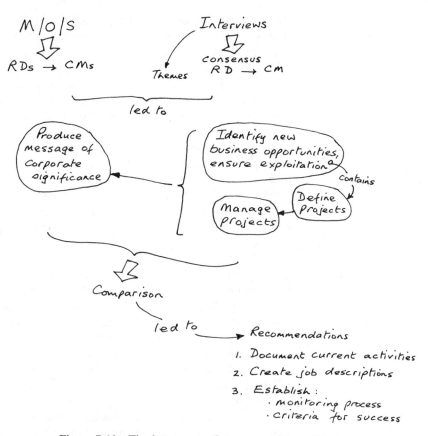

Figure 7.10 The latter part of the work in the MSS Sector

Up to the presentation to the staff meeting of the MSS Sector, the work described above was a traditional application of SSM, that is to say it used the approach as a means of doing the study. The presentation itself constituted a Stage 6 discussion of desirable and feasible change and this followed a first careful pass through Stages 1 to 5 of the methodology. Noticeable during this period was the considerable amount of 'rich picture' construction felt necessary by Cole and Hughes as, using SSM for the first time in a real situation, they strove to appreciate a very complex situation. (Not surprisingly, their pictures did not coincide, underlining Vickers's point that our previous experience gives us in-built readinesses to notice or not to notice certain features of a complex situation as significant, and to judge them by criteria developed experientially (Vickers, 1965).) Also noticeable, and perhaps of some general significance, was that the first choices of relevant systems could be seen as somewhat tortuous notions of what the analysts felt *ought* to exist if sense were to be made of a situation of great complexity. People living the situation would probably make simpler choices relevant to current issues.

Of special interest to Scholes, working again with postgraduates who had learned SSM in a classroom, having himself been exposed to it as a participant in the CCTA problem situation (Chapter 5), was the observation that initially the thinking of the apprentices was essentially SSM oriented rather than situation orientated. Everything that was done started from a reference to the methodology rather than the problem situation. It was noticeable that by the end of the extended work which Hughes did in MSS he had developed a close working relationship with members of the sector, was in fact being treated as a colleague, and was using SSM more as an internalized set of guidelines which helped the attack on a complex of problems. Perhaps we are saying here no more than that the apprentice has to go through a process in which a craft skill is absorbed and internalized before it can be truly exercised. The schoolboy batsman learning the craft of cricket thinks consciously about getting his left foot to the pitch of a good length ball, keeping his left elbow up and swinging his bat through a vertical arc. Only when he has stopped thinking consciously about these things, and has converted them into what Polanyi calls 'tacit knowledge', can he begin to be a real batsman (Polanyi, 1958, 1966). As with physical craft skill, so with an intellectual craft skill like using SSM. If the requisite balance between the action and the research is to be achieved, then the methodology itself had better be internalized as tacit knowledge. For Checkland the present study provided useful learning in the contrast between the way in which SSM was used by Scholes and by the (high quality) post-graduate students who were using it for the first time in the heat of a real problem situation. This suggested that the mode in which SSM was used was an important and neglected dimension of it.

A SYSTEMS STUDY OF PMD

As recounted earlier in this chapter, the systems study of Product Marketing Division (PMD), of which Management Support Systems Sector was a part, was one of several modules of work agreed to as part of the ICL Way programmes (see Figure 7.5). It was carried out by Elaine Cole, working on it full time as part of her Master's degree course, and by the authors who worked on it with Elaine, part time.

The Director of PMD, in agreeing that the study should be carried out, saw it as potentially contributing to improving the effectiveness of the Division in a number of ways: helping gain commitment from people in PMD to the process of change; contributing ideas on improved organization and procedures; and providing some corporate learning through its concentration on rethinking PMD and its role. It was very much in the minds of the project team throughout this piece of work that its aim was to make proposals to the Division Director and so contribute to his thinking about the future PMD and how to manage it.

As described earlier, this study was perceived as a classic application of SSM within the broader sequence of work in ICL which is described in Chapters 6, 7 and 8. Thus SSM was used to structure the study, and the account here proceeds through the stages in sequence. This is not a post-rationalization, it is how the work was done. This is of course a natural approach for someone using SSM in real life for the first time—which was Elaine Cole's position—but in this case it helped to link together this study and the parallel work in Management Support Systems Sector, previously described, as well as providing a contrast to Scholes's use of SSM in the earlier work in ICL.

Defining the Field of Study, Deciding the 'Givens'

Although this was a traditional systems study using SSM, on the pattern of the CCTA work, there were here no formal terms of reference or steering committee. The team's sense of the expected shape and climate of the study came from initial discussions with the Director and senior managers in PMD. The Director himself, without defining prohibitions, was clearly not seeking a study of the possible *structure* of PMD. The issue from his point of view was the need to make the tasks and processes of the Division more effective and efficient. Some of the senior managers in the Division did feel a need for a structural reorganization. All parties agreed that big changes in PMD could not be made in isolation because they would affect the Company as a whole through the network of transactions in which PMD was involved.

The underlying reason for the Director's reluctance to consider structural proposals became apparent only at the end of the study. It was not another example of a senior executive's reluctance to have others pontificate on organization structure, thus usurping a crucial commodity of power which directors like to retain in their own hands. In this case, as emerged later, the concern stemmed from the fact that the PMD Director knew that at Board level significant reorganization in the Company was currently being planned. Meanwhile he saw potential value in an independent study of activities and processes. (The authors' eventual involvement in the wider structural changes followed the PMD study and is the subject of Chapter 8.)

Building a Picture of the Problem Situation, Naming Relevant Systems

This study built on work done in the studies in Networked Product Line Sector, the ICL Way programme, and, in parallel, Management Support Systems Sector. (The ICL Way programme had itself provided nearly 80 interviews with 50 people in positions ranging from secretaries to the Division Director in PMD.) This meant that many previous interviews and discussions provided information and insights concerning the problem situation. Analysis of this material suggested that the main problem themes were:

- the role clarity, image and direction of the Division;
- ICL's innovative and world-wide marketing capabilities;
- the approach to personnel and management practice (e.g. appraisal and reporting arrangements);
- organizational issues concerned with matters such as matrix management, procedures and business information;
- the need for improved financial planning and control.

At a project meeting many potentially relevant 'relevant systems' were discussed. They related to such topics as: pulling together development and sales activity; mechanisms and procedures for decision making; defining new business opportunities; marketing concepts; PMD image; managing the PMD Director's office; and ICL profitability. Eleven root definitions were formulated and modelled. Half a dozen of these root definitions are listed below, and Figures 7.11 and 7.12 show two of the models produced.

(1) A PMD-owned and operated system which identifies technical developments and market needs and evaluates them so as to define new business opportunities and ensure that they can be profitably exploited.

(2) A system owned by a marketing division director within a company, operated by divisional personnel, which garners and processes ideas for new business opportunities from any source, in accordance with a perceived role that the division should be proactive as well as reactive.

(3) A system owned within the marketing division of the company, operated by division personnel, which seeks to improve the awareness and use of marketing concepts throughout the company (for the purpose of improving understanding within the company of its requirement to be market-needs-led).

(4) A PMD-owned and operated system which ensures PMD's role is appreciated by people in other divisions in a way which improves the image of PMD.

(5) A PMD-owned system which appreciates the capabilities and aspirations of people in the division and makes changes to improve their morale.

(6) A PMD-Director-owned system, manned by ICL professionals, which supports the PMD Director's role by analysing issues and relating them to a definition of PMD's role so that strategies can be presented to the PMD Director.

It may or may not be significant that during this phase of the work the more verbose root definitions were always those produced by the postgraduates who were, of course, using SSM for the first time. This may well stem from the fact that they tended to think of themselves as 'problem solvers' rather than as articulators of a process of enquiry from which purposeful action thought to be 'obviously' sensible would emerge. If you have an underlying assumption that you are 'problem solving', you are much more likely to formulate root definitions of something which 'ought to exist', since, if it existed in real life it would constitute a 'solution' to the 'problem'.

This illustrates how difficult it is in the professional executive culture of developed countries in the 1980s to get out of the habit of thinking in terms of unitary definable 'problems' which can be eliminated by 'solutions'. The language of 'problems' and

Root Definition

> A PMD-Director-owned system, staffed by ICL professionals which supports the PMD Director's role by analysing issues and relating them to a definition of PMD's role so that possible strategies can be presented to the PMD Director.

C   PMD Director

A   ICL professionals, PMD Director

T   Operation of PMD → Enhanced operation of
    Director's role   PMD Director's role

W   Staff-function support of a Director's role is worth while

O   PMD Director

E   Present structures in ICL

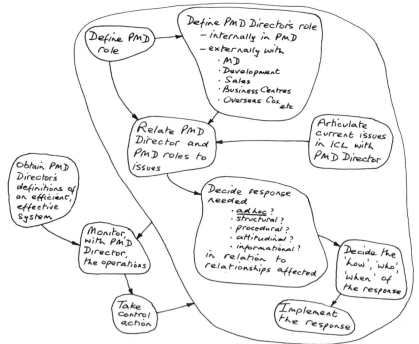

Figure 7.11 A root definition and model used in the systems study of Product Marketing Division

Root Definition

A PMD-owned and operated system which identifies technological developments and market needs and evaluates them in order to define new business opportunities and ensure that they can be profitably exploited.

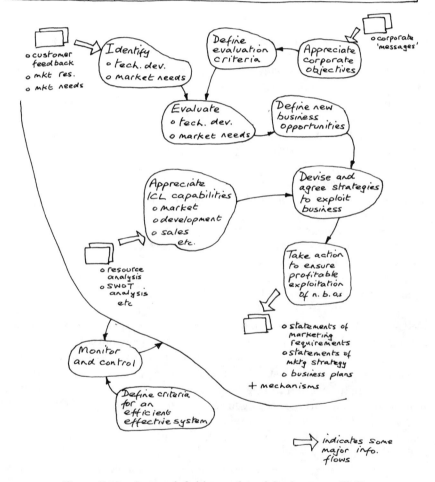

Figure 7.12 A root definition and model relevant to PMD

'solutions' which eliminate them has a powerful hold over us in this culture, even though the poverty of that language in the face of real-world multi-faceted complexity is rather blatant.

Technically within SSM, however, there is nothing wrong with root definitions of this type. As always what is important is that those formulating them should be doing so consciously, aware of their nature and pitfalls, and mixing them with other choices.

Comparing Models with Reality, Defining Feasible and Desirable Change

Since more than 100 interviews had been conducted over several months in carrying out the previous studies in ICL, the comparison between the models relevant to PMD and the real-world PMD started from the records of those interviews. From the interview notes a list of main recurring themes was built up. Each theme was annotated with the numbers of the models containing activities related to the themes (eleven models being available, of which Figures 7.11 and 7.12 show two). For example, an interview comment concerning communications would be related to models concerned with PMD links. Then, for each pairing of theme and model, the question was asked: If this conceptual model were a description of the world, how would it impinge upon the concern expressed in the theme? This comparison, carried out on paper, then guided selection of new relevant systems and led finally to the discussion which the authors had with the PMD Director.

The first overall message from the on-paper comparison was that the balance between, on the one hand, entrepreneurial flair and *ad hoc* initiative-taking and, on the other, guidelines, structures and laid-down processes was in ICL felt to be tipped very much in favour of the former. Every complex organization which wants to survive and prosper in a market has to find an accommodation between incompatibles: making sure the organization is both 'loose' enough to encourage risk-taking initiatives by its personnel and at the same time 'tight' enough through defined structures and processes to ensure that the different parts of the organization can act together as a single entity. The general feeling in ICL was that the Company was good at encouraging initiatives and independence, but could do with a few more guidelines and defined processes.

It is worth remarking on the significance of this particular comparison phase. It was important to the learning which these projects in ICL provided in two different ways. Firstly, it was of practical import-ance. It rendered *usable* all of the material from the

100-odd interviews, and made it possible to mount an argument for change which was not simply an intuitive reaction to the conversations held; it was an argument which could be explicitly retraced at any time with links to supporting evidence. Secondly, it was important methodologically. It is now possible to see (writing with hindsight) that it was a critical point in our learning from these experiences. Consider the question of what was happening, methodologically, during this comparison. It is possible to view it in two very different ways. It could, in keeping with 'classical' applications of SSM, be seen as a setting of models against perceived reality in order to ask whether the differences between models and real world were regarded as significant: would it be a good thing if the world were more, or less, like these models? Probably Elaine Cole, using SSM for the first time in real life outside the classroom, looked at it in this way. But it can also be perceived simply as a *making sense* of the interview material by examining it through the epistemology provided by the models. The authors, following the experiences of the earlier ICL studies, may have inclined more to this interpretation—it is impossible now to recall since the point was not regarded as significant at the time and was not recorded. But this comparison now seems significant in providing some of that crucial experience which led to recognition of the different modes in which SSM can be used.

Out of the comparison stage, which highlighted a relative dearth of explicit processes for decision taking, the concept emerged of treating perceived business opportunities as potential *projects* which should pull together technical and marketing skills. Such a concept might lead to ideas for change which would be feasible in an environment in which initiative-taking was prized but in which almost all those involved wanted greater clarity of purpose and less *ad hoc* responses to opportunities. This concept was developed in the way shown in Figure 7.13. A project was taken to be an operation in which a product X was delivered via a sales mode M to meet a need in a market Y, in time T, using resources R to yield a return P. The three systems of Figure 7.13 would define, manage and corporately appraise such projects. The projects could be managed at sector level by internal 'entrepreneurs', with the defining and appraising systems being manifest at Division level in a 'core' PMD of central functions. Each of the three systems was modelled, Figure 7.14

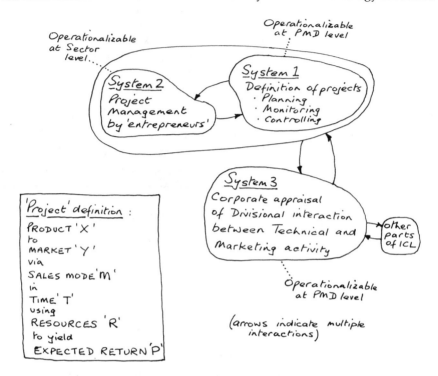

Figure 7.13 A concept of PMD based on 'projects', developed during the systems study of the Division

showing the model for the system which defines potentially profitable projects.

The concept and its models, and their contrast with existing arrangements, were used to mount an argument for a form of organizing and managing PMD which was based on projects as defined above. A presentation of this work was made to the Director of PMD, and included an exposition of the concept together with a number of suggestions for how such a concept might be realized in ICL. For example, a working group was suggested which would contain the expertise of project managers, PMD market research people and technical innovators. Such a group would define the elements and criteria for particular projects, and work under the auspices of a high-level forum combining senior managers from PMD and Technical Directorate, this forum being a possible manifestation of system 3 in Figure 7.13. A very constructive discussion with the PMD Director then developed. Although he expressed concern about some of the detail, he went along with the relevance of the concept as long as it could be realized in a way which enabled recognition of business opportunities by managers at lower levels of the Company (in what

Root Definition

A system owned and staffed within ICL which in the knowledge of ICL capabilities and market/technological opportunities defines potentially profitable projects to be carried out, and monitors and controls them, managing any arising conflicts.

C ICL

A ICL personnel

T Need for projects ⟶ need met
   to exploit opportunities

W Projects embody entrepreneurial
   enterprise and can be set
   up and run in ICL

O ICL

E ICL corporate structure; project definition

Project definition:
PRODUCT X
MARKET Y
SALES MODE M
TIME T
RESOURCES R
RETURN P

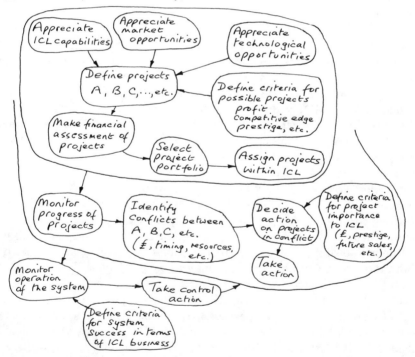

Figure 7.14 Root definition, CATWOE and model of a system to establish a managed portfolio of projects

he thought of as the Company's 'creative soup') to be captured and nurtured. He agreed with the notion of a 'core' PMD concerned with such things as overall marketing strategy and commercial policy, with projects run by entrepreneurs, but saw the suggested form of implementation as 'too loose'; structurally-institutionalized arrangements would be more acceptable! This came as a surprise to the authors, since the previous impression given was that the study should steer clear of institutional/structural change. It was at this point that some of the content of discussions which had been going on at Board level was revealed. The Managing Director and a small group of senior directors (including the Director of PMD) had for some time been developing plans for structural change in PMD, with a core PMD dealing with planning market strategy, etc., while product and industry marketing was devolved to 'Business Centres' which combined marketing and technical expertise. The work done in the systems study was usefully complementary to this thinking, and the PMD Director asked for more work to flesh out the concept of a core PMD dealing with strategic issues.

Outcomes

The study, which had started as a systems study of PMD without specific terms of reference, now became focused on helping the organization bring about a major structural change. The new work to enrich the 'core PMD' concept started with a then recently produced Mission/Objective/Strategies statement for PMD, a document which summarized the Division's main objectives and had itself been produced as part of the ICL Way programme (see the first section of this chapter). Systems to achieve PMD objectives were modelled and the models were examined to ascertain how, if they were made operational, the objectives would be achieved.

A report to the Director of PMD was written which led to the main outcome of the PMD systems study, namely a contribution to the Company's thinking concerning its prime market-oriented activities, and a role for the authors in a major new Company initiative: the commissioning of organizational units known as Business Centres. A Company team was established to think through and bring about the major change. Exactly when the PMD study should be regarded as finished and the work as part of the Business Centre Organization Team (BCOT) got underway is a moot point. The BCOT work is the subject of Chapter 8, and we have chosen to describe there in more detail the work done on the core corporate functions which would be needed in ICL to enable Business Centres to operate. That work was, indifferently, the culmination of the PMD study and/or the first work on establishing Business Centres.

Methodologically it would be very easy to describe the whole of the systems study of PMD in the language of cycling round the stages of SSM. To the authors at this stage of this sequence of studies it is probably more accurate to say that SSM had been so internalized that the way they did the work 'naturally' corresponded to the SSM cycle, even though they did not consciously and explicitly generate the expected outcome of each stage and then proceed to the next. When methodology is internalized, and becomes invisible, then, ironically, it is truly being used.

Chapter 8

The Use of Soft Systems Methodology in the Establishment of a 'Business Centre' Organization

Introduction

The previous chapter described the authors' involvement in several studies in ICL which were part of the work on the ICL Way programme in Product Marketing Division (PMD). Those studies themselves grew out of the earlier work on the Networked Product Line Sector of PMD and the work on Decision Support Systems, both described in Chapter 6. It is largely true that each study grew naturally out of the one which preceded it, and we are very aware of the extent to which it is arbitrary to describe one study as ending and a new one beginning. Real life is messier than these descriptions imply! This was especially the case in the work to be described in this chapter, which deals with the authors' participation in the establishment within ICL of an organization consisting of a set of Business Centres each of which brought together marketing and technical development expertise in order to develop and exploit particular products and markets.

Chapter 7 described a systems study of PMD which led to a report to the Division Director. Discussion of that report led directly to a request by the Director for more work on the 'core' Headquarters functions required if PMD were to fit properly into the Company as a whole. That work itself led eventually to the setting up of a Company team, in which the authors participated, to define a new Business Centre organization. This chapter describes both the work on the 'core' Headquarters functions, in Section 1, and the work of the Business Centre team in Section 2, which also includes accounts of the three different ways SSM was used in this work.

Before proceeding it is probably useful, in the interests of clarity, to provide a reminder of the sequence of studies of which the work on the new Business Centre organization was in a sense the culmination. This is done in

| Chapter 5 | 1. Review of the role of Departmental Liaison Officers in CCTA | May–Dec., Year 1 |

Chapter 6

| | 2. Study of Networked Product Line in ICL Product Marketing Division (PMD) | Nov., Year 2–Jan., Year 3 |
| | 3. What Decision Support Systems could do. | May, Year 3 |

Chapter 7

	4. The 'ICL Way' Programme in PMD	May–Oct., Year 3
	5. Study of Management Support Systems Sector	May–Oct., Year 3
	6. Systems Study of PMD	May–Oct., Year 3

| Chapter 8 | 7. The BCOT study | Nov, Year 3–April, Year 4 |

Figure 8.1 The seven studies on which the authors worked together

Figure 8.1, which records the sequence of studies which yield Chapters 5 to 8. We shall return to this sequence of studies in Chapter 10, when the learning from all of this complicated action research is discussed.

SECTION 1: 'CORE' HEADQUARTERS FUNCTIONS

Context

During the study of PMD undertaken by Elaine Cole and the authors, a concept was developed of PMD as a set of entrepreneurial projects together with a set of 'core' central functions which would service the projects and ensure that they were linked to each other (if appropriate) and to ICL activity as a whole. (This concept has been illustrated in Figures 7.13 and 7.14.)

Discussion of this with the PMD Director led to his asking for further work on the idea of a core PMD dealing with strategic issues, as mentioned in the previous chapter. This work on a core PMD, to be described in this section, is either SSM's Stage 7 of the PMD study or the initiation of a new study; we have chosen to describe it here.

> This difficulty in separating the PMD systems study from the work on core Headquarters functions, or, indeed from the subsequent Business Centre work (or from the whole sequence of work in ICL) illustrates the arbitrary nature of end points in systems studies. Taking action at Stage 7 presents new issues, for example: What action? How should it be taken? What new 'problems' does this create? In an environment in which the explicit use of SSM is accepted, this leads to further iterations round the methodological cycle. From the point of view of *action* in the real world this iteration is not problematical. However, from the point of view of research—capturing lessons from the experiences—it *is* problematical. For the purpose of reflection and providing a comprehensible account, the researcher/reporter has to decide when to stop and provide apparently static snapshots of a situation, while in real life the interacting flux of events and ideas continues to unfold. The stopping points are in the end arbitrary, and this is a permanent problem for action researchers in real situations.

Although the authors did not realize it at the time, one reason for the PMD Director's interest in the report of the PMD study was that a small group of ICL's senior directors were meeting at that time to consider the Company's infrastructure in relation to perceived external and internal pressures. They were discussing bringing together marketing units (then in PMD) and product development units (then in various development divisions) into a number of 'Business Centres' which would combine product and market responsibilities. Such a structure would clearly entail a core PMD to manage the Centres from a Company point of view. Hence the Director's interest in the similar thinking developed independently in the PMD study. He asked that further work be done to raise the thinking to Company rather than Divisional level, this work providing him with briefing for the Director-level discussions of the Business Centre idea.

Using SSM to Define 'Core' HQ Functions

The requirement was that the further systems thinking be carried out in a fortnight, alongside other work. This meant that the work had to be a thinking exercise, using the existing material from all previous work in ICL as an information base.

The methodological approach adopted is shown in Figure 8.2; again it is not chance that this maps the stages of SSM! It is a particularized version of it.

Perusal of the available interview material, the stated Mission, Objectives and Strategies for both ICL as a whole and PMD, the comments made by the PMD Director, together with the real-world knowledge gained from all the ICL work, led to a picture of the kind of company ICL aspired to be: a market-oriented company in the computer/information field which in the light of explicit business policies, exploits opportunities in the form of business projects, ensuring that resources and processes are adequate for this purpose.

With this concept as a kind of root definition, a model of necessary major subsystems was built, as shown in Figure 8.3. Then, in order to explore issues related to the notion of Business Centres, more detailed models of the subsystems were built. The process adopted was to use each of the activities in Figure 8.3 as a crude root definition, build a more detailed activity model from this direct, adding logically necessary activities as required, then work

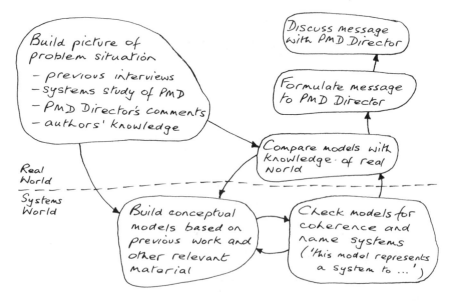

Figure 8.2 The methodological approach to the work on core HQ functions

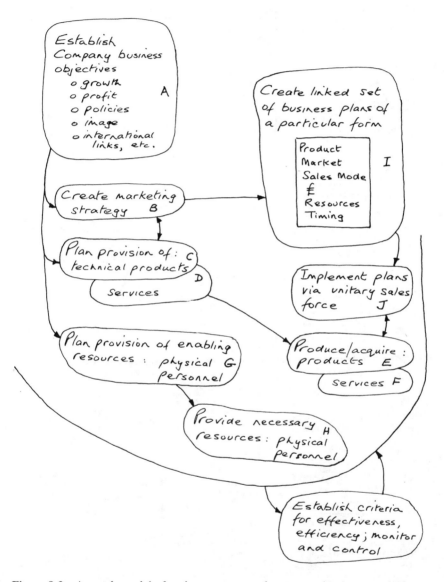

Figure 8.3 A rough model of major necessary subsystems relevant to exploring core HQ functions, based on earlier work in the PMD study

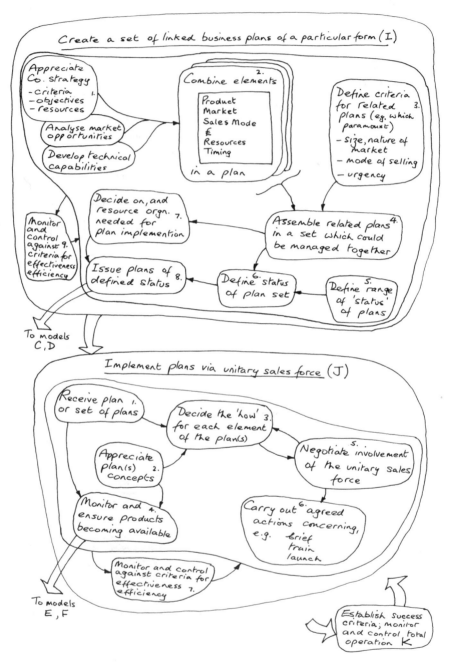

Figure 8.4 Models I and J of Figure 8.3 expanded

back to fuller root definitions by asking: What root definition would yield a model like this? As examples, Figure 8.4 shows models of subsystems I and J of Figure 8.3, their (iteratively-arrived-at) root definitions being:

System I—A Company-owned system which, within a market-oriented concept, creates a linked set of business plans of defined status (e.g.:cash cow/speculative, etc.) from information on: market analysis; technology capabilities analysis; resources; strengths and weaknesses; and Company strategy, a plan being of a particular form.

C   Company
A   planners
T   information → structured information in plan form
W   flexible rational planning is feasible and worthwhile
O   Company
E   concept of Company in main model; business centres with a unitary sales force, form of a plan.

System J—A Company-owned system which, within a market-oriented company, takes a plan or set of plans and carries out/institutes action to implement it/them via the Company's unitary sales force.

C   Company
A   Managers
T   plan(s) → implemented plan(s)
W   market-oriented plans can be implemented
O   Company
E   concept of Company in main model; unitary sales force.

It was felt that the resolution level achieved by expanding the subsystems of Figure 8.3 was enough to provide a framework for examining the Company-level activity of setting up and ensuring the proper managing of Business Centres.

> The iterative process used here was a way of quickly exploring the implications of the Business Centre concept. There was a conscious decision to use SSM as an overall framework for this part of the work (Figure 8.2) but the main thought in the authors' minds was not 'We are using SSM' but 'We are exploring the Business Centre concept using the form of SSM as a convenient guarantee of coherence'. This was not a systems study in the full sense. Only one theme was taken as the notional Stage 2, namely that in line with the PMD Director's closely defined requirement for this particular work to be done within two weeks.

Comparison between the more detailed models and what currently existed in ICL highlighted some activities which did not exist and a few which existed but were weak in some way. In general a number of activities which did exist were weak in the absence of a group function concerned with business objectives and policies. And many activities concerned with success criteria and monitoring/control were lacking at both corporate and individual business levels.

At the more detailed level of subsystem B of Figure 8.3 (provision of marketing strategy) the purposeful model which was built provided an image for *the operation* of a Business Centre. This entailed such things as deciding how to deal with each element in a plan, negotiating arrangements with the sales force, ensuring that the flow of required products was achievable, etc. Here many of the crucial activities did not exist in the real world and/or would need to be defined for future Business Centres. All this suggested an important lesson, namely that a concept of *the operation of a Business Centre* was likely to be much more important than *the definition of possible groupings* although the latter had been the issue of most concern to the group of directors discussing this issue. This is not at all surprising since the definition of possible Centres will have sharp implications for dispositions of managerial power, against which considerations of management procedures will seem rather abstract. Nevertheless, lack of attention to operational processes, especially at Company level, could cause the whole initiative to fail regardless of the structural definition of what Centres to create.

These themes emerging from the comparison were made the basis of the message from this work presented to the Director of PMD. In summary, this argued firstly for initial attention to be paid to *Company-level* processes, if Business Centres, when defined, were to be successful. Secondly, attention was drawn to the need for procedural change if Centres (however defined) were to be well managed. Finally, the need for attitudinal change was argued to be a vital wide-ranging issue. (The attitude of the sales force towards Business Centres, for example, would be a critical determinant of the success of the new structure.) Insights into desirable attitudinal change could in principle be derived from the now-available process models by asking: What real-world attitudes, on the part of which individuals and groups would be necessary for the modelled processes to work well? A programme to encourage those attitudes could then be designed.

Specifically it was suggested to the PMD Director that the minimum action necessary to plan and implement worthwhile change would be to put in place the missing *corporate* functions, and thus establish the necessary supporting procedures for Business Centres. A possible way to do this would be to set up one or two small top-level committees with terms of reference built round the missing functions, and provide the staff resource needed to service those committees with information (such as possible management processes for Business Centres and an appraisal of lessons learned from operating one or more 'pilot' Centres).

Outcomes

Two weeks after the work just described had been discussed with its sponsor, the Director of PMD, a meeting of the Company's directors decided that a number of Business Centres would be established forthwith, and an announcement was made to that effect, without going into organizational details. Business Centres would be grouped into Business Divisions. It was announced at the same time that the PMD Director would take up a new post as Business Director of Office Systems Division, and a new Director of PMD was appointed.

As he moved to his new role, the former PMD Director felt that much work remained to be done to make the Business Centre organization successful. He suggested that a presentation should be made to the Managing Director based on the systems study of PMD and the subsequent work on core functions at Company level.

A presentation was prepared by the authors and given by Scholes a few weeks later. The main conclusions from the systems study of PMD (Chapter 7) were outlined. That study did not provide a *design* as such, but could contribute to the change process in a way which would encourage initiative. If one accepted that there was an imbalance in ICL between initiative, entrepreneurial spirit and *ad hoc* action on one side, and guidelines, structures and processes on the other, then provision of a clearer framework (but *not* the imposition of a detailed design) could enhance initiative taking and release the energy of individuals more fruitfully. Such a framework would be built around the core thinking from the systems study of PMD, namely that exploiting business opportunities could be seen as defining, mounting and monitoring business 'projects' which linked technical and marketing skills. In the work already done, that idea had been pursued to define what would be needed at corporate, PMD and Business Centre level if the new organization were to be successful. It was suggested, for example, that at corporate level several activities needed strengthening: definition (and timely redefinition) of group business strategy and policies; effort to ensure that business plans linked coherently and were adequately resourced; positive management of conflicts between business areas; definition of success criteria for different levels of activity and control action based on them.

A possible approach to the management and operation of Business Centres was outlined, based on the crude model of Figure 8.5, and a number of incidental issues which had been identified during the authors' work were

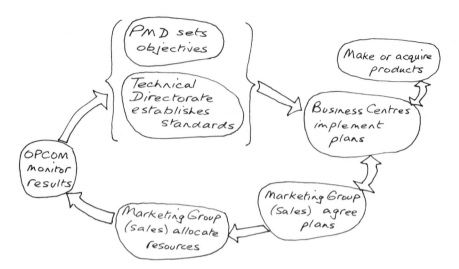

Figure 8.5 A crude model of the Business Centre context used for presentational purposes

raised: the criteria for setting up (and closing down) Business Centres; and the linking mechanisms through which Centres would interact with Sales, manufacturing, and other Business Centres.

The overall message of the presentation was that the declaration of the new structure needed to be backed and supported by a specific programme of work to make the concept into a working reality. The idea of Business Centres was not sufficiently strong or well defined that a reorganization based on it could be left to take its own course.

Following this presentation to the Managing Director it was decided at Board level that a Company team would be set up to take this thinking further. The team would be required to report to the Directors' Operating Committee (OPCOM) with proposals for action required to strengthen group functions and establish Business Centre procedures. Sponsored by the new Director of PMD, it comprised the Manager of Organization Development and the Manager of Manpower Resourcing, both from Group Personnel, the Financial Controller and Scholes from PMD, with Checkland acting as adviser to the team. The work of this team is described in the next section.

Action research is far from being laboratory experimentation, and it is impossible to know to what extent the PMD systems study and the subsequent work contributed to the actual changes which took place in ICL. A practical criterion for the authors was that the 'end' of each piece of work generated requests for follow-up work. Their *feeling* about the work, which cannot of course be proved right or wrong, is that the systems work caught a tide of changing thinking in ICL at that time. The systems-based work was perceived as relevant, and was thus able to contribute to the current, because it helped to clarify and structure the ideas which were emerging, and seemed to do so in a holistic rather than a one-dimensional way. For the authors, SSM was increasingly perceived not only as 'a way to do studies' but as 'a way to think about complexity'.

SECTION 2: THE WORK OF THE BUSINESS CENTRE ORGANIZATION TEAM (BCOT)

Context

The work of the Business Centre Organization Team (BCOT) began at a meeting at which the team and the sponsor (the new PMD Director) agreed terms of reference. These stated that the team would identify the organization required to support Business Centres, determine the procedures necessary for the effective management and operation of them, and make recommendations on organization and procedures to OPCOM (the Directors' Operating Committee) within four months. Scholes would work full time on the BCOT study, the other members part time. It was agreed that the team would build upon the PMD systems study and the subsequent work on core Headquarters activities described above.

A Company announcement made the existence of BCOT and its terms of reference official, and sought the cooperation of all of the Company's Divisions in the exercise.

The situation at this point presented the authors with an interesting dilemma. Obviously it would be perfectly possible to carry out the BCOT study using SSM. Given the prior work, this would probably be an economical thing to do. On the other hand, why should the team members be expected to immerse themselves in a particular approach with which they were largely unfamiliar? Would they in fact have much patience with time spent discussing the approach to be taken rather than getting down to grappling with the content of what was seen to be an important development for ICL? Busy managers

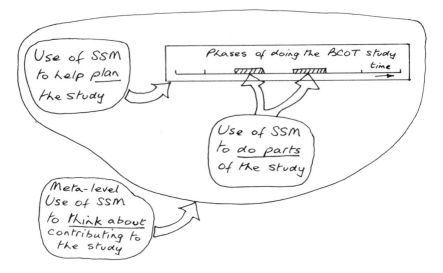

Figure 8.6   Three uses of SSM in the BCOT study

often regard anything other than dealing with the substantive content of problems as a waste of time— they are wrong, but the feeling is common. Also relevant to this dilemma was the fact that the learning from all the ICL work so far had weaned the authors away from thinking of SSM only in terms of 'a methodology for carrying out special studies'.

It was decided between the authors not to try to persuade BCOT to accept SSM (or anything else) as the study methodology; rather we would use it to help us carry out our part of the work, and we would try to learn from the experience.

In the event, SSM was used in three different ways during the few months of the BCOT study: to help *plan* the project, to *do* parts of it, and *to structure the authors' thinking* about it, as illustrated in Figure 8.6. These three uses of SSM will be described later in this chapter. First we give an outline of the BCOT work in order that the subsequent discussion of methodology will make sense.

The Work of the Team (BCOT)

At its first meeting BCOT agreed on the need for a project plan which Scholes should produce. The authors used SSM to help do this, and a plan was accepted at the next meeting and then implemented over the period of the study. It consisted of several phases:

(1) *Information collection and familiarization.* Individual team members carried out interviews with senior people inside and outside ICL, and collected information from Company documents. External work was commissioned to find out how other comparable companies faced the issues which Business Centres were to address, this work including a survey of the then-burgeoning literature on 'strategic business units' (an idea developed by McKinsey and Company in its work with GE in the 1960s) and 'portfolio planning'. (For examples of this literature, see Hall, 1978; Kruel and Glenney, 1983; Bettis and Hall, 1983; Gluck, 1986.)

All interview notes, documents and relevant articles were copied around the team, which met weekly to take stock of progress.

(2) *Analysis/model building/synthesis.* This took the form of a three-day workshop two months after the start of the work. The team developed core concepts and themes relating to corporate support and control of Business Centres, and the criteria and processes needed to set up and manage them.

(3) *Testing out the concepts.* Individual interviews with senior managers and directors tested the ideas from the workshop. The team also ran two half-day workshops for the recently-named Business Centre Managers at which they contributed to further development of the team's thinking.

(4) *Agreeing proposed changes.* At a two-day workshop the overall content of the team's report to OPCOM was agreed.

(5) *Reporting to OPCOM.* The report was drafted and circulated to members of OPCOM 3½ months after the start of the study. OPCOM discussed the report with two team members present, including Scholes. The report's recommendations were accepted and it was agreed to implement them. This led to another phase of BCOT activity being added to the original plan, namely implementation.

(6) *Implementation.* Some actions were assigned to BCOT, some to individuals and groups. Overall the team was expected to monitor the progress of implementation and keep OPCOM informed. This phase of activity continued for a year from the start of the study.

About 40 people were interviewed during the four months of the work leading to the first OPCOM report. Overall the interviews suggested that senior managers had no shared view of what Business Centres could or should be, and there were conflicting views on some important aspects of the concept. The following quotations from interviews illustrate some of these concerns:

> The main motivation behind Business Centres was to break down unwieldy development units.
> The key issue for BCs is to be customer-led and remain close to the customer.
> Autonomy? An old chestnut; they're not.
> BCs should be highly autonomous.
> Profit accountability is with the sales force, it won't be in BCs.
> BCs will create incredible management problems at the interfaces.

Enough diversity of view emerged to convince BCOT that their work was very necessary!

It was agreed in the team that its report to OPCOM needed to address a number of concerns: clarification of the background thinking leading to the Business Centre 'solution'; the role of a Centre and hence the processes for managing and judging its performance; future implications of Business Centre organization for ICL.

Before, and at, the three-day workshop the authors developed models relevant to each of these headings (e.g. a system to define and set up a Centre) and these were used to explore the interdependence of Centres and their links with corporate functions. The recommendations of the report itself, all accepted by OPCOM, covered:

- suggested arrangements to increase employee understanding of the Business Centres;
- the need to provide a corporate infrastructure;
- proposals to clarify the Centre role and improve the management of Centres;
- suggestions to ensure effective linking of Business Centres and functional units;
- arrangements to help the transition to Business Centres;
- proposals for appropriate management training and development.

It was the work which flowed from acceptance of these proposals which kept BCOT in being for a year in all.

We now turn to the ways in which SSM was used in the work just summarized.

Use of SSM in Planning the Work of BCOT

BCOT began its work by identifying the issues it was to tackle and the constraints which had to be taken as given. It was accepted that the Managing Director sought a transition to a Business Centre organization as soon as possible, that the Centres should not compete with each other (which implied a clearer product/market segment definition than existed at that time) and that as a matter of policy ICL products would be sold through the unitary sales force acting on behalf of the Company as a whole. It had also been determined that Centres would have to develop products which met pre-defined technical standards (ensuring that compatible products formed a 'Networked Product Line'). Finally there was a need to link with other initiatives, such as an exercise underway to define financial reporting arrangements for Business Centres.

BCOT asked Scholes to define a plan for its work, recognizing that team members would be available only part time for this work. The team expected some kind of detailed schedule, with tasks set against a calendar.

Producing a study plan for the situation described above was treated as a 'problem situation', in SSM language, and a number of relevant systems were named as follows:

(1) A system to service the Directors' Group overseeing the project.
(2) A system to help strengthen Group functions and establish procedures for Business Centres.
(3) A system to establish the basis for building and rebuilding necessary structures and procedures with regard to Business Centres.
(4) A system to use the existing systems study of PMD to help examine Company-level issues of current concern.
(5) A system to bring together all current organization development activities.

Root definitions were formulated and models built for each of these notional systems, Figure 8.7 illustrating the definition, CATWOE analysis and model for system 1.

Such models could not of course be compared against real-world manifestations, there being none in this 'green field'. Rather, the models were a source of coherent questions for BCOT to address. These were put to the team in a draft paper to its second meeting, in lieu of the expected plan. At that meeting much progress was made. For example, how the activities of Figure 8.7 would be done, when, and by whom were all agreed. Out of that meeting came agreement that the study activities would consist of: servicing OPCOM; bringing together existing material and activities; examining Company-level organization issues; establishing what needed to be done to strengthen corporate functions and set up procedures; establishing what ongoing (institutionalized) arrangements were needed for building and rebuilding structures and procedures as the operation of the 'new' ICL evolved.

It was accepted that these topics constituted the project plan, and that weekly meetings would monitor progress under these headings, modifying

Root Definition

> A Directors-Group-owned system manned by a staff support team which appreciates the needs of the Directors' Group and services it with the necessary information, suggestions and procedures

C Directors' Group

A BCOT

T DG needs : info.   DG needs met
       Suggestions $\longrightarrow$
       procedures

W There is need for staff support of DG at a time of reorganization

O Directors' Group

E Involvement of BCOT team

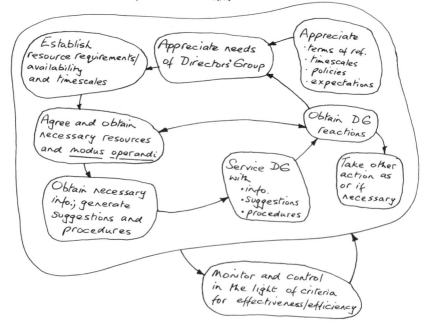

Figure 8.7 A relevant system used in the work to plan the BCOT study.

them if necessary. The agreed specific activities, individual responsibilities and timings were collated by Scholes and structured into an evolving plan. In this way the team remained harmoniously leaderless, and the study remained problem oriented rather than calendar oriented, though timing was never neglected given the amount to be done with mainly part-time effort.

This use of SSM was stimulated by the realization that the production of a project plan of the kind expected was itself problematical. It is easy to map what was done on to the stages of the methodology: models were defined, built and brought back to the problem situation. The discussion at the second BCOT meeting constituted Stages 5 and 6, with the agreement on how to conduct the study, together with the initial assignment of tasks to individuals, constituting a definition of 'desirable and feasible' activities—with 'activities' here replacing 'changes', in a green field.

However, although this is how the thinking was in fact structured (with the definitions formulated and modelled in less than a day) the methodology itself was not exposed to the team. They focused on the content to which they could all relate and to which they could contribute.

This experience convinced the authors that although SSM was not being used to *do* the BCOT study, it could nevertheless make a contribution.

Use of SSM in Doing Parts of the Work

Although not the methodology for the study as a whole, there were several parts of the project in which the use of SSM made a specific contribution, for example: in examining interface issues between Business Centres and Sales; in considering the management of organizational change in ICL; and in the example described below, which derives from the three-day workshop which formed the *analysis/model building/synthesis* phase of the study.

At a meeting with another team member, namely the Manager of Organization Development, the outline for the team's three-day workshop shown in Figure 8.8 was drawn up. This was circulated to the team members and agreed by them to provide an acceptable starting point.

At the workshop the team first discussed all the available relevant information, from documents, interviews, and the commissioned work on the experience in other companies with Strategic Business Units (SBUs). It was quickly obvious that the entities now beginning to be so casually referred to as 'ICL

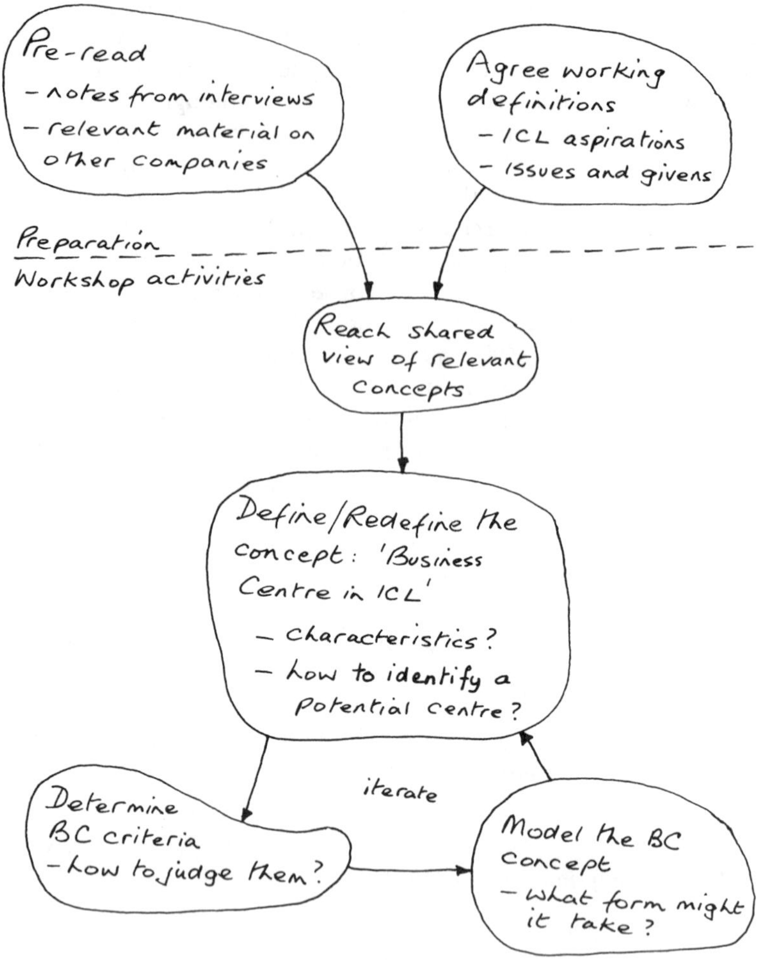

Figure 8.8 An outline of the BCOT three-day workshop programme

Business Centres' could not be easily or uniformly described. They did not have the same typical characteristics as SBUs elsewhere, and there were significant differences from one Centre to another—depending, for example, on whether they were concerned with vertical markets such as retailing or were primarily based on products, such as mainframes. Perceptions of the new concept were very different among ICL senior managers, too.

The authors saw this as an ill-defined 'problem situation', and, with the team's agreement, took it through a process of enquiry aimed at achieving conceptual clarity and agreed actions. The process used was SSM but this was

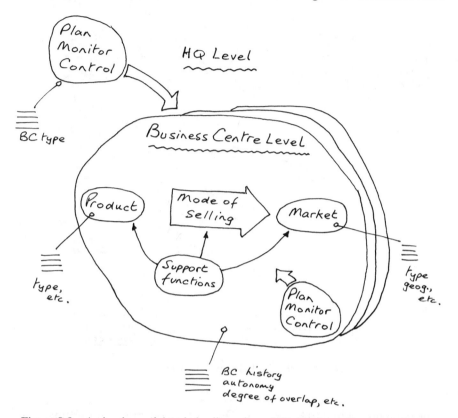

Figure 8.9 A simple model to help discussion of the Business Centre concept

not mentioned unless the team itself wanted at any point to shift discussion from problem content to the process being followed.

The starting point was a felt need for a way of describing Business Centres in simple communicable terms which would allow the variety of the real world to be expressed. The model in Figure 8.9 (deriving from the earlier systems study of PMD) was offered as one which allowed Centres to be described and differentiated in terms of characteristics such as product, market, degree of autonomy, etc. Setting that model in a wider context produced Figure 8.10, which accepted several ICL givens. Discussion of these two models yielded an agreed list of problem themes to be explored further:

- BC planning, development, marketing, and management;
- Corporate business strategy;
- Setting up BCs;
- Managing organizational change.

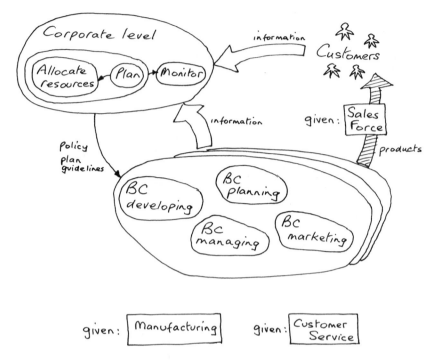

Figure 8.10 An ICL Business Centre in context

The team now split into two groups. One worked on criteria for setting up and judging Business Centres. The other, consisting of Checkland, Scholes and one other team member, continued to develop the themes outlined above in a series of conceptual models. In principle this activity was equivalent to Stages 3 and 4 of SSM although no formal root definitions and CATWOE analyses were produced. Figures 8.11 and 8.12 show two of the models produced.

These models were discussed by the full team when it reconvened. Members examined to what extent the prototype Business Centres gradually being established concurrently with the BCOT study, met, or were likely to meet, the logic of the models. During this comparison the work of the other subgroup, on criteria, was fed into the discussion. Team members contributed from their own considerable experience and knowledge of the real-world situation, and developed a concept of the likely linkages between Business Centres and other parts of ICL and an account of the transactions across those links. Also during this discussion matrices based on the idea of product life cycles were developed as a means of characterizing particular centres by product, market, mode of selling and the combination of these factors.

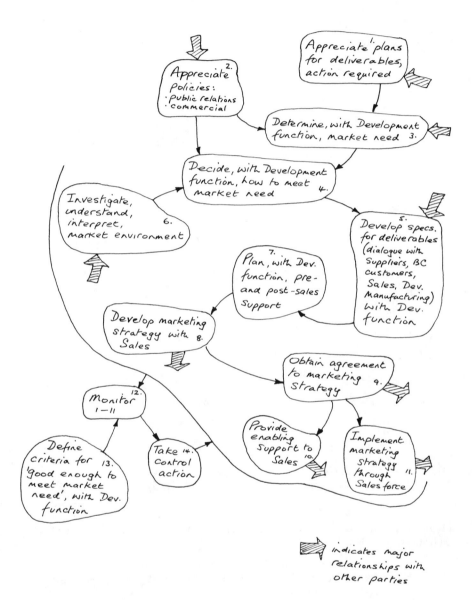

Figure 8.11 Business Centre marketing—an informal model (no RD, CATWOE) built during the BCOT workshop

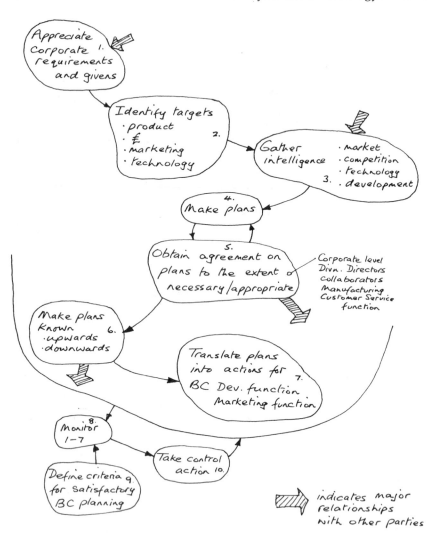

Figure 8.12 Business Centre planning—an informal model (no RD, CATWOE) built during the BCOT workshop

From this work emerged a description of ICL's Business Centres in terms of:

- development and marketing;
- responsibility for specific deliverables (products, services);
- definition of market segment;
- definition of competition;
- measurability in financial terms;

- relationships with other parts of the Company;
- operation within imposed financial and planning constraints (ie degree of autonomy).

This work enabled the team to identify an agenda for discussion with its sponsor, the new Director of PMD. That discussion, which took place at the end of the workshop (and incidentally represented Stage 6 of this use of SSM), addressed a thesis which the team had built collaboratively during their three days together. Its essence was as follows.

The label 'ICL Business Centre' describes different kinds of entity: a new venture group, a product development unit, an 'independent' or 'strategic' business unit, or a specialist marketing/sales group. Constraints on Centres are such that most, as envisaged, are really product development or product introduction units. The relationship between Centres in adding value to each other's products needed to be better defined. Finally, since Business Centres would interact with their markets via the Company's unitary sales force, and were to an extent interdependent, a significant level of corporate control through well-defined procedures was required.

These points were discussed at length with the team's sponsor, who expressed his judgement of the political acceptability of some of the messages emerging. It was agreed that a further round of interviews and discussions with senior managers and directors was needed, and that the workshop had usefully defined a menu for that phase.

Thus did the workshop, using SSM in veiled form, define the agenda for the phase *Testing out the concepts* described earlier, this being one of three uses of SSM to carry out parts of the BCOT project.

It is to be hoped that the reader, having got this far, can see clearly this 'undeclared' use of SSM at the BCOT first workshop. A messy problem situation was addressed, relevant systems of purposeful activity were named, activity models were built, though the discussion of all this was in the language of the situation, not the language of SSM. It is perhaps worth remarking on the informality of this particular use of the systems methodology. The team struggled for some time to find a way of expressing the problem situation (SSM Stage 2). This was in the end done verbally and then captured in the crude models of Figures 8.9 and 8.10, which we can dignify as basic conceptual models (SSM Stage 4), though there were never *explicit* root definitions. On the other hand these models may themselves be regarded as at least surrogate root definitions for the models built subsequently, such as those illustrated in Figures 8.11 and 8.12. These were then explicitly used in a comparison (Stage 5), leading to the Stage 6 discussion with the BCOT sponsor. Stage 7 stemmed from the decision to discuss specific issues with particular directors and senior managers, and, as always, can be seen as constituting Stages 1 and 2 of the next study.

Use of SSM in Structuring the Thinking

During the sequence of systems studies described in Chapters 6, 7 and 8 the authors gradually became increasingly aware that SSM does not have to be thought of exclusively as a way of doing special projects, although it is perfectly serviceable in that cause. They became aware that as a result of having absorbed SSM to the extent that it had become tacit knowledge, they were using it flexibly, at several different levels, and on many different timescales, from an hour or two (Decisions Support Systems) to several months (the Systems Study of PMD).

Pushing this train of thought to extremes, it is clear that the least formal, least public, meta-level use of SSM would be to use it inside one's head as a taken-as-given thinking mode. The authors 'found themselves' using it in this way many times during the studies in ICL. But such uses are by defininition private, not public, and hence are not subject to any kind of scrutiny or testing by others. Such uses only become examinable when they at least produce some tangible outputs, and such a use of SSM will now be described. This is an example of the third use of SSM, as indicated at the left-hand side of Figure 8.6: SSM used not to do a project, not to do parts of a project, not to do the planning of a project, but used at a meta-level to help thinking about those subsequent tasks. This is a type of use which might in principle be entirely private and hence unexaminable by other people, as indicated above, but in the case described produced some examinable outputs.

This particular use of SSM came about after the reporting phase of the BCOT study, when the nature of the problem was changing to that of implementation. Progressing BCOT's recommendations involved Scholes and other team members, together with ICL managers from various functions, in a number of implementation projects. These covered such areas as: the creation of a corporate strategic planning process; the development of more effective interworking arrangements between Business Centres and geographically based sales operations; and the creation and establishment of processes to evaluate and implement proposals for new business opportunities. These latter processes then had to be carried out in relation to specific opportunities.

The roles and relationships within BCOT with regard to the implementation programme were far from clear. Following acceptance of the BCOT report by ICL's Operating Committee, Scholes was made 'Business Centre Development Manager' in a newly established Group Planning Directorate which encompassed the 'core' HQ functions identified in the PMD systems study and the subsequent work described earlier in this chapter. In that role Scholes was heavily involved in some implementations and retained a more general interest in the process of implementation as a continuing part-time member of BCOT.

Initial thinking suggested that the work concerning new business opportunities could be thought of as in Figure 8.13, which can itself be regarded as an expression of what for Scholes personally was 'a problem situation'. This picture suggested a notional system in which ICL as it currently existed would establish processes for dealing with new business opportunities and then operate those processes. This notional system yielded the model in Figure 8.14. Comparing this model with Scholes's knowledge of emerging roles in the new Group Planning Directorate, and with information from discussions with the Managers of Business Development and Business Planning, enabled the activities shown in Figure 8.14 to be assigned on a reasonably logical basis between these two managers and Scholes. This gave a coherent account of Scholes's role in his new appointment. He would be concerned with the activities of Figure 8.14, either doing them himself or ensuring that they were done. He would himself take responsibility for deriving and acting upon lessons drawn from those experiences, so that ICL's reaction to new business opportunities would improve over time—something ICL would look for from its 'Business Centre Development Manager'.

Although this cycle of SSM was rapid and largely mental it *is* such a cycle. The physical outputs consisting of Figures 8.13 and 8.14 represent, respectively, a combined 'problem situation' expressed and root definition, and a conceptual model. Discussion with the Managers of Business Development and Business Planning was a version of Stages 5 and 6. Stage 7 (taking action) was realized in a second cycle which will be described shortly.

Here, then is a very informal application of SSM applied not to a general problem situation which many acknowledge, but to a 'problem situation' which is itself one man's intellectual construct. The authors came to see, experientially, that such uses of SSM are both real and far removed from instances in which SSM is the adopted methodology for a highlighted study. Such uses are simultaneously unlike

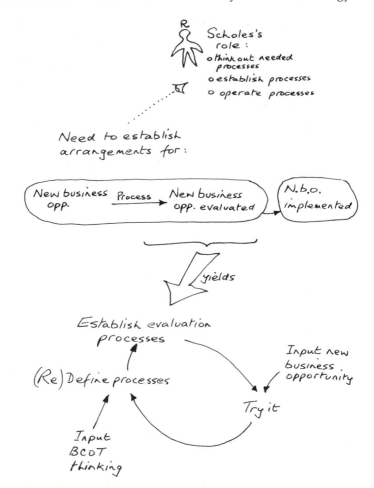

Figure 8.13 The problem situation with respect to Scholes's role in helping implement the BCOT proposals

formal systems studies and yet are recognizably SSM. That is to say, they are the enacting of a learning cycle based upon systemic constructs which are compared against a real situation in a debate about feasible and desirable action. A conventional full and formal use of SSM to do specially set up studies represents 'Mode 1' of methodology use; 'Mode 2' is the kind of informal use just described. The differences between Mode 1 and Mode 2 will be discussed in Chapter 10.

Root Definition

> A system to establish and operate processes
> for dealing with new business opportunities
> in ICL

C   JS in new role

A   JS or another manager

T   Need for establishment and operation
    of processes for dealing with n.b.os   ——→ need met

W   It is possible and useful to establish and operate
    processes for dealing with n.b.os in ICL which JS
    or other managers could operate.

O   JS

E   JS's new role, new structures in ICL

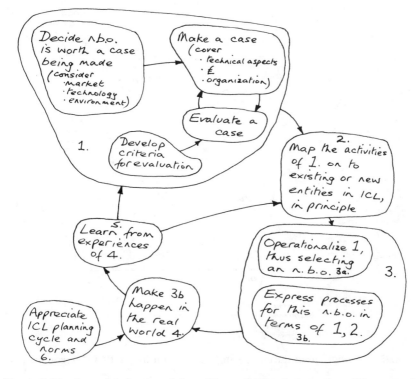

Figure 8.14 A model developed from Figure 8.13 to help explore JS's new role as Business Centre Development Manager

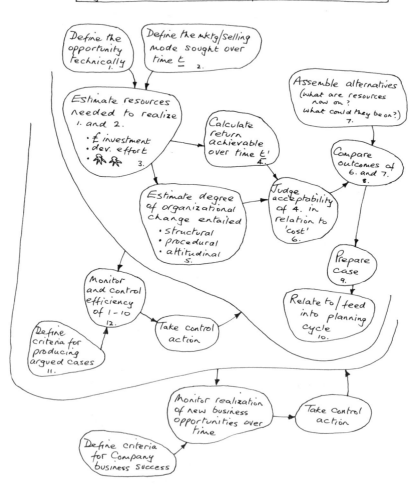

Root Definition

A central-unit-owned system which transforms a tech./mktg./£ possibility into an argued case for using resources X over time t to produce return Y over time t' in a manner achievable in terms of organization, which is a superior prospect to alternatives

Figure 8.15 A model used during the implementation of the BCOT recommendations

The end of this Mode 2 use of SSM to *think about* Scholes's new role as Business Development Manager merged immediately into a second cycle of the methodology closer to the more conventional Mode 1. The outline role for Scholes which came from the initial (Mode 2) cycle comprised Stages 1 and 2 of a new cycle in which potentially relevant systems were identified as follows:

- a system to transform an identified new business possibility into an argued business case;
- a system to ensure that new business opportunities are evaluated in terms of their business contribution, and implemented when appropriate;
- a problem-solver-owned system which takes identified new business opportunities and determines the organizational processes needed.

Figure 8.15 shows a model which was based upon the first of these.

An interesting Stage 5 comparison involved further discussion with the Director of Group Planning (previously Director of PMD) and the new Managing Director. He wanted to build ICL's organizational capability to organize resources quickly and effectively to implement business strategies, and had committed himself to a management development programme which would build a team approach to problem solving. The idea developed of linking this programme with the organizational change underway in ICL. Task forces would be created consisting of senior managers from different parts of the Company, and those teams would work on the various projects which followed from acceptance of the BCOT report. They would provide the vehicle for the development of the desired team approach. It was agreed with the Directors concerned that the task force approach would be adopted in the 'new business opportunity' projects, and two such projects were identified in discussion with the Director of Group Planning. Scholes was asked to oversee and link the total set of projects, including those concerned with new business opportunities.

These agreements constituted Stage 6 of this second cycle of SSM. The subsequent use of the methodology to plan the new business opportunity projects (among others), and to do parts of the implementation programme, constituted the action stage of this second cycle of SSM, as well as constituting further cycles of SSM application in what the authors were now thinking of as 'Mode 1'.

Conclusion

The examples of SSM in action which have been described in Chapters 6, 7 and 8 could be thought of in several ways. The language used in these three chapters has tended to treat them as *a sequence of studies*. But since each grew naturally out of those which had gone before, with Stage 7 of one study often constituting or setting up Stages 1 and 2 of the next one, it would be equally justifiable to declare something to the effect that 'SSM was used; and during the enactment of many cycles it helped to define and tackle a range of problems which arose in an organization undergoing considerable change'. In a way the choice of language is arbitrary. The important thing here is that the experience of this work in ICL expanded the authors' knowledge of SSM and in so doing modified their perceptions of it. The learning from this experience will be discussed in the final chapter, where the future directions of work with (and on) SSM will be discussed. But before that we describe briefly some contemporary current work in a major science-based company which represents SSM in action in the late 1980s. It would not have taken the form it has without the learning gained from the project (or project sequence—choose your own words) in ICL.

Chapter 9

Contemporary Soft Systems Methodology in Action: Rethinking a Service Function in the Shell Group

Introduction

Previous chapters have described SSM in action in a range of settings and in several modes, starting with that in which it was developed, namely as a methodology for carrying out a special study. Thanks to experiences like those of ICI and, especially, ICL, SSM came to be viewed more flexibly: not as methodology for a special highlighted study but simply as a way of doing managerial work, in principle at any level, with lower levels simply entailing more 'givens', more constraints.

Given the learning described in Chapters 6, 7 and 8, we may now usefully ask what a mature use of SSM looks like in the late 1980s. This chapter describes briefly some current work in the Shell Group. The work concerns a major rethinking of one of the Group's service functions, Manufacturing Function (MF), which employs several hundred people. In one sense this work is a special study rather than part of normal day-to-day work. But it is not a study carried out by a special team or task force. Rather, it is being 'orchestrated' by two Shell managers, with Checkland as an outside adviser, but is carried out by a large number of Shell managers. The work is done by all of the managers invited to take part in two-day workshops attended by twelve–fifteen of them at a time. SSM is not in any dogmatic way 'the methodology to do the study'. It is being used throughout the study, in fact, but in a late-1980s mode. It is being used not so much to plan calculative action (though that arises in the course of the work) but rather to guide and make sense of the discourse which the workshops stimulate and mediate. This is, perhaps, 'postmodern' SSM in the sense of postmodernism discussed by Cooper and Burrell (1988): SSM is being used to help 'an observer community which constructs interpretations of the world, these *interpretations* having no absolute or universal status'. New interpretations have led to new ways of structuring and managing MF, but not in any ultimate or Utopian sense. MF will no doubt have changed again by the year 2000.

Context

The Shell Group, though at core 'an oil company', is engaged in many businesses which entail obtaining raw materials from around the world and converting them into products which are sold in world markets. Its operations are fundamentally science based and, as is usual in such companies, it is essential that Shell's technology is continually being assessed and updated. It is the kind of industry in which research and technological development work now underway will, it is hoped, ensure the survival of the Group against sophisticated competitors in ten years time.

In order to help ensure this, Shell has a central staff function, Manufacturing Function (MF), which employs more than 500 people, including many engineers and scientific and technological experts. They carry out a variety of tasks: keeping up with a changing technology, and technological developments elsewhere; deciding what research and development programmes Shell needs to undertake; advising the Shell Board on its policies with regard to technology; monitoring the performance of production units; and providing instant help at the operational level to production plants around the world.

MF is the kind of central function which does not itself *directly* contribute to the Group's generation of wealth. On the other hand sophisticated companies like Shell accept that such support functions are essential for long-term survival. The permanent problem is to decide how to organize and measure the performance of such a function, since so much of its effort produces discernible results only over a period of years rather than months. It will be obvious in five and ten years time if MF is *currently* 'doing a good job', but that information may then come too late to influence events.

These thoughts were very much in the mind of Rob de Vos, head of MF, who was alert to the fact that it is necessary continually to think and rethink the possible roles for a function like MF, for which there is no simple, sharply defined 'bottom line' performance indicator. He made sure that this issue would not be forgotten or side-stepped.

It is interesting to note that at this level this is an issue which MF itself has to resolve. People who speak about those abstractions, organizations, as if they were people capable of thinking about problems and taking decisions, might imagine that 'Shell International Petroleum Company Limited' could lay down a role for MF and indicate how it will be judged. But there is no such human-like entity which is 'Shell'; the Group, like all such organizations, is an epiphenomenon. It is a concept which arises in relation to a complex of ideas and events involving a large number of people who are prepared to behave *as if* some quasi-human 'Shell Petroleum Company' actually existed. But you cannot have a conversation with an abstraction! If, in order to resolve this problem, we adopt the stance that the main Board of Shell is really the embodiment of the Shell Group, then we see at once that they cannot be expected to define tightly a role for MF and the measures by which its performance will be judged. That role and those measures will derive to a large extent from how the technology is developing, and, after all, MF are there to advise the Board on technological matters; it is their responsibility to know more about technology and its development than anyone else in the Group, including the Board. They are the experts! So there is no alternative but that MF should decide for itself how to define and play its role, and how to measure its performance, and then seek to convince the Board to which it answers that it is doing this competently. Of course, the argument just deployed could be used again to assert that MF does not exist as a quasi-human decision taker, either! But the argument grows weaker as we move to lower levels of ultimate responsibility. It would not be too unreasonable to regard MF as embodied in the Function head and his management team, since they *are* experts in the MF field. But for this assumption to be a reasonable one, it is essential that Rob de Vos and his team be engaged in a continuing discourse involving, in principle, all of MF and those with whom MF interacts.

Since becoming head of MF, de Vos had felt the need for a rethinking of the role and processes of the Function. This he discussed with Cees van der Heiden, now a senior planner in the Shell Head Office, with whom de Vos had worked earlier in their careers, and Jaap Leemhuis, now a planner in MF, who had also been with de Vos in previous jobs within Shell. Out of these discussions eventually came the idea of an 'MF reorientation', a significant rethink of MF and its role in Shell. Leemhuis would conduct the study within MF, and would work jointly with van der Heiden from the Head Office planning group. van der Heiden, though a 'Head Office man' (normally a designation with negative connotations in any large organization) was well known to, and, most important, was trusted by the senior managers in MF.

van der Heiden and Leemhuis conducted a finding-out phase which entailed 80 interviews with managers in MF, central functions and 'customers' inside and outside Shell with whom MF interacted. Findings were summarized and fed back to workshops of MF managers which helped to create a shared perception of the problems facing MF as it carried out a range of different interactions at different levels over different timescales with many different customer-clients.

van der Heiden and Leemhuis were here making use of a sophisticated approach to strategic planning which had been developed within Shell over several years, and to which they had contributed. It was an approach based upon the formal, orderly, iterative questioning of the 'vision' underlying any coherent part of Company activity which could be looked upon as a decision centre with a management team. As a vision statement emerged and was refined in discussion, the options for Shell to which it led would be questioned in terms of the many factors which could prevent the straightforward translation of vision into selected business option. Such factors as competitor activity, legislation and broader social changes would be taken into account, leading to definition and redefinition of vision and options until a plausible vision and a realizable option were agreed. A study of the efficacy of the approach had shown that there was a correlation between business success and a clear shared vision.

On the other hand, rich discussion at stimulating workshops which Shell managers are very ready to attend, does not itself necessarily lead to action being taken when busy people return to pressing immediate problems and full in-trays.

In the present case, with the nature of the MF 'vision' fundamentally in question, and the aspiration that the rethink should lead to action to relieve perceived problems, van der Heiden and Leemhuis sought an additional strategy. They asked for Checkland's collaboration in introducing SSM into the reorientation of MF.

Much thought went into deciding how to use the methodology. van der Heiden and Leemhuis did not want a study done by a small group which produced an *ex-cathedra* report. They wanted as many people as possible, including MF-ers, customers of MF, and other Shell professionals, to take part in an organized dialogue within which the MF 'vision' could be constructed and its necessary organization worked out. To the greatest extent possible, they wanted MF's 'vision' to emerge from the organization as a whole.

It was decided to run a series of two-day workshops to which groups of managers would be invited. Each workshop would have a theme to provide focus, but these would not be laid down at the start, only defined in the light of the learning which came out of the previous workshops. After workshops of this kind, there would then be several in which the participants would be Rob de Vos and his MF management team. The nature of the likely final outcome was unclear, but there was the clear aspiration that action would be taken to improve MF's effectiveness in the Shell Group.

During this sequence of workshops, SSM would be a structuring device, a way of making sense of the discourse while allowing it to be free ranging and open to many different contributions. SSM as such would not be expounded unless the workshop participants specifically asked that it should. The content of each workshop would be recorded in some detail and issued not only to participants but also to others who could thus keep in touch with the exercise. This would make 'public' not only the study itself but also the way in which SSM was giving coherence and direction to what might otherwise have been a rather chaotic 'conversation', so that that too could be challenged.

> The idea at this stage concerning methodology was to use SSM in a form closer to the 'Mode 2' described in the previous chapter, rather than in the traditional 'Mode 1', even though this was a special study of MF. The fact that the study would be, as it were, *extracted* from the work of a large number of participants at the workshops made this an appropriate aspiration. It also made it essential to pay careful attention to the process of using SSM in this study.

The Preliminary MF Mission Statement

Prior to the first workshop of the SSM-based study, and providing an intellectual base for it, we had all the material from the first stages of the traditional Shell strategic examination of MF which van der Heiden and Leemhuis had initiated earlier. Open-ended interviews and discussions with highly articulate groups of MF managers and others who interacted with MF in the course of their day-to-day work, had produced an abundance of cogent points concerning all aspects of MF, its role in Shell, how its customers perceived it and what could be done to improve it. These discussions had been captured in a series of flip-chart-like pages which collected headline-like points concerning different aspects of MF. Over 100 such pages provided a very rich picture indeed!

It is not possible easily to summarize such material, and in any case there was no intention of doing so definitively at the time. Rather, these discussions and their outcomes were a source of some broad problem themes, as well as being a repository of attitudes and ideas upon which the subsequent study could draw. As an indication of the nature of the material we may usefully indicate its scope. The collected points cover:

- role definitions for MF (11 pp.);
- the culture of MF (and Shell) (18 pp.);
- objectives of MF (14 pp.);
- MF's interfaces (12 pp.);
- technology development (10 pp.);
- service to 'customers' (10 pp.);
- project management (6 pp.);
- MF characteristics (8 pp.);
- human resource development (8 pp.);
- organization (9 pp.);
- MF reorientation (8 pp.);

The material revealed much recognition of the need for MF to reconcile irreconcilables. They must develop technology expeditiously in a very focused way, but at the same time provide a service to internal 'customers' from a very broad base. They need generalists with a broad understanding of business problems; but the value of having at the centre shrewd and experienced technological experts who have 'seen it all' was recognized—such professionals were referred to as 'the old foxes'. People at the centre need to stay in post for a considerable time to become truly expert; on the other hand people need to rotate if they are to remain in touch with real life on the refineries and other production units. Do what your customers want; but take initiatives, be proactive . . . and so on. A few quotations give the flavour of this material:

> The high degree of specialization leads to procedures and 'official positions'.
> The procedural culture leads to: 'What can I safely say?' rather than: 'What is the best solution?'
> A shift towards market pull is required.
> You cannot stop technology push in this business.
> Create *contracts* at the interface.
> Is MF a business or a cost centre?
> If you have a specific problem and know who to ask, you see MF at its best.
> The final outcome is always satisfactory, the problem is to get them moving.

Using this mass of material as a starting point, Cees van der Heiden and Jaap Leemhuis, working with Shell managers, had produced an MF technology vision statement which accepted the need for a strong technological position in a capital intensive industry, looked for increased emphasis on satisfying customer needs, not least through improvements in information technology, expressed the need for management education directed towards the middle-management-less plant, and urged a proactive MF to 'lead on the basis of a business logic'. Such statements, especially in large organizations, can always be dispatched to oblivion as being no more than flag waving in favour of motherhood and apple pie, but in this case there was a very solid compendium of Shell opinion underlying the work done so far; and this was now sufficient to make it appropriate to hold the first formal workshop of the MF reorientation study.

General Workshop 1

When an invited group of managers assembled in a hotel for a two-day workshop with the theme 'Technology Development', formal proceedings began with several hours of unstructured but intense discussion of MF and its problems. Even if Checkland, van der Heiden and Leemhuis had wanted it differently, they would have been unable to corral or direct this first morning of the workshop. Articulate people with strong views had come to the workshop (some of them from half-way round the world) in order to express some strongly held views: they were unstoppable! It took about three hours for the energy of this discussion to subside. By lunchtime it was becoming somewhat circular, with now-familiar stances being re-expressed, and a set of issues 'on the table'. At this point Leemhuis asked Checkland to make a comment on the morning, as an outsider. He suggested that the discussion—from which an outsider had learnt much—was rather like a bird's nest: having a shape but one difficult to describe clearly, and having many strands interwoven at many different levels. In order to build upon the morning discussion it was suggested that after lunch the participants should meet in small groups and should aim, in the light of the issues raised, to produce some flip charts containing possible plausible statements of MFs 'core purpose'.

By mid-afternoon a number of statements were available, typically half a dozen from a syndicate group, covering such primary tasks for MF as generating new technology-driven business opportunities, bringing added value to current production operations, developing R and D programmes and people in order to develop competitive positions, and servicing (in a problem-solving sense) manufacturing sites and operations. Checkland then demonstrated on a couple of examples how such 'core purpose' statements could be framed as statements sharp enough to enable us to derive from them the structure of activities needed to realize the purposes expressed. (In fact, of course, these were root definitions formulated using CATWOE, together with their conceptual models, but we were not thrusting the methodology at the participants, lest it get in the way of the substantive issues.) Having seen a couple of examples worked through, the syndicate groups then spent an hour or two model building.

This was not as difficult a step to accomplish as might be thought, not least because a number of these participants were engineers or scientists and so were familiar with the idea of models as intellectual constructs. The elements new to these professionals were two: using verbs as a modelling language and working from declared descriptions of purposeful 'holons'.

The decision not to describe SSM as such, and not to use its language, reflected the desire to remain both problem oriented and rooted in the world of the workshop participants. The technicalities of root definitions, CATWOE and model building were explained briefly (with special emphasis on the T of CATWOE) when a few curious participants enquired into the process being followed. But the fact that there *was* such a process was never thrust at participants, it was allowed to emerge. (It became a workshop joke among participants that Leemhuis was 'manipulating' them, but they were indulgent as long as what was happening was clearly relevant to issues they regarded as important.)

In this mode it was easy to convince people that looking at known real-world happenings in relation to the logic of the models provided both an interesting way of subjecting the familiar everyday world to scrutiny, and a way of revealing insights into possible change which might never emerge, or might not be *noticed* in unstructured discussion.

Technically the modelling could be described as 'not bad', although one group did spend a happy hour flow charting *existing procedures* for generating the annual Shell Research and Development Plan, rather than following the logic required by their root definition—only to find subsequently that such a model yielded only a comparison of X with X! Figure 9.1 illustrates three of the eight root definitions which became the basis of detailed comparisons.

The second morning of the workshop was spent comparing models with the real world of MF and its customers, using the comparisons to list a handful of 'issues highlighted'. The comparisons were carried out according to the format of Figure 9.2. The work was done in syndicate groups, and the outcomes provided the agenda for a final plenary discussion of the workshop.

The whole workshop was captured in an eighteen-page book which was subsequently sent to participants and others. It includes all the definitions,

Root Definition 1

> An MF-owned and staffed system which meets the need for a relevant technological base by managing and updating, in a cost-effective manner, the accumulating reservoir of information and the skills to enable MF to carry out its tasks

Root Definition 2

> An MF-owned and professionally-staffed system which, in the areas of Shell technology, analyses and evaluates experience from operations and from R and D results, and the experience of others, in order to identify development options at an early stage and to make proposals to Shell companies for safely maintaining and improving existing and potential processes and products to enhance their competitive position, within the resource limitations of the Group.

Root Definition 3

> A Shell-Group-owned system staffed by appropriate experts which ensures that Shell's technology (both 'doing' and 'enabling') is as good as any in the business in which Shell is, or could be, active, gives advice to the Group and companies within it, in order that Shell's technology base will be adequate for survival and enable Shell businesses to compete successfully (technically), and takes action to refurbish Shell's technological base.

Figure 9.1 Three root definitions developed and used at the first general workshop

Activity in model	Exists ?	How ?	Who ?	Good/Bad ?	Alternatives ?
Accumulate skill reservoir	Yes	Discussion and management action	MF mgmt. Shell companies Personnel	G	Contractor
Determine nature of action needed	Yes	MF/Shell Co. discussion. Various/y formal-informal	MF and Shell Co. people	G in general	No alternative
Decide Scope and depth of skill accumulation	No, not formally	—	—	B	Special exercise, task force, data base regularly updated

Figure 9.2 The format of comparison used at the first general workshop with a few examples

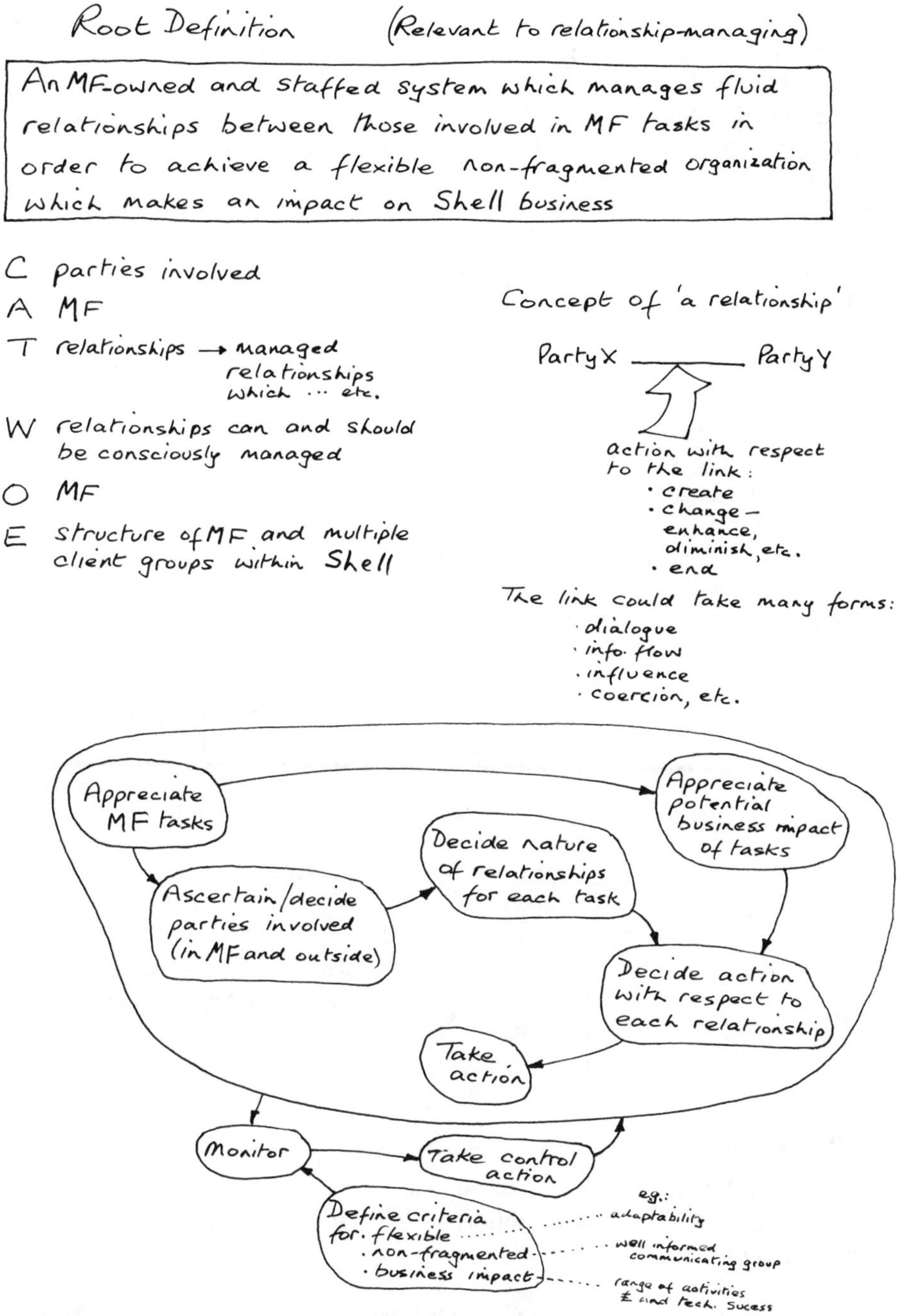
Root Definition (Relevant to relationship-managing)
An MF-owned and staffed system which manages fluid relationships between those involved in MF tasks in order to achieve a flexible non-fragmented organization which makes an impact on Shell business
C parties involved
A MF
T relationships → managed relationships which … etc.
W relationships can and should be consciously managed
O MF
E structure of MF and multiple client groups within Shell
Concept of 'a relationship'
Party X
Party Y
action with respect to the link:
· create
· change – enhance, diminish, etc.
· end
The link could take many forms:
· dialogue
· info. flow
· influence
· coercion, etc.
Appreciate MF tasks
Appreciate potential business impact of tasks
Decide nature of relationships for each task
Ascertain/decide parties involved (in MF and outside)
Decide action with respect to each relationship
Take action
Monitor
Take control action
Define criteria for · flexible · non-fragmented · business impact
eg.:
adaptability
well informed communicating group
range of activities £ and tech. Sucess

Figure 9.3 An issue-based root definition and model developed at the first workshop

models and comparisons generated, and tries to record the whole event. For example it records that one group, applying CATWOE to each activity in a model (something they themselves decided to do), found that all elements remained the same except 'C' (the activity's 'customer'—i.e. victim or beneficiary) which varied considerably. This led them to the need for a further system 'to balance and prioritize the requirements of MF's clients': this group had discovered for themselves the possibility of issue-based, as well as primary task root definitions! This led to discussion of this point and a couple of issue-based definitions were modelled by Checkland, one being shown in Figure 9.3.

The workshop report also includes letters from participants written to Leemhuis after they had had time to reflect upon the experience, and a final summary of 'Recommendations from the Workshop to MF Management'. From the 'technology development' perspective of this workshop the three main MF tasks were identified as: maintaining and updating a reservoir of 'know-how'; developing relevant R and D objectives and programmes; and creating business options on the basis of new or improved technology. The implications of each of these were discussed in some detail. Although this was a confidential document within Shell it was a 'public' record of the event, made available on what civil servants call a 'need to know' basis.

> The workshop seemed to the three 'orchestrators' of it to have remained satisfactorily problem oriented. SSM as such had remained unobtrusive to the participants but had provided Checkland, van der Heiden and Leemhuis with the feeling that they were working to a structure. Thus the resource of these managers' time could be used efficiently to steer towards the final outcome, namely the workshop document as a contribution to the overall reorientation of MF.
>
> On the basis of the learning from the first workshop it was decided to hold another one a few weeks later, this time with a focus on 'service provision' instead of 'technology development'.

General Workshop 2

The discussion at the first workshop had repeatedly hinged on the many different interactions, at many different levels, in which MF had to engage. In the words of the report from that event:

> The whole of MF seems a complicated balancing act of different activities. This includes balancing responses to demands, prioritizing and balancing provision of

services versus development work, and developing and maintaining fluid relations with a wide body of customers. ('Fluid' means continuously adapting to the
requirements of the moment.)

The first workshop had illustrated that the longer term development of
technology frequently interacted with MF's provision of service, and com-
peted for MF's resources. Answering telexes requesting help was a strong
feature of MF life, and answering them quickly and competently was a source
of esteem for MF-ers. From the point of view of an operating unit, the most
obvious and public manifestation of MF lay in its help with immediate
technical problems, rather than in its formulation of a manufacturing strategy
for the next decade, and judgements of MF were mainly in terms of MF's
ability to provide an instant useful service to operating units in the Shell
Group. It was decided to explore the provision of service in a second
workshop similar to the first.

During this period, as a facilitator of the study and its process, Checkland
had learned much about MF's operations, its culture and that of Shell
generally. He had in fact been doing Analyses Two and Three (described in
Chapter 2) asking himself what roles, norms and values were strong in MF and
Shell, and trying to elucidate the commodities of power in the Shell culture.
The evidence suggested that above all MF was an 'engineering culture', one in
which technical proficiency is prized and technical perfection is an ideal. A
powerful image for many of the MF professionals was that they were the
creators and guardians of necessary skill pools within the Group, the level of
the skills being such that they were as good as any in the world. A number of
remarks had been made about what, from the stance of a production plant,
looked like 'gold plating' by MF: 'The MF strategy seems to be to provide a
"Rolls Royce" service at a "Rolls Royce" price'. All this confirmed the value
of looking at MF's activity with the theme of service provision—whether that
service is to a production plant or to the Main Board of Shell.

The second general workshop, set up to do this, took the same form as the
first, again with about a dozen managers invited, some from MF, some from
Shell plants. In order to stimulate Shell's thinking about the provision of
service, also invited to the workshop was Richard Normann of the Service
Management Group, Paris, a consultancy which specializes in working with
service-provision businesses (although SMG would argue that *every* business
needs to think about itself and its role in terms of service provision, this
applying to manufacturing companies as well as to banks and insurance
agencies).

Normann made a presentation to the workshop which very usefully provided an alternative *Weltanschauung* to that unconsciously accepted in their
day-to-day work by many of the participants. In the concept of the Service
Management Group, if you base a business on providing a service, then you

need to bring together a number of guiding ideas: the business idea (market segment, product, production system, organization); the concept of organizational culture; and the concept of service system (service concept, delivery system, image) (Normann, 1984). In his presentation Normann argued that the preferred definition of a service is 'the total object of transaction (including intangibles)' between provider and consumer of the service, this leading to the idea that for any service business it is necessary to have:

- a service culture and dominating service idea;
- a segmented market;
- offerings which fit these two;
- a delivery system which delivers in that context; and
- an image which is maintained and improved by all these factors.

Against this framework the workshop participants, again working in small groups, looked at MF in three different service roles: as technical adviser to operations; as a designer and builder of plant; and as an enabling system for strategic decision making. Again the use of SSM was to be tacit rather than explicit, but in fact with a number of participants having read the document from the first workshop, there was this time a greater readiness to acknowledge the process being followed and to work explicitly to its guidelines. Prior to the workshop Checkland, van der Heiden and Leemhuis had prepared five root definitions and models which it was hoped would be relevant to the workshop discussions, and these were fed into the debate alongside five root definitions and models developed by syndicate groups from initial 'core purpose' statements. These latter included a system to carry out 'appraisals', a formally established procedure in Shell in which MF, jointly with production managers, reviews a production plant's performance over a stated time period in order to generate plans and standards for future performance. Figure 9.4 shows two of the root definitions developed by syndicates including one covering appraisal; Figures 9.5 and 9.6 show one of the 'readymades' produced by the workshop organizers, a system in which the service provided to the Shell Group by MF is the development of well-trained people.

Using all these models, and Normann's service business framework, the comparisons carried out on the second day of the workshop consisted of asking whether or not, and how, and to what extent, the existing MF service provisions met the logical requirements of the models and the concepts of the service framework.

The final plenary discussion brought together all the issues raised in the comparisons. Thirty points are listed in the book which records this workshop. Participants accepted ruefully that the concepts developed in their models were relatively unimaginative when set against the (to them) radical ideas in the service business framework. In general it was accepted that MF

Root Definition 1

A functional system exercising top class expertise on the processing of [raw materials X Y], providing integrated technical and economic information, such that this information will contribute to strategic decision making by the appropriate management level.

Root Definition 2

A system which periodically carries out a review, done jointly by production plant management and MF management, of the plant's past performance against previously agreed targets, and generates plans and yardsticks for future performance improvements, such that the results can be reported to both the Shell company concerned and regional coordination in order to improve the competitive performance of that company.

Figure 9.4 Two root definitions produced by syndicates at the second workshop

presently operated largely as a delivery system, giving somewhat Olympian technical advice from a position of sapiential strength. (For example, MF's appraisals were perceived by some as part of a policeman-like control process rather than as a positive service to generate mutually agreed improvements.) In order to develop MF as a service business it was accepted that significant changes would be needed: rethinking the services offered and the relationships with customers they required; developing different kinds of relationship; and redesigning at least part of the organization as a 'service system'. This implied structural, procedural and attitudinal change, an important message which was carried beyond the workshop itself in the 45-page book which recorded its work.

Methodologically SSM was used in the same way as at the first workshop. As a result of the report from that, the language of models and comparisons was becoming established in MF. Fifteen pages of report from the second workshop contain either root definitions or models. One outcome from their use

Root Definition   (Relevant to 'training')

An MF-owned and staffed system which, in response to
a continuous need for higher quality personnel for
servicing and managing the manufacturing operations
of the Shell Group, and a need for manufacturing
expertise in other functions, develops and trains
people and provides experience in a cost-effective
manner, within constraints imposed by MFs
carrying out its core tasks as service provider
and technology.

C   Those trained; through them, the Company

A   MF personnel

T   Need for trained   ⟶ Need met
     experienced people

W   Training can emerge from careful planning of
     MF work with a view to providing suitable experience

O   MF

E   MF core tasks

Concept :

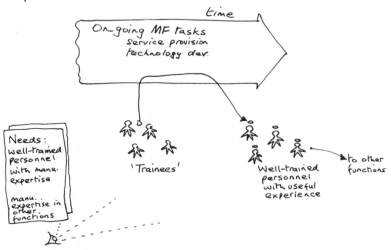

Figure 9.5 A root definition and concept used at the second workshop

Figure 9.6 A conceptual model from the root definition of Figure 9.5

was that the explicitness of root definitions and models made very stark the contrast between the rather conservative models built by syndicates and the richness of Normann's concept of a service business. That contrast could not be fudged, as might have happened in more general discussion not anchored in models and explicit comparison with the real world.

The Three Management Workshops

Rob de Vos, as head of MF, had kept in touch with the progress on what he had called MF's 'reorientation' through discussions with van der Heiden,

Leemhuis and Checkland. He agreed that the next stages of the work should continue the same pattern of two-day workshops. But this time the participants would be his senior management team from MF, and this group of six people would attend several workshops aimed at deciding how to bring about a MF reorientation. Three workshops took place on this basis over a period from June to September.

Management Workshop 1

It was first necessary to take stock of the work done so far and to reintroduce it to the MF management team (some of whom had taken part in one or other of the general workshops). As a starting point for discussion the original work on the MF vision was re-examined. It was expressed in terms of there being a challenge to MF to:

- develop, formulate, communicate, implement and review Shell's manufacturing strategy;
- service the existing structure for optimal profitability;
- create long-term competitive advantage for the Group by identifying, selecting and developing preferred manufacturing technologies.

The 'challenge' was to do these things simultaneously, within each other's context and in a focused way.

The acuteness of the problem was expressed in deliberately sharp language ('MF clients are taking steps to decrease their dependence on MF'; 'How can MF culture become less passive, less defensive?'; 'How can we *manage* dilemmas, rather than just live with them?' etc.). Also responses from the general workshops were reviewed. It was accepted that MF needed somehow to enact the purposeful activity implied by the definition shown in Figure 9.7. The three subsystems implied there (develop technology base; develop manufacturing strategy; support operations) were themselves made the subject of root definitions and were modelled. Figure 9.8 shows the definition for 'develop technology base', together with its concept expressed pictorially; Figures 9.9 and 9.10 show the definition, CATWOE, concept and model for the subsystem 'support operations'.

Much of the workshop was spent using these models as a vehicle for examining current practices, and conclusions were gathered under the headings 'providing operational support' and 'maintaining and developing a technology know-how base'.

In the case of the former there was felt to be insufficient appreciation of the competitive position of MF's customers. It was agreed that enormous quantities of data and know-how were available in MF but there were problems in processing this into knowledge useful for operational support. Requests for

Root Definition   (Relevant to MF as a whole)

A Shell-Group-owned system, staffed by business- and
technologically-experienced professionals which provides
services comprising technology development, manufacturing
strategy and operational support by accumulating, analysing,
structuring, synthesizing, developing, using and transferring
technology-based know-how in order to increase the
competitiveness of its 'customers' (corporate management,
functions, regions, Shell companies including joint ventures
and third parties) consistent with the Group's standards
and its overall interest.

'Technology' includes
                        process technology
                        engineering technology
                        operations
'Operational support' includes
                        projects
                        audits, appraisals
                        operational advice
                        manpower provision

Figure 9.7   A root definition which summarizes concepts from the general workshops

help from MF came in addressed to a wide range of people, and responses
were not necessarily shared: knowledge gained in one situation was not
automatically transferred to similar situations in other Shell operating companies. We needed to analyse how customer participation could be achieved
in the kind of transactions required in a service business.

As regards maintaining and developing a skill pool, MF was felt to be good
at this in the sense of having available much high quality technical expertise.
But although MF was seen to be very active in process engineering, it was
much less active in such areas as maintenance, manning, management systems, logistics and the modern use of information technology. A specifically
contentious point was whether MF could be expected to give Shell companies
objective advice on technology which it would be possible to acquire from
companies outside the Shell Group. Occasionally such technology would be
in competition with something developed through MF's work within the Shell
Group. How could MF be neutral in such situations? In general there was felt
to be insufficient clarity of expression in the strategic technology vision and
insufficient sharing of it between MF and those with whom the function had
service transactions.

Root Definition   (Relevant to 'Develop Technology Base')

An MF-owned system of professionals which, on the basis
of information derived from R and D results, operational
experience and an appreciation of Shell and non-Shell
technologies, identifies, selects and develops a
portfolio of technologies and sponsors an R and D
programme against a background of both competitors'
effort and the defined strategic mission as a reference,
with the objective of creating a long-term competitive
advantage for the Group's manufacturing companies.

Concept:

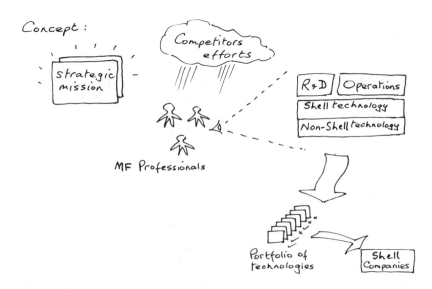

Figure 9.8 Root definition and concept of a system to develop Shell's technology base

Root Definition   (Relevant to Supporting Operations)

> An MF-owned system of experienced professionals which both pro-actively and on demand provides Shell companies, joint ventures, third parties and groups within Shell with relevant operational service support and advice, exploiting pooled know-how and specifically generated data in order to enhance the competitiveness of its 'customers'.

C   Shell companies, joint ventures, Shell groups, third parties

A   Experienced professionals

T   Need for operational ⟶ Need met, pro-actively
       Service support   and on demand

W   Operational support can be achieved via a professional group organized to supply it

O   MF

E   Structures within Shell

Concept :

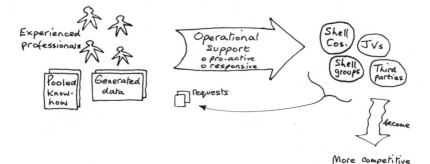

Figure 9.9 Root definition, CATWOE and concept of a Shell system to support operations

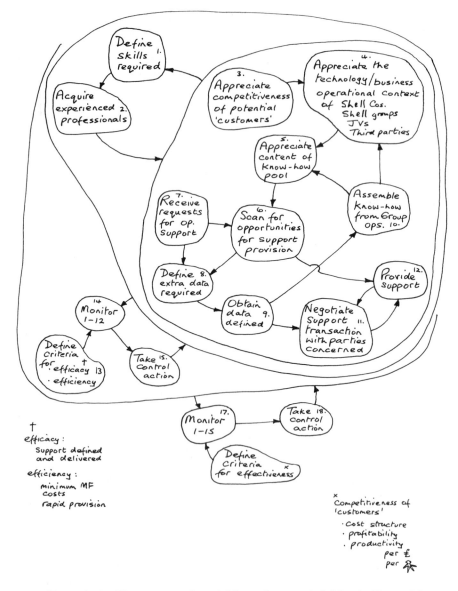

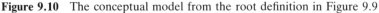

Figure 9.10 The conceptual model from the root definition in Figure 9.9

The workshop was recorded in a 31-page book which left participants thoughtful. They were very willing to take part in another management-team workshop in two months' time, this one to focus again on MF as a service business.

Management Workshop 2

Like the first management workshop, the second took place in a hotel away from the constant interruptions which normally characterize a manager's day. This one was to focus again on the implications of looking at MF as if it were a service business, rather than simply a source of technical advice, and again Richard Normann was present.

Since most of the management group had not been at the 'service provision' workshop, much of these two days was spent in discussing the nature of service provision as it could be applied to MF working within the Shell Group. Normann emphasized that if a service provider is to think wholly within the 'service business' concept, then a crucial question is not, Who is my customer?, but rather, Who is my customer's customer? This emphasizes that the purpose of the service provision is to enable the recipient of it successfully to meet *his* customers needs. Focusing on *this* transaction enables the provider and the recipient to decide jointly upon the nature of the service needed. (This concept is illustrated in Figure 9.11.) Translated into Shell, this clearly implied a particular kind of relationship between MF as provider (A in Figure 9.11) and Shell refineries and production plants as recipients (B in Figure 9.11), a relationship very different from that which had developed historically between MF and the plants. It also implied a rethink of MF's relation with its other customers such as the Shell Main Board, and regional operations.

In order to make the service provision concept sharp and usable, an activity model was built of a system in which parties A and B of Figure 9.11 work jointly to enhance B's capabilities and commercial success. The root definition and model are shown in Figure 9.12. This model was helpful in structuring the discussion, and enabled many different possible versions of MF's service provision to be examined, covering day-to-day support, technology development and the generation of long-term strategic options for Shell. Many specific examples sprang to mind and were examined against the framework of the model—for example the situation, not unknown, in which MF builds an alliance with a particular production plant in order to counter some proposed course of action by the plant's local corporate management.

With hindsight, Checkland, van der Heiden, and Leemhuis feel that in the whole sequence of work which this study represents, the production of the model of Figure 9.12 at the second management workshop was the crux of the study. After this the thinking of those involved was crucially changed; the

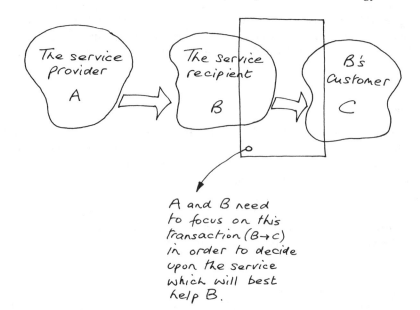

A and B need
to focus on this
transaction (B→C)
in order to decide
upon the service
which will best
help B.

Figure 9.11 An important part of the 'service business' concept (*source*: Normann, 1987)

model summarized and encapsulated a shift in the 'appreciative settings' (Vickers, 1965) of those involved. The model expresses compellingly the concept of service provision thought about in the way discussed in two workshops: namely, *co-production of added value* by the provider of a service and its recipient, and a provider who asks: *Who is my customer's customer?* The model expresses how this might be made a reality, and it is difficult for any MF-er to look at Figure 9.12 without immediately mentally comparing it with his or her own interactions with customer/clients. To the reader Figure 9.12 may look innocuous enough; but it provided a watershed in the thinking in this study. After it had been produced (and it was developed out of participants' versions of such a model) there was little doubt that the study would lead to action being taken.

The discussion at this second workshop of de Vos's management team was intense: new conceptualizations of MF were emerging which were at the same time exciting and—to some—threatening. No conclusions were reached on the spot. At the end of the event, nine tired people agreed that at a third workshop in a month's time the work from the whole 'reorientation' project would be pulled together and decisions reached. Rob de Vos ended the workshop by saying that at the final meeting he wanted each participant to come prepared to present his ideas on possible new structures for MF, and its processes, which would at the same time embody the insights from the whole

Root Definition   (Relevant to service transactions)

> A system owned and staffed by service supplier and
> client—within an environment of access, open
> communication and transparency— which, by joint
> examination, seeks to match the capabilities of party
> A (supplier) with the potential for opportunities of party
> B (recipient) to maximize B's success, thereby
> simultaneously increasing the experience and income of A.

Figure 9.12 A root definition and model of a system for the joint operation of a service transaction between two parties (developed from Figure 9.11)

study so far and yet be (politically) realistic in the real world of MF. Each participant was asked to produce several options.

In order to help with this 'homework', Checkland, van der Heiden and Leemhuis did some work to help structure the task and to sum up the study so far at the conceptual level. This led to a letter to the participants which also went into the report recording the second management workshop, this time a book of 37 pages. (This was a superbly documented study!)

The summary from the facilitators mounted the following argument, here drastically condensed.

(1) The work has tried to separate the everyday world and the conceptual world, cycling between the two using models of purposeful systems to interrogate the real world and gain insight. (Models from the first two general workshops were listed.)

(2) The elements: technology development, strategy formulation and operational support can all be described in terms of a service business concept, but enacting this would entail rethinking the services and making new logical 'bundles' of them, taking account of segmentation of customers, as well as reshaping the 'delivery system'.

(3) We can express the accepted basic concept of MF in terms of Figure 9.13, which consists of the three subsystems together with an added system to manage the three. This 'system to manage' is itself modelled in Figures 9.14 and 9.15. This gives the picture of Figure 9.16. If we use a service business concept, then MF's task is to enact transactions with a range of customers, from Main Board to individuals on production plants. The range of transactions covers different types of transaction with different time spans, from day-to-day help (hours) through technological projects (one to two years) and strategies (three-year cycle) to generation of options for the long-term future (up to 20 years).

(4) The above account is mainly located in what we might call 'the world of business logic'. We now need to move to 'the world of management', defining the actual structures and processes which would best enable MF to do its job, covering customers, tasks, skills/knowledge and time span. At the next workshop the suggested options for MF can be tested against the models as shown in Figure 9.17.

At this stage of the work an interesting encouragement came from an unusual source. A senior manager in the Shell Group, deputed by the Board to make an appraisal of MF (rather in the same mode that MF formally appraises production plants) declined to do so when he discovered that MF were thoroughly rethinking their own role. He wrote to Rob de Vos,

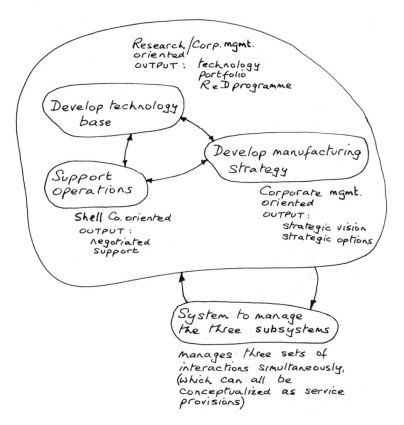

Figure 9.13 The basic concept of MF as it emerged in the workshop

cautiously but encouragingly:

> I think this kind of organisational conceptualising must be a crucial stage in your thinking and one which should be given time and the benefit of creative thinking free from the shackles of the past, despite its significant contribution to the Shell Group. Your future challenges are well defined, and imaginative solutions to meet them will determine your future success.

Management Workshop 3

The third meeting of the MF management team got straight down to a highly specific task: hearing expositions from each participant of his or her options for the structure and processes of MF, and testing them to see whether and how they embodied the logic of the models in a service-business mode.

Root Definition   (*Relevant to managing MF*)

> An MF-owned and staffed system which, in the light of discussion of Shell Group expectations of its technology function (discussions which help to create those expectations) obtains and allocates resources among the three subsystems (technology; strategy; support for operations), seeking through dialogue an appropriate balance between them, and ensuring their efficacy, efficiency, effectiveness.

C   Shell Group, the three subsystems

A   MF management

T   need for a managed   ⟶   need met
     technology function

W   This kind of learning management is possible and desirable where there is no simple 'bottom line' to help judgements

O   MF management

E   Existence of MF as an entity; the three subsystems

Concept:

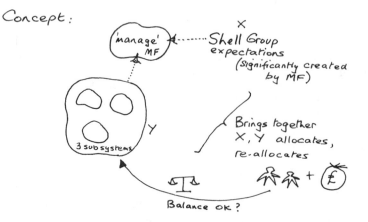

Figure 9.14   Root definition, CATWOE and concept of a system to manage MF

Figure 9.15 The conceptual model from the root definition in Figure 9.14

Many possible structures and accompanying procedures were presented, including such possibilities as:

- a 'shareholder'-oriented group and a customer-oriented group together with skill pools of engineering expertise;
- splitting MF on regional lines, each unit servicing a different part of the Shell Group;
- setting up a service and development group together with a 'standards' group as guardians of technical excellence;

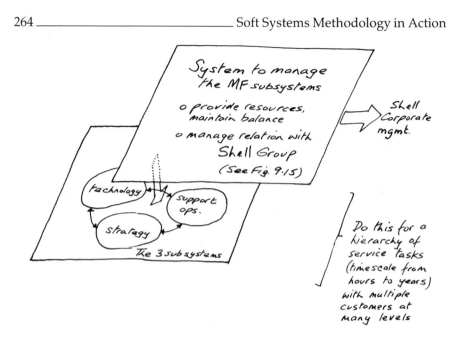

Figure 9.16 The concept 'managing MF' developed at the workshops

- sticking with the existing structure, reinterpreted as a service business;
- moving to a totally commercial service organization, owned by Shell companies;
- having a service group and a development unit, with changing task forces to ensure an input from technology providers to service providers.

Appraisal of such options as these from the management team made some use of pre-prepared forms which listed 23 crucial activities derived from the models of the four subsystems of Figure 9.13. We could in principle enquire for each of these activities whether the option under scrutiny catered adequately for it, in principle, recording +, 0 or − according to whether in the team's view the activity would be done better, the same, or worse in the examined option. In fact it was necessary to give three answers for each question, separating technology development, service provision and strategy definition! This sounds extremely tedious, and was found to be so. The forms (as in Figure 9.18) helped the discussion to re-find its feet every so often, rather than provided its structure. Nevertheless, three options were examined in detail using this approach.

The management workshop did not itself reach a new organization structure within its two days. Rather it provided the material (in another, this time 43-page book!) from which a new organization structure and its associated

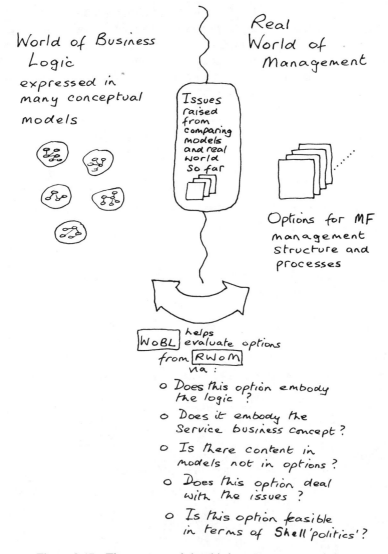

World of Business
Logic
expressed in
many conceptual
models

Real
World of
Management

Issues
raised
from
comparing
models
and real
world
so far

Options for MF
management
structure and
processes

WoBL helps
evaluate options
from RWoM
via :

o Does this option embody
the logic ?

o Does it embody the
service business concept?

o Is there content in
models not in options ?

o Does this option deal
with the issues ?

o Is this option feasible
in terms of Shell 'politics'?

Figure 9.17 The concept of the third management workshop

procedures were selected in a process involving de Vos, Leemhuis and members of the MF management team. This process was itself spread over the next month or so, and involved much consultation as implications were worked out and examined.

Most important, the discussions at the third management workshop convinced all who took part that naming a new *structure* is only a first (and relatively easy) step in achieving significant organizational change. Much effort needs to be devoted to the processes which go on within the structure. In addition, and equally important, at a meta-level to this it is extremely important consciously to think out and articulate the *process* by which the new structures and processes will be broadcast, explained, argued, internalized, implemented, made to work.

OPTION No.

Does this option, in principle, enable us to carry out the following activities better, the same, or worse than at present? (Answer for Strategy, Technology Development and Service provision to ops.)

1. Measuring the efficacy, efficiency and effectiveness of MF as a whole .

      S _ _ _ _ _
      TD _ _ _ _ _
      SP _ _ _ _ _

2. Allocating resources over the three major tasks.

      S _ _ _ _ _
      TD _ _ _ _ _
      SP _ _ _ _ _

3. Deciding what resources MF should employ.

      S _ _ _ _
      TD _ _ _ _
      SP _ _ _ _

etc.
[23 activities in all derived
from the conceptual models
relevant to MF]

Figure 9.18 The structure of the form used to evaluate suggested options for MF structure and processes from the management team

Outcomes

Following the work described above Rob de Vos addressed a meeting of more than 500 members of MF and indicated that the mysterious reorientation they had no doubt heard rumours about was reaching a stage at which their involvement was requested. They would soon be taking part in the implementation of a new MF with changed structure and new procedures, a process to which their contributions were sought. An organization structure which could embody the new conceptualization of MF was selected by the management team; it consists of organizational units which reflect fairly closely the models described above: human resource development and planning, services, technology development, and strategy development, all perceived in terms of running a service business (see Figure 9.19). At another meeting of the whole function de Vos presented the new structure and gave out a 21-page book which described it and the principles upon which it was based.

Now began a carefully planned process spread over several months of filling in the details within this structure, and defining its processes with the participation of large numbers of MF personnel. Three groups of 40 MF-ers, chosen from across the organization at random, with many levels represented, were invited to seminars at which the principles of the reorganization

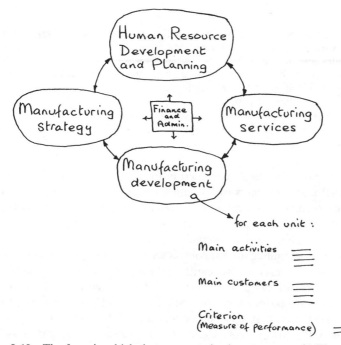

Figure 9.19 The form in which the new organization structure of MF was expressed

were explained. Participants were invited 'to express their concerns . . . to generate ideas and suggestions for the further development of new ways in which we can conduct our activities and organise ourselves' (de Vos). The introductions from speakers and the subsequent discussion at the seminars triggered 2355 ideas/suggestions; Leemhuis and van der Heiden subsequently analysed and grouped these and re-presented them back to every member of MF in a 65-page book structured according to the new organizational units.

'Luncheon information sessions' then carried the message to groups of people who had missed the seminar selection, these being addressed by members of the management team. In the end all 600 members of MF had the chance to contribute ideas and suggestions through one kind of meeting or the other.

The new MF began operation some eight months after the third management workshop. Figure 9.20 summarizes the whole process which had led to this outcome.

It is worth dwelling briefly on the new organization itself in order to illustrate how a felt-tò-be-practical structure embodies the thinking developed in the workshops.

The book describing the selected organization contains no hierarchical structure diagram of the traditional kind. Its introduction sets the tone:

> The attached deals with managing the MF tasks, managing the relationships with the various customers and customer groups and managing the internal relationships and interactions.
>
> The diagrams embody activities based on groupings consistent with the task, geography and skill-organizing principles and are at this stage unspecific about seniority and numbers of people.
>
> It is proposed to engage a wide body of MF managers and staff in a process to determine the detail of the substructures and processes of interaction.

Each organizational unit is described in terms of the major *activities* of its sub-units; the main *customers* of the unit are listed, and the key *criterion* for measuring performance is defined. For example the Service Unit has five sub-units: operational services; projects; process and engineering services; service business management; and information services. Projects sub-unit is concerned with project management, contractor assessment, construction, commissioning, contract service agreements and project management administration. The main customers of the Service Unit are listed as Shell companies, joint ventures, third parties and other Shell sectors. The overall criterion for this unit is that the cost of it should be less than the benefits to the financial 'bottom line' of the customer—a clear reflection of the 'service business' idea.

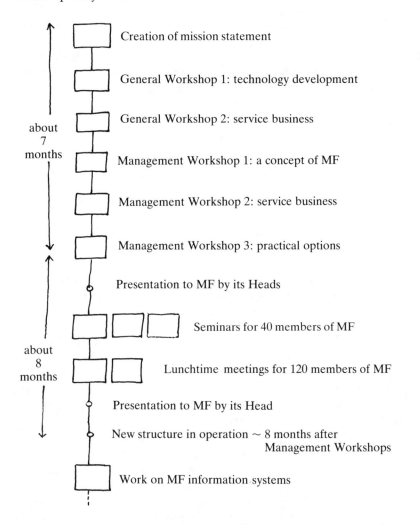

Creation of mission statement

General Workshop 1: technology development

General Workshop 2: service business

about
7
months

Management Workshop 1: a concept of MF

Management Workshop 2: service business

Management Workshop 3: practical options

Presentation to MF by its Heads

Seminars for 40 members of MF

about
8
months

Lunchtime meetings for 120 members of MF

Presentation to MF by its Head

New structure in operation ~ 8 months after
                              Management Workshops

Work on MF information systems

Figure 9.20 The process of the reorientation of MF

The document then discusses the major processes and the roles involved in them. For example:

> Task forces are drawn from Services and Development to move from feasibility study to project specification. Since this activity takes place on behalf of the customer, the Account Manager should be heading the task forces under 'normal, core business' circumstances.

These Account Managers look after the interests of a group of Shell production plants as far as service provision is concerned. Such a manager is a focal entry point for requests and is proactive in linking customers to MF.

Thus does the account of the new MF try to graft practicality on to the idea of carrying out purposeful activity conceptualized according to a particular *Weltanschauung*. At the workshop seminars held for 40 people at a time, the participants were asked at the end to each indicate a reorientation objective which he or she could pursue personally. Suggestions were very varied, including such ideas as: providing crisp job descriptions; accepting that customer problems are more important than central office solutions; using less telex, more user-friendly fax; bringing in people who have not attended the seminar; ensuring that there is humour in MF. These people seem to have heard the message.

Work currently underway, again using SSM, is directed to answering the question: What information systems does the new MF need? It is interesting to note that that question is being answered only now, after all the thinking about the nature of MF. This is in keeping with the idea that *information systems serve activity systems*, and can be designed only when questions concerning the nature of an organization and its task have been answered (Checkland, 1988d). If the information system question had been posed too early it is now clear that MF would have put much misdirected effort into developing information systems to support the creation and preservation of skill pools, that being the concept which, in the learning process of the study was replaced by the 'service business' concept. As it is, the information strategy for MF now being created participatively by the MF managers will define the information systems needed to support the new MF.

Although the reorientation of MF was a special study, rather than the everyday activity of that function, it was a study carried out through the contributions of a very large number of people in Shell. There was no special team or task force to do the work. The role of Checkland, Leemhuis and van der Heiden was to 'orchestrate' or 'facilitate', and their use of SSM was not as 'the methodology to do the study'. SSM was used in a way closer to the 'Mode 2' described in the previous chapter. SSM was seen as an intellectual structure which could help the 'orchestrators' to keep a grip on what could very easily have become a chaotic rambling debate. Although the stages of SSM could be mapped on to happenings in the reorientation of MF (Management Workshop 3, for example, could be regarded very clearly as an example of Stages 5 and 6 in action) there was never any feeling that the *structure* of SSM came first. An internalized SSM was there to be used when appropriate to make sense of the complex happenings throughout the study, so that the study as a whole represented a process of learning. The role of SSM in a study of the kind is to provide a ready-made structure for the reflections which enable learning to be recognized and extracted.

_____ Part III

Learning from Real-World Research

Chapter 10

Gathering and Learning the Lessons

Introduction

At the start of this book two quotations set its themes and tone. Chester Himes, clearly a believer in the 'cock-up' rather than the 'conspiracy' theory of history, shrewdly notes that when people complain that things always go wrong they are not usually moaning, they are expressing a positive belief that people are not predictable automata. And Barbara Trapido, pointing out that in spite of our ultimate autonomy we are prisoners of our origins and upbringing, nevertheless adopts a spunky tone which indicates that we don't *have* to follow the script prepared by our aunts and grandmothers.

These quotations convey, for the authors, a good sense of the felt texture of involvement in human affairs—something which is always part logical, part irrational, part farce and part tragedy, with human affairs themselves always, in the end, unpredictable. Nevertheless, this book is written in the belief that the rich problematical pageant of human affairs can be improved by some structured thinking. Furthermore, we have evidence that thinking can be developed around systems ideas, in particular in the form in which they are used in SSM, experience having shown the value and transferability of that approach.

This does not imply, however, that there is any such thing as an unchanging SSM which can be applied time after time like using a template. The account of the developed form of SSM, in Chapter 2, as well as the accounts of it in action in subsequent chapters, should have made clear that SSM not only develops and changes but also gets used in different ways by different users in different circumstances. Every use of it can be seen—if the users are sufficiently alert—as *research into its use*. Reflection can lead to the recognition of lessons which mould further use of the approach, and a long steady series of experiences can in principle yield lessons of some general validity.

The ten studies described here are representative of the larger set of experiences of using SSM which the authors have accumulated in the last decade, and the selection has been made of those examples which best illustrate the more general lessons. It is these general lessons which will be discussed in this final chapter. Much specific and detailed learning has been indicated in Chapters 3 to 9 or subsumed in Chapter 2. Here attention is on the broader patterns of learning which a long sequence of experiences can reveal. These concern: *the mode of use* of SSM; the question of a *constitutive definition* of the approach; and the question of how *the system to use* SSM might be conceptualized in a way which derives from all the experiences, guides any particular study, and yet does not constrain or inhibit the user.

This chapter is intended to demonstrate an acute case of the kind of reflection which Schon (1983) advocates in *The Reflective Practitioner*. In that book its author seeks to demonstrate the inadequacy of the dominating idea of 'Technical Rationality':

> the view of professional knowledge which has most powerfully shaped both our thinking about the professions and the institutional relations of research, education and practice.

The 'Technical Rationality' view is that

> professional activity consists in instrumental problem solving made rigorous by the application of scientific theory and technique.
>
> (Schon, 1983, p. 21)

Unfortunately, although there is a 'high, hard ground where practitioners can make effective use of research-based theory and technique', there is also a swamp lower down in which lie the 'confusing "messes" incapable of technical solution'; and it is in the swamp that we find 'the problems of greatest human concern' (p. 42). Schon points out the emergence of many attempts to turn 'soft' problems into 'hard' ones via computerized models, and rightly points out that in spite of successes in 'undemanding areas' such as inventory control and logistics, the algorithms

> ... have generally failed to yield effective results in the more complex, less clearly defined problems of business management ...
>
> (Schon, 1983, p. 44)

Schon wishes to abandon the idea of the manager as technician, since there is a widely held belief that managers

> learn to be effective not primarily through the study of theory and technique but through long and varied practice in the analysis of business problems, which builds up a generic essentially unanalysable capacity for problem solving.
>
> (Schon, 1983, p. 240)

He advocates that more attention be paid to the *reflection in action* which he sees as characterizing what managers (and other professionals) actually do in practising their craft.

The present authors recognize and endorse Schon's analysis; but they part company with him when he imagines that, in the field of management, the choice is between the algorithmic management science approach and the case study approach in which, somewhat forlornly, the classroom tries to provide artificial vicarious experience of managing by talking about it through case studies.

The overall argument developed in *Systems Thinking, Systems Practice* (1981) and continued in the present book is that the putative 'unanalysable capacity for problem solving' *can in fact be analysed*, and can be translated into methodology which *embodies* reflection in action.

The methodology is system-thinking-based, but the trick is to *re-cast* the idea of 'a systems approach' into the form which underlies SSM. Systemicity is shifted from the world to the process of enquiry into the world: 'the system' is no longer some part of the world which is to be engineered or optimized, 'the system' is the process of enquiry itself. If *that* is formulated as a learning system which the practitioner consciously enacts in one of many possible ways (a number of which have been illustrated in earlier chapters), then the reflection in action becomes analysable. Furthermore, the structuring which derives from consciously enacting the system of enquiry enables apparently disparate studies to be examined as a group through the epistemology which SSM provides. Out of that more tranquil reflection, after the event, general lessons may be extracted. This chapter is an example of that higher level reflection on the reflection in action which characterizes uses of SSM.

The Experiences Revisited

In arguing the general lessons which the studies described illustrate, it will be useful briefly to summarize each of the studies previously described.

Information and Library Services (ILSD) in ICI Organics (Chapter 3)

Set up by a manager who understood SSM as being most powerful when used by participants in a problem situation, the study was carried out by three managers in ILSD, with some methodological help provided by outsiders. Four fairly formal cycles of the methodology were followed; the first, based on a very rudimentary model, was used as a way of structuring the finding-out phase. In practical terms the study led to a new conceptualization of ILSD, and helped persuade the Company to devote new resources to it. Methodologically, the 'arms-length' role occupied by the methodology experts suggested

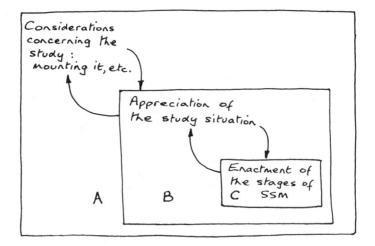

Figure 10.1 The basic structure of a system to use SSM

the basic structure for 'a system to use SSM' which is shown again in Figure 10.1. This way of thinking about using SSM is developed further in this chapter.

A Study in a Community Medicine Department in a Health Authority (Chapter 4)

This was a study in which SSM was used *by outside researcher/consultants* to explore the role of a Department of Community Medicine in a District Health Authority. In this sense it is the most 'traditional' study in this book. However, the managers concerned were interested in the approach to the study as well as its content, and did not blench when faced with some fairly elaborate conceptual models. The study provided a way of evaluating any health care project, and this had been tested in three then-current projects in the Health District in question. Methodologically the study was part of the development of Analyses Two and Three, and also provided evidence that so far as SSM is concerned, there is no *methodological* difference between using it in industry and in the National Health Service.

A Study in a Government Agency (CCTA) (Chapter 5)

This was a highlighted study carried out by a team consisting of two insiders (civil servants) and three outsiders. A 'final report' was a required outcome, and the 'political' requirement for a very large programme of interviews meant that much attention had to be paid to arranging and carrying out the study coherently (i.e. to areas A and B in Figure 10.1). External constraints led to area C consisting of one careful pass through the seven stages of SSM. A Steering Committee always saw its focus of concern as the acceptability of the final report, rather than the content of the study. Ultimately, the study contributed to the rethinking of the role of the Central Computer and Telecommunication Agency by a new Director for whom some further work was done. Methodologically, the cultural differences between industry and Whitehall meant that aspects of the study beyond the application of the methodology (i.e. A and B in Figure 10.1) required much more attention than was usually the case in industry. Finally, it did not occur to the team that SSM could conceivably be anything other than 'the methodology to do the study'. The significance of this was appreciated not at the time but later, with the help of the hindsight provided by the work in ICL which followed the CCTA study!

The Studies in ICL (Chapters 6, 7, 8)

The uses of SSM in ICL described in Chapters 6, 7 and 8 cover a range of different engagements with problem situations of various kinds. The study of the Networked Product Line Sector of Product Marketing Division and the brief study for the marketing manager responsible for Decision Support Systems, both represented Scholes as a manager making use of SSM not to 'do a systems study' but to carry out his normal day-to-day work. This led to new thinking about SSM which will be discussed below. Those two experiences permeated all the subsequent work done in ICL, whether it consisted of further managerial day-to-day use of SSM (as in the 'ICL Way' programme) or more formal highlighted studies (as in the studies of Management Support Systems Sector and Product Marketing Division). When the authors found themselves part of the Business Centre Organization Team they were ready to use SSM much more flexibly than they had in the CCTA study. Outcomes in the ICL studies were many and various: structures for programmes of work, help in carrying out such programmes, contributions to decisions about what ICL should do and how it should organize itself, new ways of conceptualizing parts of the Company and its activities. In general, these can all be described as contributions to the coherent enactment of ICL's appreciative system (Vickers, 1965; Checkland and Casar, 1986).

Reorientation of the Manufacturing Function in Shell (Chapter 9)

The 'reorientation' of MF (Shell's name for the study) entailed a sophisticated use of SSM in the sense that it was carried out by a changing group of Shell managers who each took part in some of the two-day workshops. A wider group still—all 600 members of MF—had the opportunity to make their contributions. SSM was never mentioned unless the participants wanted to discuss the process aspect of the study, but SSM provided the structure for the study and also its technical content, which was methodologically far from naive. This work, in fact, illustrates both careful attention being paid to areas A and B of Figure 10.1 and a technically sophisticated enactment of the stages of SSM (area C in the figure). Outcomes included a new structure for MF with new processes for running it, and the establishment of a concept of the Function from which the required strategy for creating its appropriate information systems was defined and is being implemented. Appropriate information systems could not have been defined without the prior work on the core purposes, structure and processes of MF.

SSM in 'Highlighted Studies' and in 'Managing': Mode 1 and Mode 2

Although the significance of it was not realized at the time, the position of the group which originally developed SSM inevitably had a strong influence both on the form SSM took and, probably more important, on how the developers thought about it. They were located in a postgraduate department in a university; their research object (messy real-world problem situations) was mainly located in companies, institutions and Government departments outside the university. In order simply to get access to the research object it was necessary for 'outsiders' to enter other people's problem situations; and this was sometimes done in a consultancy mode in order to ensure that the situations entered were *real* problem situations. This reinforced, both for the university-based researchers and for those in whose situations they worked, an unexamined (but limiting) model of 'intervention'. According to that model there are four elements: outsiders enter problem situations; do work in it (or on it); write report; and depart. Not surprisingly, SSM emerged in the form of a methodology for tackling ill-structured problem situations. This was normally described as a seven-stage sequence, and could, indeed, be used in this way even though sophisticated users approached the use of the methodology more flexibly.

This unquestioned model can be seen reflected in, if never *exactly* followed in, some of the studies described in earlier chapters. Phil Belshaw in ICI Organics wanted a highlighted study carried out, and freed an internal team a day a week to do it. The study was to use SSM as a stage-by-stage process, and did so, with the role of the outsiders that of advisers to the team on SSM. In East Berkshire Health Authority, Dr Jeremy Cobb wanted a special study done but was interested in methodology as well as substantive content, so that the project notes described both the work done and the approach taken. Nevertheless the role of Checkland, Caiger and Martin in that Health Authority was to do a study using SSM. In CCTA a joint insider–outsider team carried out a study which its initiator Gerald Watson, then Director of the Agency, did not constrain methodologically. But the effect of the outlook and attitudes of the Steering Committee was to drive the study towards being a single careful pass through the stages of SSM, leading to a final report.

It was in the ICL work that scales began to fall from the authors' eyes. The existence and significance of taken-as-given assumptions about the methodology became apparent. Scholes, as 'Business Planning and Control Manager' in a sector of Product Marketing Division, and having been a member of the team working on the CCTA study, began using SSM in his day-to-day work. Working together at this time, the authors came to see a real difference, not merely a semantic conceit, between 'using SSM to do a study' and 'doing work using SSM'.

The essence of this difference emerged as the difference between, on the one hand, mentally starting from SSM, using it to structure what is done, and, on the other, mentally starting from what is to be done and mapping it on to SSM, or making sense of it through SSM.

It was very useful at this point in the thinking to be able to include in the ICL work alongside Scholes's managerial work, some studies of a more conventional kind, namely the work on Management Support Systems and on Product Marketing Division. These were worked on in teams which included Hughes and Cole, who were using SSM in a real-world situation for the first time; and it was interesting to note that their thinking, like that of the authors a few months previously, saw SSM unproblematically as a methodology for doing systems studies. This is probably how most of those readers of this book who have previously heard about SSM will have been thinking about it.

Later in the ICL work the authors, in ways described in Chapter 8, found it perfectly possible to map the work done on, for example, the core Headquarters functions and in the Business Centre organization, on to the stages of SSM. But, as is remarked in Chapter 8, exactly when one stage began and ended, or when the PMD study should be regarded as 'finished' and the work as part of the BCOT team as 'starting' was a moot point. SSM was seen to be being used in an internalized form.

These experiences led to the recognition of a spectrum of (in-principle) use of SSM from, on the one hand, a formal stage-by-stage application of the methodology (let us call it Mode 1) to, on the other, internal mental use of it as a thinking mode (which we will call Mode 2). We are now in a position to formalize this development in a form which enables it to become part of the epistemology of SSM development. (In fact, doing that has enabled a number of subsequent experiences to underline the value of the Mode 1/Mode 2 distinction.)

To express the difference between Mode 1 and Mode 2 we may usefully borrow from Vickers his much-used metaphor of the inextricable 'two-stranded rope' of events and ideas, unfolding through time, which constitutes experienced daily life for human beings. Given that image, expressed in

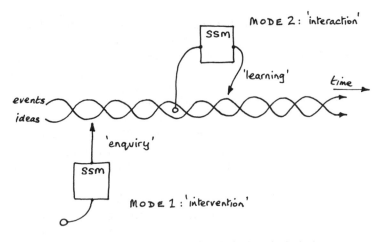

Figure 10.2   SSM in use in Mode 1 and Mode 2

Figure 10.2, we can notionally, at one extreme, operate a Mode 1 use of SSM to investigate from outside a part of the flux using SSM to structure enquiry. At the other extreme, in full Mode 2 use of SSM, we start *inside* the flux but may use SSM to make sense of the experience (Scholes, 1987). Here we are not 'operating the stages' of SSM but are using it to provide a coherent way of describing the would-be problem-solving involvement in the flux. This is also shown in Figure 10.2, where the two modes are described as *intervention* (Mode 1) and *interaction* (Mode 2). This distinction was made spontaneously by Sue Holwell and Paul Jackson, working recently with Checkland in Group Planning in Shell, in a study which was, methodologically, nearer to Mode 2 than Mode 1. And Sheila Challender, using SSM in a study of District Nursing services in a local Health Authority, was thinking on the same lines when she declared 'The worst thing about SSM is the seven-stage model'!

Mode 1 and Mode 2 were described above as 'in-principle' ways of using SSM. The careful wording was deliberately chosen to indicate that both modes are in fact 'ideal types' in terms of which SSM use may be described, rather than descriptions of actual uses. Most uses will be somewhere between the two, and Figure 10.3 subjectively places the studies described in this book between the two poles. (Most commentary on SSM, especially that from those with not much experiential knowledge of it, assumes only the possibility of Mode 1, and the approach is comfortably, but inaccurately, pigeon-holed as a well-defined seven-stage approach to problems of a complex, fuzzy, ill-structured kind.)

The Mode 1/Mode 2 distinction has been found to be of practical importance in aiding clear thinking about application of the methodology. Intellectually, we can pin down the distinction by making use of a general model of

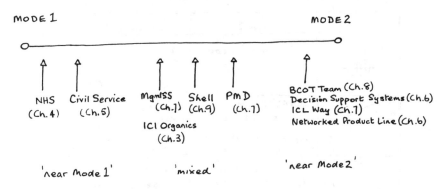

Figure 10.3 The use of SSM in ten studies described here, subjectively assessed on a spectrum from 'ideal-type' Mode 1 to 'ideal-type' Mode 2

any consciously scholarly activity in which a framework of ideas is embodied in a methodology which is then used to examine some area of application (Checkland, 1985a). In terms of this conceptualization (Figure 10.4) the ideal type Mode 1 of SSM uses a framework of systems ideas embodied in SSM as expressed either in Figure 2.5 (the 'seven stages' model) or Figure 2.6 (the 'logic-driven and cultural streams of enquiry' version) in order to enquire into and improve some part of the real world. The ideal type Mode 2, however, takes SSM *itself* as its framework of ideas, takes as its methodology conscious reflection upon interactions with the flux of events and ideas, and takes as its focus of enquiry the process of learning one's way to purposeful improvement of problem situations. This sharply delineates Mode 2 as a meta-level use of SSM compared with Mode 1, and both modes can be discerned in serious uses of the approach. This is clear in a case like the reorientation of Shell's Manufacturing Function (Chapter 9). Within the study's workshops, models were built and compared with the real world in a debate about change. This felt close to Mode 1, though the language of SSM was not used. *Between*

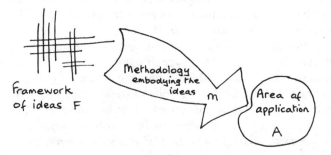

Figure 10.4 A basic conceptualization of intellectual work

	Mode 1	Mode 2
 ![grid sketch] F	Systems ideas 1.	SSM as in Fig. 2·5 (7 stages) or Fig. 2·6 (2 streams) 1.
 ![arrow sketch] m	SSM as in Fig. 2·5 (7 stages) or Fig. 2·6 (2 streams) 2. (intervention)	Reflection upon the everyday flux of events and ideas, using SSM to make sense of it. (interaction) 2.
 ![blob sketch] A	Some part of the real world e.g. NHS, a company, the civil service, etc. 3.	The learning of whoever does 2. above 3.

Figure 10.5 Mode 1 and Mode 2 use of SSM defined in terms of Figure 10.4

workshops however, the digestion and reflection upon the happenings—in order to define the next thrust of the study—was much closer to Mode 2.

Figure 10.5 shows Modes 1 and 2 defined according to the conceptualization of Figure 10.4. In fact, the reflective practitioner will always make sure that any serious use of SSM contains elements of both modes.

A Constitutive Definition of SSM

In the innocent salad days of its early development, given its emergence as a re-cast systems engineering, SSM was automatically thought of as a methodology for problem solving. We would now say that it was automatically thought of in Mode 1 terms. Experience quickly showed that the language of 'problems' and 'solutions which eliminate problems', though a useful epistemology in the classrooms of management schools, is too poverty stricken to cope semantically with the complex mix of actions, norms, standards and judgements which constitute the flux of everyday experience. (Much textbook management science has stayed with this language of 'problems' and 'solutions' long after this language has overstayed its welcome.) Serious users of SSM, however, learned quickly never to speak of 'problems', only of 'problem situations', and to use SSM flexibly, not necessarily in sequence from Stage 1 to Stage 7. Nevertheless, in spite of such softenings, SSM was always perceived as an organized use of systems ideas in a methodology for learning one's way to purposeful action to improve a problem situation.

One feature never in doubt was the fact that SSM is *methodology* (the *logos* of method, the principles of method) rather than technique, or method. This means that it will never be independent of the user of it, as is technique. This is something very clear to anyone who uses the approach, and Atkinson (1987) has demonstrated it more rigorously.

It was SSM's status as methodology which caused Naughton (1977) to raise the important question of the sharp definition of it. His motivation was pedagogical rather than substantive. He wished to be able to teach SSM coherently to Open University students on systems courses. Nevertheless his contribution raises a very interesting substantive question concerning SSM. How can we know that what someone may *claim* to be a use of SSM is legitimately so described? He suggested that there were 'Constitutive Rules' which 'must be obeyed if one is to be said to be carrying out a particular kind of enquiry at all', and 'Strategic Rules' which are more personal, which 'help one to select from among the basic rules . . . those which are "good", "better" or "best" '. Strategsic Rules include the use of suh options as doing the situation exploration through Analyses One, Two and Three, or conducting the comparison stage by writing scenarios based on conceptual models, to compare with historical happenings. As methodology rather than technique, SSM leaves open many choices of this kind.

It is the Constitutive Rules which are of greater interest since they answer the stark question: What is SSM? If there are no such rules then in what sense can SSM be said to exist?

In 1981 Checkland gave a modified version of Naughton's Constitutive Rules. They declare that SSM is 'a seven-stage process' and include: defined outputs from each stage of SSM (such as root definitions formulated using CATWOE, at Stage 3); conceptual models checked against root definitions and the 'formal system' model (which has now been put to rest in Chapter 2); models derived only from root definitions; and models not treated as systems to be engineered except as an occasional special case. These are clearly redolent of the language of Mode 1, and demonstrate that that was how SSM was interpreted in the early 1980s. There is no room for Mode 2 in these supposed 'Constitutive Rules'! But another decade of development has shown the value of uses of SSM well towards the Mode 2 end of the spectrum. The Constitutive Rules need to be rethought.

We have to accept at the start of such rethinking that the extreme ideal-type Mode 2, as a purely internal mental process, is publicly untouchable by testing against Constitutive Rules of any kind. If someone declares they have used SSM in that extreme version, we might conclude from observed outcomes that it must have been used incompetently. But we could not *prove* one way or the other that it had or had not been used, since what people say and claim may not actually reflect what is going on inside their heads! But such an extreme version of Mode 2 is of little interest; what we need are redefined Constitutive Rules covering typical uses of SSM which would now, at least for reasonably sophisticated users, be a mix of use in Mode 1 and Mode 2.

The recognition of the potential spectrum covering the mode of use of SSM, defined by the extremes of Mode 1 and Mode 2, might at first seem to militate against there being Constitutive Rules for SSM, at least towards the Mode 2 end of the spectrum. It has already been conceded that the extreme case of Mode 2—in which use of SSM is entirely a mental act of structured thinking—will escape any external examination, against Constitutive Rules or anything else. Nevertheless there will be family resemblances between all explicit uses of SSM and it is useful to redefine these resemblances by a rewriting of the original Constitutive Rules, this time to cover all forms of use other than 'extreme Mode 2', rather than only Mode 1. The purpose of doing this is not so that miscreants can be struck off some imaginary list of SSM users! It is to enable coherent critical debate about the experience of using SSM to take place.

It is perhaps superfluous to point out that the new Constitutive Rules will themselves be an 'ideal type' construct, its purpose being, like that of Weber's original ideal types, to enable a particular kind of discourse to take place, rather than to pigeon-hole part of the real world.

What follows is an account of the new Constitutive Rules of SSM based upon all the experiences which underlie the writing of this book. They are written in the form of an account of the family resemblances which characterize the whole spectrum of SSM use.

(1) SSM is a structured way of thinking which focuses on some real-world situation perceived as problematical. The aim is always to bring about what will be seen as improvements in the situation, and this is true whether or not the work done is part of normal day-to-day managerial work (defining 'managerial' in the broad sense discussed in Chapter 1) or a special highlighted study.

(2) SSM's structured thinking is based on systems ideas, and its whole process has yielded an explicit epistemology. Any account of work which lays claim to being SSM-based *must be expressible in terms of that epistemology* whether or not SSM language was used as the work was done. The epistemology is summarized in Table 10.1. ('Expressible in terms of' does not mean that the whole process has to be followed each time SSM is used. But whatever gets done must be describable using the language of Table 10.1, regardless of the scope of it.)

(3) The full claim "SSM was used' (implying some version of the approach as a whole) ought to refer only to instances in which the following guidelines were followed.

(a) There is no automatic assumption that the real world is systemic. If part of the real world is taken to be a system to be engineered, then that is by conscious choice.

(b) Careful distinction is made between unreflecting involvement in the everyday world (the unfolding flux of events and ideas) and conscious systems thinking *about* the real world. The SSM user is always conscious of moving from one world to the other, and will do so many times in using the approach.

(c)   In the systems thinking phases, holons are constructed. (These will usually take the form of purposeful 'human activity systems' which embody the four basic ideas: emergent properties, layered structure, processes of communication and control.)

(d)   The holons are used to enquire into, or interrogate the real world in order to articulate a dialogue, discourse or debate aimed at defining changes deemed desirable and feasible.

(4)   Since SSM can be used in many different ways in different situations, and will in any case be interpreted somewhat differently by each user, any potential use of it ought to be characterized by conscious thought about how to adapt it to a particular situation.

(5)   Finally, and again because SSM is methodology, not technique, every use of it will potentially yield methodological lessons in addition to those about the situation of concern. The methodological lessons may be about SSM's framework of ideas, or its processes, or the way it was used, or all of these. The potential lessons will always be there, awaiting extraction by conscious reflection on the experience of use.

These five statements, together with Table 10.1, define SSM sufficiently for its use to be discussed coherently. This may be illustrated by some examples from published accounts of SSM or work said to be based upon it.

In the journal *Management Services*, Patching (1987) gives an account of work in Essex County Council in which SSM was used 'to clarify the organizational context of potential IT applications'. This article builds upon an earlier one in the same journal which introduced SSM to practitioners of 'management services' (McLoughlin, 1986). In giving an account of the use of SSM as a suitable strategy for the appropriate introduction of information technology, Patching describes rich-picture building and then writes:

> The root definition is a statement of what the system shown in the rich picture is designed to achieve, based either on the issues which have been uncovered, or on the 'primary task' of the organisation. (p. 17)

Now it is often useful to seek primary task root definitions on which to base analysis of fundamental information flows. This has been done in the continuation of the work in Shell's Manufacturing Function, for example. But Patching's focus on *only one* root definition, and the assertion that the rich picture (singular) represents a *system*, rather than a situation, shows that, in terms of the three Constitutive Rules above, what is here being described is a

Table 10.1 SSM's epistemology: the language through which its process makes sense

	Soft Systems Methodology
Real world	The unfolding interacting flux of events and ideas experienced as everyday life.
Systems thinking world	The world in which conscious reflection on the 'real world' using systems ideas takes place.
Problem situation	A real-world situation in which there is a sense of unease, a feeling that things could be better than they are, or some perceived problem requiring attention.
Analyses One, Two, Three	*Analysis One*: examination of the intervention or interaction in terms of the roles; 'client' (caused the study to take place), 'problem solver' (undertakes the enquiry) and 'problem owner' (plausible roles from which the situation can be viewed, chosen by the 'problem solver').
	Analysis Two: examination of the social (cultural) characteristics of the problem situation via interacting roles (social positions), norms (expected behaviour in roles) and values (by which role-holders are judged).
	Analysis Three: examination of the power-related (political) aspects of the problem situation via elucidation of the 'commodities' of power in the situation.
Rich pictures	Pictorial/diagrammatic representations of the situation's entities (structures), processes, relationships and issues.
Root definitions	Concise verbal definitions expressing the nature of purposeful activity systems regarded as relevant to exploring the problem situation. A full RD would take the form: do X by Y in order to achieve Z.
CATWOE	Elements considered in formulating root definitions. The core is expressed in T (transformation of some entity into a changed form of that entity) according to a declared *Weltanschauung*, W. C (customers): victims or beneficiaries of T. A (actors): those who carry out the activities. O (owner): the person or group who could abolish the system. E: (the environmental constraints which the sytem takes as given).
The 5Es	Criteria by which T would be judged: Efficacy (does the means work?); Efficiency (are minimum resources used?); Effectiveness (does the T help the attainment of longer term goals related to O's expectations?); Ethicality (is T a moral thing to do?); Elegance (is T aesthetically pleasing?).

Table 10.1 (*contd*)

Soft Systems Methodology	
Conceptual model	The structured set of activities necessary to realize the root definition and CATWOE, consisting of an operational subsystem and a monitoring and control subsystem based on the Es.
Comparison	Setting the conceptual models against the perceived real world in order to generate debate about perceptions of it and changes to it which would be regarded as beneficial.
Desirable and feasible changes	Possible changes which are (systemically) desirable on the basis of the learned relevance of the relevant systems, and (culturally) feasible for the people in the situation at this time.
Action	Real-world action (as opposed to activity in conceptual models) to improve the problem situation as a result of operation of the learning cycle for which this epistemology provides a language.
Use of SSM	
The system to use SSM	The language and structure of Figure 10.8 provides an epistemology which makes sense of the process of using SSM.

variant of SSM with a strong flavour of hard systems thinking. In fact, the work in Essex County Council was a field study within a European Community 'ESPRIT' project, and the account of *Functional Analysis of Office Requirements* which came out of that work (Schafer *et al.*, 1988) uses SSM in a mode compatible with the Constitutive Rules, as this sentence relevant to Patching's remark illustrates:

> The models generated in the systemic analysis are used in the client organization to initiate debate with people in the problem situation about what to do (Schafer *et al.*, 1988, p. 34)

Enough has been said to illustrate how the new Constitutive Rules could lead to coherent debate about a use of SSM such as Patching describes.

A second example is provided by Watson and Smith's review of applications of SSM in Australia (1988). They describe briefly eighteen applications of SSM and discuss four in detail. A feature of their examples is the recurrence of the phrase 'implicit root definitions used'. Now the work in ICL (Chapters 6, 7, 8) included the occasional model built from a root definition which the authors expressed only verbally, but this is rare, at least in European use of SSM. Why antipodean users should do this more often is an intriguing question! It is clear from the Constitutive Rules, however, that such a departure from the usual form of SSM will make more difficult an explicit declaration and discussion of the lines of argument which in the end deem certain changes to a situation to be both desirable and feasible.

A final example is provided by the study of policing vice in the West End of London mentioned in Chapter 2. In their account of the work, Flood and Gaisford (1989) suggest that it is difficult to answer the question: Who is the client? because

> . . . if we *select* a client are we not promoting this personage to a most relevant
> position (but on what grounds?)
> (Flood and Gaisford, 1989 p. 13; present authors' italics).

Clearly the notion 'client' is here being used in its everyday language sense, not in the exact sense in which the term is used in SSM's Analysis One (Checkland, 1981). And it is not being distinguished from 'problem owner'. (In Analysis One, 'client' is whoever caused the study to happen—and there is always a real-world person or group in this role. It is never necessary to 'select' a client. It is clear from their account that in this instance the 'client' initiating the study of West End vice by Flood and Gaisford was either Superintendent Gaisford or his superiors.) The point being made here is that the new Constitutive Rules provide a sharply defined framework for clarifying issues of this kind, so that critical debate of substantive issues concerning SSM can be based upon clear concepts and a tightly defined epistemology. That is their function: to introduce some of the kind of rigour normal in natural science into this more difficult field of applied social science.

The System to Use SSM

During the early years of SSM's development, most of the attention was inevitably directed to its content as an alternative to systems engineering, an alternative which was turning out to be radically different from its parent. The concept of 'human activity system' emerged and was refined. Ways of naming such notional systems in root definitions were created. Methods of model building using verbs were developed; CATWOE was proposed and tested against 50 used root definitions; four ways of comparing models with perceptions of the real world emerged and were given further tests . . . and so on. All this work put the emphasis on SSM seen as a describable approach to working in messy problem situations to bring about improvement.

When SSM was then described as a seven-stage enquiring or learning system, the standard form of expression of it was the seven-stage model of Figure 2.5. It is not by chance that this model is of a connected set of *entities* rather than *activities*. Those who developed SSM were very conscious of its status as mouldable methodology rather than rigid technique, and they wished to leave 'how, exactly, to do it' as a strategic choice for the user to make. Gradually, however, as experience has built up, the degree of attention given to 'how to use SSM' has increased (see also Davies, 1989 and Ledington, 1989.) Much has been learned in this area of concern, and it is now possible to elaborate that learning in an account of a notional 'system to use SSM' without reducing the approach to technique.

When the work in ICI Organics described in Chapter 3 was set up, Phil Belshaw stated unequivocally: 'I wish the task force to carry out the study, rather than, say, bringing in the Department of Systems at Lancaster University to carry it out on our behalf.' This forced Checkland and Perring to think rather carefully about how SSM was to be used, and led to the conceptualization of 'using SSM' shown in Figure 10.1. Other experiences added their contributions. In the CCTA study (Chapter 5) the fact that the team was physically dispersed, contained only one full-time member, and had to carry out a very large number of interviews, again forced much attention to be paid to the way of carrying out the study. And once the authors' work in ICL was underway, what was being done was so different from previous experiences that new thoughts about how to apply SSM were inevitably on the agenda.

We now gather those and other experiences together in order to expand upon Figure 10.1. This is done on the assumption that what we are describing is a full use of the whole of SSM, rather than a partial use which might consist of, say, spending an hour or so examining some real-world action in the form of a number of transformation processes based on declared *Weltanschauungen* and defining the '3Es' for each. Such partial uses can be very helpful, but for the sake of completeness it is best to focus on a complete use of the methodological cycle, from 'finding out' to 'taking action' in a problem situation.

Starting with area C in Figure 10.1, we have 'the stages of SSM'. The common experience here is to formulate some potentially relevant systems based on both initial finding out about the problem situation and doing Analysis One. Building models and doing some initial comparisons is contingent upon this, but is also itself a source of better rich pictures and more relevant systems. A cycle consisting of picturing/selecting/modelling/comparing/picturing normally ensues, of which the product is possible changes to make in the situation. We emerge from the cycle as the conviction

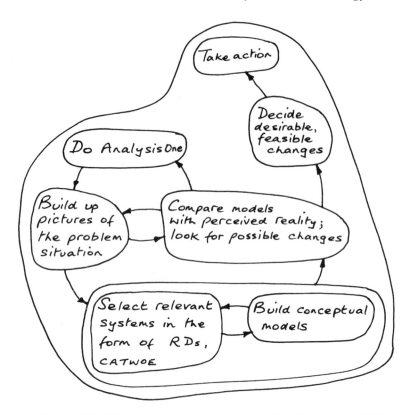

Figure 10.6 The common dynamics of operating the stages of SSM

develops that particular changes are both desirable and feasible. These common dynamics are illustrated in Figure 10.6. Note that this figure also includes the possibility that models may precede root definitions, as long as the outcome is a defensible pairing of definition and model.

Although Figure 10.6 has been described as if it could be initiated in a vacuum, it will of course actually be contingent upon a greater appreciation of the problem situation than is provided by Analysis One. Study of the situation in which SSM is being used, by means of Analyses Two and Three, will affect the activities of Figure 10.6 just as carrying them out will feed those analyses. (This is the mutually contingent relationship between B and C in Figure 10.1.)

Exactly similar thinking applies to the link between the appreciation of the situation via Analyses Two and Three (area B) and area A, which concerns mounting the study (though 'mounting', of course, here covers everything from setting a team to work on a highlighted study, to the case in which a manager is thinking about a current issue by means of SSM). B will be

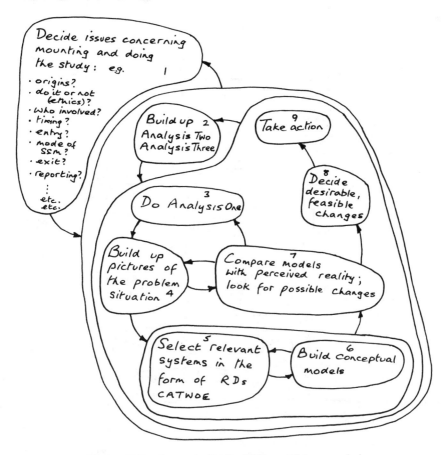

Figure 10.7   Areas A, B, C of Figure 10.1 expanded

contingent upon A and vice versa, in the same kind of mutually-creating relationship as that between B and C. This gives the picture shown in Figure 10.7.

Now since we have used the phrase '*the system* to use SSM' it behoves us to add the monitoring and control activities by means of which the system being assembled could in principle adjust and adapt in the course of a study. Figure 10.8 adds these considerations and names the criteria by which system performance would be judged. Finally, Activities 13, 14, 15 and 16 ensure that learning from use can affect a particular study and lead to improved practice in future. These activities ensure that a use of SSM is also the operation of a system which learns about the methodology through action.

In fact Figure 10.8 is a tighter version, based on explicit systems thinking, of the model of an appreciative system which Checkland and Casar (1986)

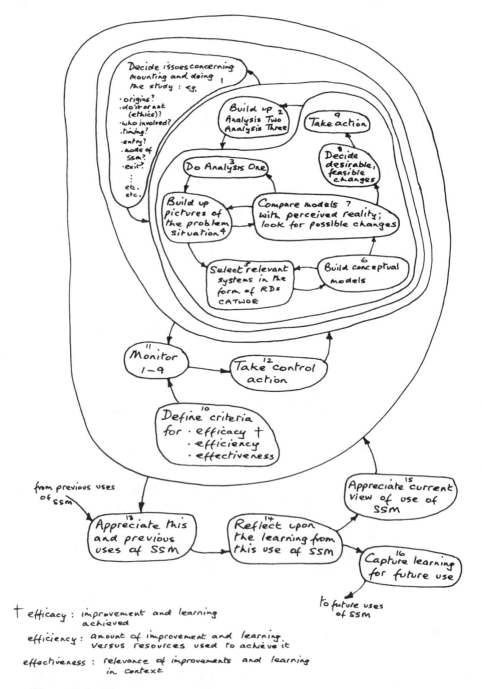

Figure 10.8 The system to use SSM which maps all the experiences of using it

built from a careful reading of Vickers. No doubt that model helped to form the mental processes which have produced this system. So also, no doubt, did the recognition by Checkland and Davies (1986), following Fairtlough (1982), that *Weltanschauung* is used in more than one sense in SSM. Referring to Figure 10.8, what Checkland and Davies call 'W_1', is used in model building in Activities 5 and 6. This is the *Weltanschauung* of CATWOE. Activity 2 implies exploration of another *Weltanschauung*, their 'W_2', the one which in a particular situation makes certain notional system 'relevant'. Finally, Activity 1 relates to Checkland and Davies's 'W_3', the *Weltanschauung* behind the perceived social reality of the situation in which the study is mounted.

In other words the model of the 'system to use SSM' stems not only from experiences such as that in ICI Organics but also from other then-current strands of thinking in the development of SSM. It is not plucked out of the air.

Since Figure 10.8 is a model of a system to use SSM, and is one which uses SSM's own modelling method, it is appropriate to name the root definition and CATWOE compatible with it in a defensible pairing. This is done in Figure 10.9.

The model of the system to use SSM in Figure 10.8 captures the full range of those experiences of using the approach which have followed the whole methodological cycle. As with all such models, the weighting of different activities in terms of both the amount of time and the degree of attention given to them will vary considerably from one usage to another. Sometimes Activities 1 and 2 may take up most of the time and energy, with modelling and comparing being done very quickly; in more formal studies, with Activity 1 done according to local bureaucratic norms, most time may be spent on the activities which comprise moving through the stages of SSM (Activities 3–9).

With this in mind, all of the authors' experiences can be mapped on to Figure 10.8. In the study in East Berkshire Health Authority, for example, Activity 1 was fairly straightforward. Three outsiders, Caiger, Checkland and Martin, were to carry out a systems study in a department of community medicine. SSM was to be used flexibly but in a manner close to the 'Mode 1' end of the spectrum in Figure 10.3 (though that concept did not exist at the time). The client, Dr Cobb, was interested in the methodology as well as the outcome, and reporting was to cover both. All this was settled very quickly. Activity 2 consisted of serious attempts to carry out Analyses Two and Three at a time when these were first being developed. These informed the work on the stages of SSM (Activities 3–9 in Figure 10.8) and were themselves informed by that work. The model building was that reported in Chapter 4; it led to the way of describing community medicine initiatives which was tested in three then-current projects in East Berkshire, and to the framework for evaluating the community medicine department itself.

In the reorientation of MF described in the previous chapter, the balance of effort among the activities of the 'system to use SSM' was rather different.

Root Definition

> A system to use SSM (as defined by the new Constitutive Rules) in a real-world problem situation in order to take action in the situation and to learn from the experience

CATWOE :

C   'Client' (Anal. 2), participants, those affected by outcomes

A   'Would-be problem solvers' (Anal. 2)

T   Intention to use SSM in a real-world problem situation in order to take action and to learn   →   Intention met by use of SSM

W   Use of SSM can yield learning ; the systemicity of SSM's approach can be modelled by using SSM's modelling technique

O   'Would-be problem solvers'; 'client'; participants; those outside these groups with power in the situation

E   new Constitutive Rules

Figure 10.9 The root definition and CATWOE defensible against the model in Figure 10.8

Activity 2 was done formally only by Checkland, who throughout the study tested its outcome (namely his growing understanding of the Shell culture) in conversations with workshop participants. Activities 3–8 consisted of the work done in the sequence of workshops, which entailed more than twenty significant models being built and compared with real-world action. One of those models, the issue-based model of a system to enact a service-providing transaction between MF and a 'customer', can with hindsight be seen as providing a trigger which shifted the thinking of MF-ers. This was the hinge on which the project thinking turned. But the strongest impression the study leaves with Checkland, an outsider brought in not to do the study but to help mount it, is of the care devoted to Activity 1. In retrospect this was very worth while. The two-day workshop format; the selection of a theme for the next workshop only after learning from all those already held; the decision not to use the language of SSM unless the participants raised it; the involvement of Rob de Vos and his management team in three successive workshops; the thorough documentation of each workshop in reports circulated more widely than workshop participants; the later 'seminars for 40' and the 'lunchtime meetings for 120'; all these features were the product, at points throughout the study, of Activity 1. They set its tone and enabled the reorientation of MF to emerge as a genuine co-production by the 600 members of that Function.

In contrast, any of the 'near-Mode 2' uses of SSM in ICL give a very different weight to the activities in the 'system to use SSM'. Although it would be possible to identify the manifestations of Activity 1 in, say, the study for the manager responsible for the Decision Support Systems Sector of Product Marketing Division, they did not occupy more than a minute or two, and Activity 2 was taken to be the existing appreciation of ICL culture by Scholes and his colleague. Focus was immediately on some models and their comparison with the real situation. Activities 3 to 8 were done in two short meetings, and Activity 9 was manifest in subsequent actions taken in ICL. But these are an observer's conceptualizations. To Scholes and his colleague it was simply a matter of tackling an urgent problem using some relevant ideas.

Similarly, in the 'near-Mode 2' work on the 'ICL Way' programme described in Chapter 7, 'thinking and action flowed together' and the epistemology of SSM was used for sense-making. Here most of the effort went into ensuring that the 'ICL Way' programme in Product Marketing Division itself was implemented as a learning system, so that the main emphasis was on Activities 1, 10, 11 and 12 of Figure 10.8.

These examples could be extended to cover every use of SSM in which the authors have been involved. The model of the 'system to use SSM' in Figure 10.8 has shown its value as an epistemology relevant to using SSM, hence its addition to Table 10.1. Moreover, in studies done since the ones described here, it has shown its value in directing thought as studies are initiated and carried out. But it is important to keep in mind that its role is to supply mental props and guidelines. The model itself is an ideal type. Its role is similar to that of a cookery book for chefs. Earnest trainees may thumb it nervously and follow it doggedly, while for the experts it provides helpful guidelines which become internalized.

Conclusions

I

During the work in the Product Marketing Division of ICL described in Chapter 7, Elaine Cole surprised the authors one day by remarking 'I've just realized you never use SSM!' Since the authors felt that 'using SSM' was exactly what they were doing in a succession of problem situations in ICL, they expressed some astonishment at Elaine's remark. She went on to say that her perception was that what was being done was 'research on SSM' rather than routine 'use of SSM'. This is an interesting distinction. On reflection, it is probably the case that whenever a reasonably sophisticated body of knowledge and associated methodology has been *internalized*, then any 'use' can be described as 'research'. The exercise of expertise can probably best be thought of as researching the potential relation between a body of knowledge and methodology on the one hand and a potential application on the other. (In this sense so-called 'expert systems' which try to capture in the machine a set of inference rules which they imagine experts follow in their work are on entirely the wrong track.) Being expert consists, rather, of doing *unprogrammed* research on the linking of a particular body of knowledge and methodology to a specific instance of its use (Johnson, 1984). Every sophisticated use of methodology *needs to be* research on its use in a particular context.

In practical terms, Elaine's astute remark probably arose from her observation of, and participation in, a crucial transition in the development of SSM. Although it was not sharply realized at the time, the Mode 1–Mode 2 distinction was just then emerging as meaningful.

In the whole development of SSM a number of such transitions can, with hindsight, be recognized. Examining the development and dissemination of SSM, Forbes (1989) finds several discernible phases. Initially systems engineering was 'softened' by the introduction of the idea of root definitions and conceptual models which referred to concepts relevant to a problem situation rather than being would-be descriptions of it. The 1972 paper (Checkland, 1972) marks this phase, the apotheosis of which was Naughton's (1977) formulation of the original Constitutive Rules. Meanwhile a wider range of application areas and, no doubt, more sophisticated use of the ideas in SSM led to a phase in which the focus was on the approach as an articulation of the social process Vickers (1965) calls 'appreciation'. The work in ICI described in Chapter 3 illustrates this phase. In a third phase, covered in this book, a mature methodology is defined by the Mode 1–Mode 2 distinction, the new Constitutive Rules and the system to use SSM shown in Figure 10.8. (Forbes, 1989, finds the general understanding of SSM stuck in the early 1970s, but we may hope that this book does something about that!)

This history has yielded an SSM defined by the new Constitutive Rules and the system to use the methodology as in Figure 10.8, its manifestations in application covering the spectrum defined by Mode 1 and Mode 2.

That is as simple an account of the nature of SSM as we can concoct, given the range of experience it has to envelop. It is necessarily more complicated than the account of SSM usually given to students in the classrooms of the management schools. There the white lie is that SSM is a seven-stage problem-solving methodology applicable to problems of a certain kind, namely messy, ill-structured ones. We can forgive the pedagogues' simplification; after all, the true complexity of the real world *has* to be simplified for classroom consumption. But in the real world outside classrooms, if we want seriously to bring about improvements, we need to work with the richer account of SSM given above.

II

It is an irony of work of the kind described here that the obvious question posed by the very existence of an approach to bringing about improvements in problem situations cannot be answered! When people hear for the first time about SSM, or systems engineering, or cognitive mapping, or any such methodology, they naturally ask: Is it any good? Does it work? And they are very frustrated when they develop an understanding of the nature of methodology sufficient to realize that the question is unanswerable, that methodology is in fact *undecidable*. This was pointed out in relation to systems-based methodology in the early days of development (Checkland, 1972) but the point requires continual reiteration in the face of the misguided zeal of academic comparers of methodologies and students with dissertations to write. (Academics find this a problem, but not managers; the latter simply find their way to approaches which work for them in their context.)

The point was bluntly put in 1972:

> if a reader tells the author 'I have used your methodology and it works', the author will have to reply 'How do you know that better results might not have been obtained by an *ad hoc* approach?' If the assertion is: 'The methodology does not work', the author can reply, ungraciously but with logic, 'How do you know the poor results were not due simply to your incompetence in using the methodology?'
>
> (Checkland, 1972, p. 114)

Given the nature of methodology as a set of principles of method, the inseparability of the methodology from the use made of it by a particular user in a particular context will always mean that SSM itself, as a description on paper according to the new Constitutive Rules, will forever remain undecidable. The transformation of the Manufacturing Function of Shell described in Chapter 9 was brought about using SSM, and the work reached an outcome satisfactory to those who felt the need for a root-and-branch rethink of that function. But there is in principle no way in which it could be proved or disproved that this was the *best* way to do it, or that a *more competent* use of SSM would have achieved the result more quickly.

Alas for the academics, that is the nature of any methodology, and its recalcitrant nature is even greater, for SSM, now that that approach can adapt chameleon-like within the spectrum defined by the extreme special cases of Mode 1 and Mode 2.

III

An approach with claims to any kind of general applicability will necessarily have to be able to exhibit considerable variety if it is to match the variety with which real life surprises us. SSM and the ways of using it now show a wide range of forms, but its essence and its deep values are simple and straightforward.

As regards the latter, it would be difficult to develop much feel for the approach if you did not accept its prime implicit value, namely that learning is an ultimate good and that in human situations learning is never complete. SSM is not an approach for dogmatists. Its essence is also simple. It organizes coherently what we do whenever we relate mentally to the world outside ourselves. It is simply an organized version of doing purposeful 'thinking'!

When we think purposefully we always start by an act of selection. A general perception of reality leads to selection of some subject of discourse. This selection creates a figure–ground (F–G) relationship, as in Figure 10.10. We then initiate discourse by constructing sentences in which we say something about F: we *predicate* F; that is the form our language takes. But by using a facility and skill which escapes cats and cuckoos, we as human beings can always formulate many alternative predicates. Discourse then consists of *comparing* predicates either with each other or with perceived reality or both. These comparisons create arguments in relation to evidence, and underpin decisions to act in particular ways. In these terms, SSM simply makes use of the notion of holons (normally, in SSM, the holons we call 'human activity systems') and goes about constructing holons and comparing them with each other and with perceived reality. This is a more organized and formal version of *what we do anyway* when we think purposefully, as Figure 10.10 argues. Thus you cannot help but use the *form* of SSM whenever you do serious organized thinking; or, put the other way round, SSM simply articulates that kind of thinking in a particular way using the notion of holon. That is why, once the epistemology of SSM is grasped, using it seems so natural. That is why SSM can be internalized so easily, making 'Mode-2-like' uses of it seem the most natural thing in the world.

The question which remains for any would-be users is that of convincing themselves that the disciplined use of purposeful holons will in fact improve a process which they will be enacting in any case.

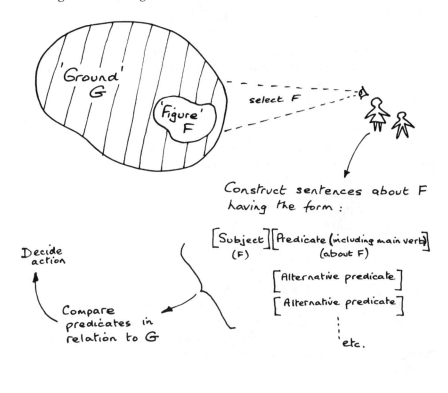

'Ground'
G

'Figure'
F

select F

Construct sentences about F
having the form :

$$\left[\begin{array}{c}\text{Subject}\\ \text{(F)}\end{array}\right]\left[\begin{array}{c}\text{Predicate (including main verb)}\\ \text{(about F)}\end{array}\right]$$

Decide
action

$$\left[\text{Alternative predicate}\right]$$

$$\left[\text{Alternative predicate}\right]$$

Compare
predicates in
relation to G

'etc.

Any purposeful thinking	SSM
Perceive R	Perceive problem situations
Select F	Select relevant systems expressed as root definitions
Predicate F	Build conceptual models
Compare predicates	Compare models and perceived situation
Decide action	Decide action

Figure 10.10 The relation between SSM and any purposeful thinking

IV

If SSM, like all other methodology, is undecidable, and if would-be users ultimately have to convince *themselves* that it might be helpful for them, what can we say to such potential users as they come to it for the first time?

We can say that structured thinking using holons is not a bad idea, in the sense that it allows arguments to be made explicit so that, for example, abandoning an intellectual cul-de-sac can be done consciously, and steps retraced. This brings a little of the rigour to the kind of thinking entailed in natural science into applied social science. And the use of the epistemology of SSM can help the creation of *shared* appreciations or the recognition that endemic conflicts have to be *accommodated*.

We can say that the way SSM is adopted and adapted by any users in a particular situation will not be exactly like other users' versions: all users must find a version with which they are comfortable, given their own cast of mind. And we can say that you should have nothing at all to do with it if you know what is the case, follow some rigid dogma, are incapable of questioning your own *Weltanschauungen*, cannot see the world through the eyes of another.

Finally, it seems relevant to the authors to remark that the very best uses of SSM seem always to exhibit a certain dash, a lightfootedness, a deft charm. In this sense the role of the approach is akin to that of the cavalry in nineteenth-century war: it can add a certain tone to what might otherwise be a vulgar brawl.

Appendix

Information Systems and Systems Thinking: Time to Unite?

(The Second Annual Rank Xerox Lecture: see note on page 314)
Peter Checkland

The notion of 'information system' is examined particularly from the perspective that those concerned with such 'systems' might learn something from the whole body of ideas concerned with the notion 'system'. The first section examines briefly the possible approach to understanding information systems through the notion of information theory. The second section examines the fundamentals of systems thinking as developed in the 1950s and 1960s, and reviews new developments during the 1970s and 1980s, examining their application to information systems; the third section reviews the implications of these developments for work on information systems in the future.

INTRODUCTION

Human beings appear to be uniquely capable of attributing meaning to what they perceive. We can obtain the *data* which is the position of the hands on a clock and convert it into the *information* that we are late for an appointment or that we have time for another cup of tea. This transformation of data into information by the attribution of meaning makes the study of information a very broad and hybrid field; but it is a very central one in understanding the nature of the complex culture which will be created by a meaning-endowing animal like *homo sapiens*, at once autonomous and gregarious.

My concern here is the special nature of organized attempts to provide information; that is to say with 'information systems'. As with the parent field of information studies itself, the topic of information systems is a rather chaotic hybrid field. It is one which has been dominated by a particular *means* —the computer—and by much conceptual confusion which stems from giving a means the status of an end. The confusion has been compounded from the start by the unfortunate anthropomorphic language used about computers. Both of the pioneers von Neumann and Turing, for example, used the

303

unjustifiable metaphor 'memory' in relation to computers. They could, perhaps, have justified the alternative metaphor 'storage'. As computers manipulate stored tokens for data, they certainly do not exercise 'memory' in the human sense of that word!

Ironically, the use of anthropomorphic language has gone along with a relative neglect of the fact that an information system as an organized attempt at meaning-creation-from-data *by an institution* will have social implications at least as important as the technical ones. Many of the technically expert computer professionals vigorously resist being drawn into the processes of social change inevitably entailed in introducing information systems in an organization.

Add to this a rapidity of technological development which has had the effect of largely handing over the technology to potential users, via micro-computers and application packages, side stepping the technical experts, and we have a recipe for a very chaotic field indeed.

The whole field of information systems (like that of information studies more generally) lacks its Newton to bring it conceptual clarity, in spite of the ubiquity of the equipment and the packages, and it is still worthwhile trying to examine the field at a fundamental level in order to get it into a perspective which could provide both theoretical and practical help, and could also help definition of useful future work in the field.

INFORMATION SYSTEMS: AN APPROACH THROUGH INFORMATION THEORY?

It might be thought that the obvious approach to an understanding of information systems would be through the 'information theory' developed in the 1940s. At that time the statistician Fisher, the mathematician Wiener and the communications engineer Shannon were all working independently on the idea of a quantitative statistically-based theory of information transmission. This built upon the work of communication engineers in the 1920s, and the conceptualization which emerged clearly stemmed from that source: a 'message' could be encoded into a 'signal'; the signal could be transmitted via a 'channel'; a 'decoder' could recover the message. This is clearly based on the engineers' view of the world, and *from the point of view of engineers* the interesting questions concern measuring what is transmitted and its degree of distortion. The triumph of the theory was to define a quantitative measure of what is transmitted; but the pioneers then made a disastrous mistake. They referred to what was transmitted as 'information'. What should have been called 'signal transmissions theory' was unfortunately called 'information theory', as if a coded signal and information were synonymous. At least some of the pioneers themselves recognized this, Weaver writing in 1948 that

> The concept of information developed in (Shannon's) theory at first seems disappointing and bizarre—disappointing because it has nothing to do with meaning, and bizarre because it deals not with a single message but rather with the statistical nature of a whole ensemble of messages . . .
>
> (in *The Mathematical Theory of Communication*, 1949)

What is 'disappointing and bizarre' is that a so-called Theory of Information makes no distinctions which derive from the meaning content of what is transmitted. To the communications engineer acting as such, there is no fundamental difference between 'I have just changed my socks' and 'I have just pressed the nuclear button'!

In order to meet the expectations raised by the phrase 'information theory' we need what has been called 'semantic information theory': a theory concerned with how transmitted symbols convey the desired meaning and how the significance of the meaning can be measured in context. Progress has been slow here since the 1950s, but there is promise in the kind of work being done by Ronald Stamper and his colleagues at LSE: an approach to information systems based upon semantic analysis of a particular task and situation, an approach which may be able to ground itself on the work of modern philosophers on the theory of 'speech acts'.[1] This is an approach to be welcomed, linking as it does to the everyday connotations of 'information' as implying the generation of meaning; but it is not yet the case that information theory can provide an intellectual base for dispelling the confusion in the field of information systems.

INFORMATION SYSTEMS: AN APPROACH THROUGH SYSTEMS THINKING?

To many people a computerized information system is the very paradigm of what they mean by 'system', so it is perhaps particularly surprising that work on information systems has paid little heed to the general development of ideas of 'system'.

I shall argue that work on information systems has in fact tacitly followed the systems thinking of the 1950s and 1960s; that *that* version of systems thinking (which is *systematic*) is complementary to more recent developments in the 1970s and 1980s of systems thinking, which are *systemic*; and that these more recent developments are relevant to current problems in work on information systems.

In order to mount this argument I shall drastically summarize the development of systems thinking using the conceptualization of Figure 10A.1, in which a human observer tries to make sense of his or her perceived reality, *PR*, by means of some intellectual concepts used in some mental processes or methodology, *M*.[2] 'System' is one of the concepts used in this process, but

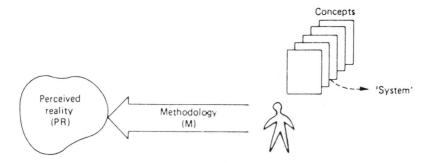

Figure 10A.1 The human observer uses concepts (including 'system') in a methodology, *M*, to explore perceived reality, *PR*

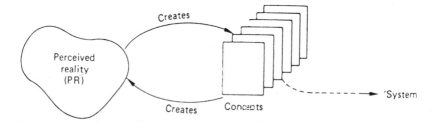

Figure 10A.2 The mutually creating relationship between perceived reality, PR, and intellectual concepts (which include 'system')

considerable confusion is caused by the fact that the word is used not only as the name of an abstract concept which the observer tries to map onto perceived reality, but also as a label word for things in the world—as when we casually refer to 'the education system' or 'the legal system'. The confusion is of course not helped by the fact that the ultimate source of the 'concepts' in Figure 10A.1 is (can only be) perceived reality; that is the ultimate source of the concepts through which we try to make sense of perceived reality in a never-ending cyclic process of learning as in Figure 10A.2. Perceived reality and the intellectual concepts steadily create each other. 'System' seems on present experience to be a useful epistemological device in that process.

The 1950s and 1960s: 'hard' systems thinking

In the late 1940s Ludwig von Bertalanffy, an organismic biologist, suggested that the ideas about organisms as whole entities, which he and fellow non-reductionist biologists had developed, could be generalized to refer to wholes of any kind called 'systems'. (It is unfortunate that he chose a word for 'concept of a whole' which was already familiar in everyday language, rather than constructing a neologism such as Koestler's later 'holon': that would have avoided the fundamental current confusion between 'system' as concept and as object.)

The idea of an autonomous whole was developed and applied in many fields, and the process of Figure 10A.2 generated much knowledge in different areas, including biology, ecology, engineering, economics, geography, sociology, organization theory, and many others. Out of this work, in the 1950s and 1960s there emerged a particular notion of 'a systems approach' which is still in the late 1980s the conventional wisdom for all who do not keep up with the systems literature and developments in the field. This conventional wisdom can be expressed as taking a particular view of Figure 10A.1, namely that:

PR is systemic, while
M can be systematic

Systems are assumed to exist in the world, and it is further assumed that *M* can be a *systematic* appraisal of alternatives, followed by choice of that which will best achieve explicit objectives.

This is the version of a systems approach which underlies and informs Systems Engineering, classical Operational Research, and RAND Corporation-style Systems Analysis. All assume that what 'the system' is is not problematical, that the system's objectives can be defined, and that alternative means of achieving them can be modelled and compared using some declared criteria, enabling a suitable selection to be made of the most desirable form of the system. This can then be implemented and monitored. This is 'hard' systems thinking, a means–ends schema which assumes that problems can be perceived as a search for an efficient means of achieving declared objectives or meeting declared needs.[3]

Now, when we examine the structure of thinking behind the ubiquitous accounts of the creation of computerized information systems via organized projects, the same ideas are revealed to be at work. The conventional wisdom here takes as given the concept of 'project life cycle', the project being that of analysing the information requirements of some organization, department or section, designing, constructing and implementing a computer system to provide those requirements, and monitoring its operation. Miles analyses numerous accounts of this process in the literature and shows how thoroughly they map the 'hard' systems paradigm.[4] From the 1960s to the 1980s, in such accounts, organizations are conceptualized as goal-seeking machines and information systems are there to enable the information needs associated with organizational goals to be met.

The adoption of this approach was probably historically inevitable, given that the early computers were large machines requiring their special air-conditioned locations and the recruitment of a new organizational priesthood to operate them. They were also expensive, and their acquisition had to be treated as what it was: a significant capital investment. So the initial importation of thinking from the world of engineering projects was not foolish. But the computer technology has changed dramatically over the last two decades, while the thinking which structures its acquisition and use has moved only slowly. In the most recent book in this area on my shelves, a text used for teaching young Californians 'software engineering', the structure of the thinking is entirely that of the means–ends framework of 'hard' systems thinking:

> A software engineering project begins with expectations that are not now being satisfied. An expectation involves a perceived need and a perceived means that may satisfy that need . . . A need seeking a means is called a problem.[5]

In the late 1980s the changes in the technology make problematical the automatic use of the means–end schema and the concept of 'project to realize the system'. But there is a second reason for trying to rethink this approach to computer systems analysis and design. In the 1960s the adoption of the standard assumption from management science that organizations could be treated as if they were instrumentalities, goal-seeking machines, seemed not unreasonable. But in the 1980s such an assumption seems increasingly dubious. Why not treat organizations as if they were not goal-seeking machines, but discourses, cultures, tribes, political battlegrounds, quasi-families, or communication and task networks? In management science itself phenomenology and hermeneutics have begun to creep in where once an automatic functionalism held sway.[6]

These changes have been signalled by the newer developments in systems thinking, in which the observer's ability to attribute meaning to what is perceived is treated as a prime element. Perhaps those developments might be relevant to improving both the thinking and the practice entailed in information system creation?

The 1970s and 1980s: 'soft' systems thinking

In order to understand the recent developments in system thinking it is useful to start from the most fundamental image or metaphor embodied in it. This is the notion of an autonomous entity in an environment which may be able to survive as the environment delivers shocks to it. Survival will be possible in principle *if* the entity has available processes of communication and control, the latter in the control engineers' sense of the word. The entity may itself contain similar entities, and may itself be part of a larger similar entity in a layered, or hierarchical, structure. The most fundamental system idea is that the entity as a whole has so-called 'emergent properties', properties which are properties of the whole and are *meaningful* only at the level of the whole. (The vehicular properties of a bicycle are emergent properties of a particular whole entity structured in a certain way: the braking system would be a sub whole within it, also possessing *its* emergent properties.) The four most fundamental system ideas are thus emergence, hierarchy, communication and control: they are the ideas needed to describe the core system metaphor.[7]

These basic concepts have been elaborated in many fields in the work of the systems movement, from engineering to politics, from biology to social work. Ideas of 'natural systems' (which might map onto frogs, fishes and foxgloves) and 'designed systems' (which might map onto files and fire engines) have been elaborated and used to give systemic insights in many different areas of work.

From the early 1970s the author and his colleagues sought a better approach to tackling the messy ill-structured problems which characterize human affairs. The research strategy was to take systems engineering (the 'hard' systems paradigm) as given and try to use it in unsuitably 'soft' problem situations. Systems engineering failed in such circumstances and had to be reconstructed. The outcome of a decade of the research was 'Soft Systems Methodology' (SSM).[8]

The crucial move in the research was to add to the notions of 'natural' and 'designed' systems the idea that a set of activities so linked as to form a purposeful whole could itself be regarded as a kind of system, a 'human activity system'. Then came the hard-won learning that such systems could be adequately clearly described only in relation to a particular world view, or *Weltanschauung*: purposeful activity which one observer perceives as 'a terrorist system' is to another observer 'a freedom-fighting system'. Meaning attribution is crucial, so that it is essential to declare a world view when giving an account of any purposeful activity.

Methods for naming and modelling human activity systems (according to declared *Weltanschauungen*) were developed, and SSM developed as a process in which such models could be compared with real-world action in a problem situation in order to structure a debate about change. Note that the models are not models *of* real-world activity, they are models *relevant to* debating it. The modern form of SSM is shown in Figure 10A.3. In the right-hand stream of analysis the debate about desirable and feasible change is structured by building models and comparing them with real-world action. In the process the participants *learn* their way to which 'relevant' systems are truly relevant in the social context. In the left-hand stream of analysis the problem *situation* (the notion of 'the problem' having been abandoned) is explored as a culture, with social and political characteristics being

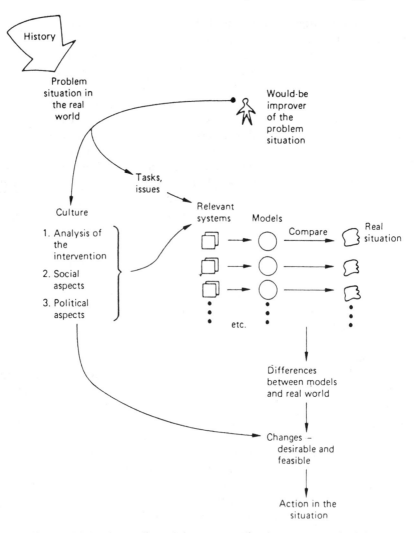

Figure 10A.3   An outline of the process of soft systems methodology

examined.[9] This feeds both the choice of relevant systems and the debate about change.

At an early stage of the work it was realized that activity models could be transformed into information flow models by asking of each activity in a model of a purposeful system: What information is required in principle to do this activity, in what form, with what frequency, from what source? And: What information is generated by doing the activity?[10] Thus the debate in SSM could be information-focused as well as (or instead of) activity-focused.

It thus provided in principle an approach which could be elaborated into a methodology aimed at information system provision, and has been developed and exploited in that way.[11]

Figure 10A.3, then, can be seen as a description of SSM as a learning system: a process which learns its way to the meanings which characterize an organization. In this process an organization is perceived as entailing readinesses on the part of its members to conceptualize it and its internal and external relationships in a particular way. Those *readinesses* are in a real sense the condition for the existence of that abstraction: the organization. Of course, they change through time, sometimes incrementally, sometimes in a revolutionary way, as perceptions and membership change.

Returning now to Figure 10A.1, we can define SSM in a way which sharply denotes the construction of a systems methodology to replace systems engineering in situations in which systems and objectives could not be taken as given but were themselves problematical. In the language of Figure 10A.1, SSM as an approach to messy ill-defined problem situations embodies the view that

> *PR* is problematical, while
> *M* can be systemic.

This is a very different perspective on Figure 10A.1: systemicity is no longer assumed to be in the outside world (which is regarded as problematical): it is in the process of inquiry. This is a fundamental shift, a shift from the idea of optimizing to the idea of learning the meanings by which people sharing a human situation seek to make sense of it. The significance of this shift is sometimes obscured because of the complication that SSM is in fact doubly systemic. It is, as a whole, *a learning system*; and it is one which happens to make use of *system models* (though other kinds of models could in principle be inserted). The important point is that, in using SSM, we must never lose sight of the fact that the models are *not* would-be descriptions of parts of the world. They are abstract logical machines for pursuing a purpose, defined in terms of declared world views, which can generate insightful debate when set against actual would-be purposeful action in the real world.

The view of social reality implied both by the form of SSM and by the way it is used is that it is the ever changing outcome of a social process in which human beings continually negotiate and renegotiate, and so construct with others their perceptions and interpretations of the world outside themselves and the rules for coping with it. These rules are never fixed once and for all.

In a recent major use of SSM in a multinational science-based company, the study rethought the role structure and processes of the Company function which ensured that their science-based technology was at least as good as that of the competition (and would still be in five and ten years time).

The function traditionally enacted its role both by keeping up to date with developing technologies in a scientific sense and by working with plants on day-to-day and longer-term problems. Major learning occurred for a number of the professional managers who took part in the study over the course of it. One group gained the realization that they had been conceptualizing their role in terms of the need to create, support and preserve professional skill pools within the Company. This is a plausible and defensible view, but an alternative perception developed convincingly during the course of the debate initiated at the comparison stage of SSM. This was the view that it would be useful to conceptualize the function's role as that of engaging in transactions, technologically based, with second parties, transactions whose added value had necessarily to be *co-produced* by the two partners involved in the transactions. This change of perception would of course change totally the nature of the information systems needed to support the operations of the function. It is rare that there is an adequate, single, unitary view of an organization so delineated and accepted that information system provision can be based upon it without further thought.

THE IMPLICATIONS OF THE NEW SYSTEMS THINKING FOR THE PROVISION OF INFORMATION SYSTEMS IN ORGANIZATIONS

It now seems that, in the future, the computer project managed through a 'project life cycle' will increasingly become the occasional special case in which some uncontentious and relatively mechanical administrative procedures are computerized. Where perceptions and meanings, and hence tasks, are more problematical, the 'project' approach needs to be complemented by a process for the continuous rethinking of organizational tasks and processes, together with the rethinking of the enabling information flows. As an organization steadily learns and reconstitutes itself, in a changing environment, rethinking the meanings it attributes to its world, it will need to have available a process by which the nature and context of the required information flows can be rethought and reprovided. This will be a social and political, as well as a technical, process, and, as Keen notes, researchers in information systems will increasingly need a social and political perspective.[12]

A 'process' rather than a 'project' approach to the creation of information systems can in fact be seen in recent developments such as prototyping,[13] and in the 'information resource centre' approach to placing computing in the hands of the end user.[14] In the former approach a rough speedily-assembled version of the required system can enable the designers and users to *learn* their way to an adequate arrangement. In the latter we have an attempt to overcome perceived deficiences in the traditional process of system development. It is an approach driven, in part, by technology changes, in which the user interacts with a corporate information resource. With the help of software support tools (and the occasional professional!) the user creates his or her own information system.

Both of these approaches attempt to move beyond the traditional project approach, and are usefully complementary to it. However, in both cases, as with all problems of information provision, the first stages are nothing at all to do with data, hardware, or software. They concern perceptions and politics, the interpretations of their world by the organizations in question; they concern the meanings attributed to the flux of events and ideas through which the organization lives. Analysis has to start there, and has to accept that the meanings (and hence the conversions of data into information) will remain static only for the most basic mechanical processes—such as, say, logging the receipt of mail in a mail-order company.

It is in this situation that the new systems thinking can provide something useful in the field of information provision, providing as it does both a social stream of analysis (the left-hand side of Figure 10A.3) and a logic-driven stream (the right-hand side) in which are developed models of purposeful activity systems. When iterations of the process of Figure 10A.3 produce models which are *widely agreed* to be *relevant* in a company situation, then such consensus activity models can be converted into information flow models and the more traditional methods of information system design can be initiated.

In this SSM offers a process through which an organization can continually reflect upon its aspirations and tasks, thus continually reviewing its information strategy.

At a less global level, too, the inquiring process of SSM can make a contribution. Suppose for example that an organization wishes within an information strategy to pursue the 'information resource centre' approach. In the language of SSM a relevant purposeful 'human activity system' would be 'a system to establish, develop and maintain an information resource enabling users to develop information systems suited to their needs'. Another might be 'a system to appraise users' experience of working with the information resource centre'. Using the methods of SSM such activity systems could be modelled and the models used in a design mode to ensure that processes were institutionalized by means of which the organization would continue to learn from its flow of experience. Both prototyping and the information centre approach imply the need for such built-in learning if the same problems are not to be tackled again and again in isolation.

CONCLUSION

In summary, it has been argued that the field of information systems has rather surprisingly neglected systems thinking as an underpinning to both its theoretical and practical concerns. Pragmatically, and unconsciously, the field adopted the (systematic) systems thinking of the 1950s and 1960s. Since then the information system environment, the information processing technology and systems thinking have all changed. The new (systemic) systems thinking of the 1970s and 1980s is complementary to that of the earlier decades. Its process orientation offers help with some of the crux problems of information provision in organizations in the 1990s.

NOTE

This is the text of the Second Annual Rank Xerox Lecture of the Worshipful Company of chartered Secretaries and Administrators. The Lecture is funded by Rank Xerox and delivered at the City University under the auspices of the Department of Information Science. This lecture was delivered on 2 March 1988 by Professor P. B. Checkland, University of Lancaster. The author is grateful to Butterworth and Co., publishers of *International Journal of Information Management* in which the lecture was published (Vol. 8, 1988, 239–248), for permission to reprint it here.

REFERENCES

[1]SEARLE, J. (1969). *Speech Acts*, Cambridge: University Press.
[2]CHECKLAND, P. B. (1984). Systems thinking in management: The development of soft system methodology and its implications for social science. In H. ULRICH and G. J. B. PROBST (eds), *Self-Organisation and Management of Social Systems*, 94–104. Berlin: Springer-Verlag. CHECKLAND, P. B. (1988). Images of systems and the systems image. Presidential Address to ISGSR, Budapest, June 1987. *Journal of Applied Systems Analysis*, *15*, 37–42.
[3]CHECKLAND, P. B. (1981). *Systems Thinking, Systems Practice*, 757–767. Chichester: John Wiley. CHECKLAND, P. B. (1985). From optimizing to learning: A development of systems thinking for the 1990s. *Journal of the Operational Research Society*, *36*, (No. 9), 757–767.
[4]MILES, R. K. (1985). Computer systems analysis: The constraint of the 'hard' systems paradigm. *Journal of Applied Systems Analysis*, *12*, 55–65.
[5]STEWART, D. V. (1987). *Software Engineering with Systems Analysis and Design*. Monterey: Brooks/Cole Publishing Co.
[6]CHECKLAND (1981). *Op cit.*, Ref 3. BLUNDEN, M. (1985). Vickers' contribution to management thinking. *Journal of Applied Systems Analysis*, *12*, 107–112.
[7]CHECKLAND (1981). *Op cit.*, Ref 3.
[8]CHECKLAND (1981). *Op cit.*, Ref 3.
[9]CHECKLAND, P. B. (1986). The politics of practice. International Roundtable on the Art and Science of Systems Practice, IIASA, November 1986.
[10]CHECKLAND, P. B. and GRIFFIN, R. (1970). Information systems: A systems view. *Journal of Systems Engineering*, *1* No. 2), 29–42.
[11]WILSON, B. (1984). *Systems: Concepts, Methodologies, Applications*. Chichester: John Wiley.

[12]KEEN, P. G. W. (1981). Information systems and organisational change. *Communications of the ACM*, *24* (No. 1), 24–33.

[13]BRITTAN, J. N. G. (1980). Design for a changing environment. *The Computer Journal*, *23*, (No. 1), 13–19. DEARNLEY, P. A. and MAYHEW, P. J. (1983). In favour of system prototypes and their integration into the systems development cycle. *The Computer Journal*, *26* (No. 1), 481–484.

[14]HEAD, R. V. (1985). Information Resource Centre: A new force in end-user computing. *Journal of Systems Management*, *36* (No. 2), 24–29.

Bibliography

Acheson, D. (1988). Public Health in England: the report of the Committee of Inquiry into the future development of the Public Health Function. Cm 289, HMSO.

Ahituv, N., and Neumann, S. (1982). Decision making and the value of information, in Galliers, R. D., q.v.

Albert, K. J. (Ed.) (1983). *The Strategic Management Handbook*, McGraw-Hill, New York.

Atkinson, C. J. (1984). Metaphor and Systemic Praxis, PhD Dissertation, University of Lancaster.

Atkinson, C. J. (1987). Towards a plurality of soft systems methodology, *Journal of Applied Systems Analysis*, **13**, 19–31.

Atkinson, C. J. (1989). Ethic: a lost dimension in soft systems practice, *Journal of Applied Systems Analysis*, **16**, 43–53.

Atkinson, C. J., and Checkland, P. B. (1988). Extending the metaphor 'System', *Human Relations*, **41**(10), 709–725.

Bemelmans, Th. M. A. (Ed.) (1984). *Beyond Productivity: Information Systems development for Organizational Effectiveness*, North Holland, Amsterdam.

Bertalanffy, L. von (1968). *General System Theory*, Braziller, New York.

Bettis, R. A., and Hall, W. K. (1983). The business portfolio approach—where it falls down in practice, *Long Range Planning*, **16**(2), 95–104.

Blackett, P. M. S. (1962). *Studies of War: Nuclear and Conventional*, Oliver and Boyd, Edinburgh.

Blair, R. N., and Whitston, C. W. (1971). *Elements of Industrial Systems Engineering*, Prentice Hall, Englewood Cliffs, NJ.

Blondel, J. (1978). *Thinking Politically*, Pelican Books, Harmondsworth.

Blunden, M. (1985). Vickers' contribution to management thinking, *Journal of Applied Systems Analysis*, **12**, 107–112.

Boland, R. J., and Hirschheim, R. A. (Eds) (1987). *Critical Issues in Information Systems Research*, John Wiley & Sons, Chichester.

Bowden, D. and Gumpert, R. (1988). Quality versus quantity in medicine, *Royal Society of Arts Journal*, **CXXXVI**, April, 333–346.

Bulow, I. von (1989). The bounding of a problem situation and the concept of a system's boundary in soft systems methodology, *Journal of Applied Systems Analysis*, **16**, 35–41.

Caws, P. (1988). *Structuralism: The Art of the Intelligible*, Humanities Press International Inc., Atlantic Highlands, NJ.

CCTA (1989). *'Compact' Manual*, Version 1.1, No. 1, Central Computer and Telecommunication Agency, Norwich.

Checkland, P. B. (1972). Towards a systems-based methodology for real-world problem solving, *Journal of Systems Engineering*, **3**(2), 87–116.

Checkland, P. B. (1975). The development of systems thinking by systems practice—a methodology from an action research programme, in Trappl and Hanika. q.v.

Checkland, P. B. (1980). The systems movement and the 'failure' of management science, *Cybernetics and Systems*, **11**, 317–324.

Checkland, P. B. (1981). *Systems Thinking, Systems Practice*, John Wiley & Sons, Chichester.

Checkland, P. B. (1983a). OR and the systems movement: mappings and conflicts *Journal of the Operational Research Society*, **34**(8), 661–675.

Checkland, P. B. (1983b). Editorial introduction to Vickers, G., Human systems are different: two chapters from an unpublished work. *Journal of Applied Systems Analysis*, **10**, 3. The chapters were subsequently published in Vickers, G., *Human Systems are Different*, Harper and Row, London, 1983.

Checkland, P. B. (1984). Systems thinking in management: the development of soft systems methodology and its implications for social science, in Ulrich and Probst, q.v.

Checkland, P. B. (1985a). From optimizing to learning: a development of systems thinking for the 1990s, *Journal of the Operational Research Society*, **36**(9), 757–767.

Checkland, P. B. (1985b). Achieving desirable and feasible change: an application of soft systems methodology, *Journal of the Operational Research Society*, **36**, 821–831.

Checkland, P. B. (1986). The politics of practice, IIASA International Roundtable on 'The Art and Science of Systems Practice', International Institute for Systems Analysis, Laxenburg, Austria.

Checkland, P. B. (1988a). Soft systems methodology: an overview, *Journal of Applied Systems Analysis*, **15**, 27–30.

Checkland, P. B. (1988b). The case for 'holon', *Systems Practice*, **1**(3), 235–238.

Checkland, P. B. (1988c). Images of systems and the systems image. Presidential Address to Annual Meeting of The International Society for General Systems Research, Budapest, *Journal of Applied Systems Analysis*, **15**, 37–42.

Checkland, P. B. (1988d). Information systems and systems thinking: time to unite? *International Journal of Information Management*, **8**, 239–248.

Checkland, P. B. (1989). An application of soft systems methodology, in Rosenhead, J., q.v.

Checkland, P. B., and Casar, A. (1986). Vickers' concept of an appreciative system: a systemic account, *Journal of Applied Systems Analysis*, **13**, 3–17.

Checkland, P. B., and Davies, L. (1986). The use of the term *Weltanschauung* in soft systems methodology, *Journal of Applied Systems Analysis*, **13**, 109–115.

Checkland, P. B., and Griffin, R. (1970). Management information systems: a systemic view, *Journal of Systems Engineering*, **1**(2), 29–42.

Checkland, P. B., and Wilson, B. (1980). Primary task and issue-based root definitions in systems studies, *Journal of Applied Systems Analysis*, **7**, 51–54.

Chestnut, H. (1967). *Systems Engineering Methods*, John Wiley & Sons, New York.

Ciborra, C. U. (1984). Management information systems: a contractual view, in Bemelmans, Th. M. A., q.v.

Ciborra, C. U. (1987). Research agenda for a transaction costs approach to information systems, in Boland, R. J. and Hirschheim, R. A., q.v.

Cooper, R. and Burrell, G. (1988). Modernism, postmodernism and organizational analysis: an introduction, *Organization Studies*, **9**(1), 91–112.

Crick, B. (1962). *In Defence of Politics*, Weidenfeld and Nicolson, London.

Dahl, R. A. (1970). *Modern Political Analysis* (2nd edn), Prentice-Hall, Englewood Cliffs, NJ.

Davies, B. M. (1984). *Community Health and Social Services* (4th edn), Hodder and Stoughton, London.

Davies, L. (1989). The Cultural Aspects of Intervention with Soft Systems Methodology, PhD Dissertation, University of Lancaster.

Davies, L., and Ledington, P. W. J. (1987). *Creativity and Metaphor in Soft Systems Methodology*, Paper submitted to 31st Annual Meeting of the International Society for General Systems Research, Budapest, June 1987, Vol 1, pp. 176–184.

Davies, L., and Wood-Harper, A. T. (1989). Information systems development: theoretical frameworks, *Journal of Applied Systems Analysis*, **16**, 61–73.

Demerath, N. J., and Peterson, R. A. (Eds) (1967). *Systems, Change and Conflict*, Free Press, New York.

Easton, D., (1961). *A Systems Analysis of Political Life*, John Wiley & Sons, New York.

Eden, C., Jones, S., and Sims, D. (1983). *Messing About in Problems*, Pergamon Press, Oxford.

Espejo, R., and Harnden, R. (1989). *The Viable Systems Model*, John Wiley & Sons, Chichester.

Fairtlough, G. (1982). A note on the use of the term 'Weltanschauung' in Checkland's *Systems Thinking, Systems Practice, Journal of Applied Systems Analysis*, **9**, 131, 132.

Flood, R., and Gaisford, P. (1989). Policing vice in the West End, *OR Insight*, **2**(3), 10–15.

Forbes, P. (1989). The Development and Dissemination of Soft Systems Methodology, PhD Dissertation, University of Lancaster.

Forbes, P., and Checkland, P. B. (1987). *Monitoring and Control in Systems Models*, Internal Discussion Paper 3/87, Department of Systems and Information Management, University of Lancaster.

Fuller, R. B. (1983). *Critical Path*, Hutchinson, London.

Galliers, R. D. (Ed.) (1987) *Information Analysis: Selected Readings*, Addison-Wesley, Wokingham.

Galliers, R. D., Whittaker, B. D., Clegg, J. D., and Mouthon, M. (1981). Improving employment prospects for mentally handicapped people in Camden: a systems study, *Journal of Applied Systems Analysis*, **8**, 101–114.

Gardner, J. R., Rachlin, R., and Sweeney, H. W. A. (Eds) (1986). *Handbook of Strategic Planning*, John Wiley & Sons, New York.

Gerard, R. W. (1964). Entitation, animorgs and other systems, in Mesarovic, q.v.

Gilmore, T., Krantz, J., Ramirez, R. (1985). Action based modes of inquiry and the host-researcher relationship, *Consultation*, **5**(3), 160–176.

Gluck, F. W. (1986). Strategic management: an overview, in Gardner *et al*, q.v.

Goddard, A. (1989), Are the '3Es' enough? *OR Insight*, **2**(3), 16–19.

Government White Paper (1989). Working for Patients. Cm 555, HMSO.

Gray, W., and Rizzo, N. D. (Eds) (1973). *Unity through Diversity: a Festschrift for Ludwig von Bertalanffy*, Gordon and Breach, New York.

Griffiths, R. (1983). NHS Management Inquiry—a report to the Secretary of State for Social Services. HMSO.

Griffiths, R. (1987). Radcliffe–Maud memorial lecture, quoted in *NHS Management Bulletin*, No. 7, August, DHSS.

Hall, A. D. (1962). *A Methodology for Systems Engineering*, Van Nostrand, Princeton, NJ.

Hall, W. K. (1978). SBUs: hot new topic in the management of diversification, *Business Horizons*, February, 17–25.

Hirschheim, R. A. (1985). *Office Automation: A Social and Organizational Perspective*, John Wiley & Sons, Chichester.

HMSO (1984). *Review of the Central Computer and Telecommunication Agency CCTA, IT Series: Information Technology in the Civil Service*, No. 8, July.

HMSO (1985). *Central Computer and Telecommunication Agency Progress Report CCTA, IT Series: Information Technology in the Civil Service*, No 11, August.

Huhne, C. (1989). *The Guardian*, 15.2.89, quoting J. Wells, Cambridge University.

Hult, M., and Lennung, S. (1980). Towards a definition of action research: a note and a bibliography, *Journal of Management Studies*, **17**(2), 241–250.

Hunter, R. B. (1972). Report of the Working Party on Medical Administrators (DHSS), HMSO.

Jacob, F. (1974). *The Logic of Living Systems*, Allen Lane, London.

Johnson, P. E. (1984). The expert mind: a new challenge for the information scientist, in Bemelmans Th. M. A., q.v.

Jones, G. (1989). Letter to *OR Newsletter*, April, 18.

Jordan, N. (1965). *Themes in Speculative Psychology*, Tavistock, London.

Keen, P. G. W., and Scott Morton, M. S. (1978). *Decision Support Systems: An Organizational Perspective*, Addison-Wesley, Reading, Mass.

Koestler, A. (1967). *The Ghost in the Machine*, Hutchinson, London.

Koestler, A. (1978). *Janus: A Summing Up*, Hutchinson, London.

Kruel, W. J., and Glenny, L. H. (1983). Business unit strategy, in Albert, K. J., q.v.

Land, F., and Hirschheim, R. A. (1983). Participative systems design: rationale, tools and techniques, *Journal of Applied Systems Analysis*, **10**, 91–107.

Lasswell, D., and Kaplan, A. (1950). *Power and Society*, Yale University Press.

Ledington, P. W. J. (1989). Intervening in Organisational Conversations using Soft Systems Methodology, PhD Dissertation, University of Lancaster.

Lord, W. T. (1989). Letter to *OR Newsletter*, April, 18.

Lyytinen, K. (1988). Stakeholders, information system failures and soft systems methodology: an assessment, *Journal of Applied Systems Analysis*, **15**, 61–81.

McConnell, M. (1988). *Challenger: A Major Malfunction*, Unwin Hyman, London.

McLoughlin, P. (1986). Soft systems methodology—its role in management services, *Management Services*, August, 16–20.

Machol, R. E. (1980). Comment on Eden and Jones 'Publish or perish', *Journal of the Operational Society*, **31**, 1109–1110.

Maddison, R. N. (1983). *Information Systems Methodologies*, British Computer Society and Wiley-Heyden.

Mathiassen, L., and Nielsen, P. A. (1989). Soft systems and hard contradictions: approaching the reality of information systems in organisations, *Journal of Applied Systems Analysis*, **16**, 75–88.

Merton, R. K. (1957). Manifest and latent functions, in Demerath and Peterson, q.v.

Mesarovic, M. D. (1964). *Views on General Systems Theory: Proceedings of the 2nd Systems Symposium at Case Institute*, John Wiley & Sons, New York.

Miles, R. K. (1985). Computer systems analysis: the constraints of the hard systems paradigm, *Journal of Applied Systems Analysis*, **11** 55–65.

Miles, R. K. (1986). Author's response—Soft approaches can be deceptive: a reply to Veryard, *Journal of Applied Systems Analysis*, **13**, 95–96.

Miles, R. K. (1987). The Soft Systems Methodology: A Practicable Framework for Computer Systems Analysis, PhD Dissertation, University of Lancaster.

Miller, G. A. (1968). *The Psychology of Communication*, Allen Lane, the Penguin Press, London.

Miller, J. D. B. (1962). *The Nature of Politics*, Butterworth, London.

Mumford, E. (1983). *Designing Human Systems*, Manchester Business School.

Mumford, E., Hirschheim, R. A., Fitzgerald, G., and Wood-Harper, T. A. (Eds) (1985). *Research Methods in Information Systems*, North Holland, Amsterdam.
Naughton, J. (1977). *The Checkland Methodology: A Reader's Guide* (2nd edn), Open University Systems Group, Milton Keynes.
Normann, R. (1984). *Service Management: Strategy and Leadership in Service Businesses*, John Wiley & Sons, Chichester.
Normann, R. (1987). Private communication to P. B. Checkland.
Passos, J. (1976). In Warley, Zuzich, Zajkowski and Zagornik, q.v., p. 196.
Patching, D. (1987). Soft systems methodology and information technology, *Management Services*, August, 16–19.
Polyani, M. (1958). *Personal Knowledge*, Routledge and Kegan Paul, London.
Polyani, M. (1966). *The Tacit Dimension*, Doubleday, New York.
Popper, K. R. (1963). *Conjectures and Refutations: The Growth of Scientific Knowledge*, Routledge and Kegan Paul, London (revised edn, 1972).
Pruzan, P. (1988). Systemic OR and operational systems science, *European Journal of Operational Research*, **37** 34–41.
Richie, B. D. (1987). The pyramid of Cheops: using metaphor in information systems design, *International CIS Journal*, **1**(1), 31–38.
Rivett, P. (1983). A world in which nothing ever happens twice, *Journal of the Operational Research Society*, 34(8), 677–683.
Rivett, P. (1989). Letter to *OR Newsletter*, March, 17.
Rodriguez-Ulloa, R. A. (1988). The problem-solving system: another problem content system, *Systems Practice*, **1**(3), 243–257.
Rosenhead, J. (1989). *Rational Analysis for a Problematical World*, John Wiley & Sons, Chichester.
Ryan, B. J. (1973). In *Journal of Nursing Administration*, **3**, 301.
Schafer, G. (1988). *Functional Analysis of Office Requirements: A Multiperspective Approach* (with contributions by Hirschheim, Harper, Hansjee, Domke, Bjorn-Andersen), John Wiley & Sons, Chichester.
Scholes, J. (1987). Extending the Application of Soft Systems Methodology, PhD Dissertation, University of Lancaster.
Schon, D. A. (1983). *The Reflective Practitioner: How Professionals Think in Action*, Temple Smith, London.
Schweder, R. A., and LeVine, R. A. (1984). *Culture Theory: Essays on Mind, Self and Emotion*, Cambridge University Press, Cambridge.
Seedhouse, D. (1988). *Ethics: The Heart of Health Care*, John Wiley & Sons, Chichester.
Shils, E. A., and Finch, H. A. (Eds) (1949). *Max Weber's Methodology of the Social Sciences*, Free Press, New York.
Simon, H. A. (1960). *The New Science of Management Decision*, Harper & Row, New York.
Simon, H. A. (1977). *The New Science of Management Decision* (revised edn), Prentice Hall, Englewood Cliffs, NJ.
Simon, H. A., and Newall, A. (1972). *Human Decision Making*, Prentice Hall, Englewood Cliffs, NJ.
Smith, G. W. (1984). Towards an organisation theory for the NHS, *Health Services Manpower Review*, **10**(3), 3–7.
Smyth, D. S., and Checkland, P. B. (1976). Using a systems approach: the structure of root definitions. *Journal of Applied Systems Analysis*, **5**(1), 75–83.
Stowell, F. (1989). Change, Organisation, Power and the Metaphor 'Commodity', PhD Dissertation, University of Lancaster.

Susman, G., and Evered, R. D. (1978). An assessment of the scientific merits of action research. *Administrative Science Quarterly*, **23** (December), 582–603.

Symons, V., and Walsham, G. (1988). The Evaluation of information systems: a critique, *Journal of Applied Systems Analysis*, **15**, 119–132.

Todd, A. R. (1968). Report of the Royal Commission on Medical Education 1965–68. Cmnd 3569, HMSO.

Trappl, R., and Hanika, F. de P. (Eds) (1975). *Progress in Cybernetics and Systems Research*, Vol. II, Hemisphere Publications, Washington.

Tusa, A., and Tusa, J. (1988). *The Berlin Blockade*, Hodder and Stoughton, London.

Ulrich, H., and Probst, G. J. B. (Eds) (1984). *Self-Organization and the Management of Social Systems*, Springer-Verlag, Berlin.

Varela, F. J. (1984). Two principles of self-organization, in Ulrich and Probst, q.v.

Veryard, R. (1986). Computer systems analysis, models and prototypes: a reply to R. K. Miles, *Journal of Applied Systems Analysis*, **13**, 89–93.

Vickers, G. (1965). *The Art of Judgement*, Chapman and Hall, London. (Reprinted 1983 Harper and Row, London.

Wang, Ming and Smith, G. W. (1988). Modelling CIM systems; Part 1: methodologies, *Computer-integrated Manufacturing Systems*, **1**(1), 13–187.

Warley, H. H., Zuzich, A., Zajkowski, M., and Zagornik, A. D. (Eds) (1976). *Health Research: The Systems Approach*, Springer, New York.

Warmington, A. (1980). Action research: its method and its implications, *Journal of Applied Systems Analysis*, **7**, 23–39.

Watson, R. and Smith, R. (1988). Applications of the Lancaster soft systems methodology in Australia, *Journal of Applied Systems Analysis*, **15**, 3–26.

Weber, M. (1904). Objectivity in social science and social policy, in Shils and Finch, q.v.

Wilson, B. (1984). *Systems: Concepts Methodologies and Applications*, John Wiley & Sons, Chichester.

Wood, P. (1985). Private communication to P. B. Checkland.

Woodburn, I. (1985). Some developments in the building of conceptual models, *Journal of Applied Systems Analysis*, **12**, 101–106.

Ziman, J. M. (1968). *Public Knowledge: An Essay Concerning the Social Dimension of Science*, Cambridge University Press, Cambridge.

Author Index

Subject Index

Soft Systems Methodology: a 30-year retrospective

JOHN WILEY & SONS, LTD

Chichester · New York · Weinheim · Brisbane · Singapore · Toronto

Other Wiley Editorial Offices

John Wiley & Sons, Inc., 605 Third Avenue,
New York, NY 10158-0012, USA

WILEY-VCH Verlag GmbH, Pappelallee 3,
D-69469 Weinheim, Germany

Jacaranda Wiley Ltd, 33 Park Road, Milton,
Queensland 4064, Australia

John Wiley & Sons (Asia) Pte Ltd, 2 Clementi Loop #02-01,
Jin Xing Distripark, Singapore 129809

John Wiley & Sons (Canada) Ltd, 22 Worcester Road,
Rexdale, Ontario M9W 1L1, Canada

Typeset in 10/12 pt Times by C.K.M. Typesetting, Salisbury, Wiltshire.
Printed and bound in Great Britain by Biddles Ltd, Guildford and King's Lynn.
This book is printed on acid-free paper responsibly manufactured from sustainable forestation,
for which at least two trees are planted for each one used.